Emergence, Mind,
and Consciousness

Emergence, Mind, and Consciousness

A Bio-Inspired Design for a Conscious Agent

GARY A. LUCAS

iUniverse, Inc.
Bloomington

Emergence, Mind, and Consciousness
A Bio-Inspired Design for a Conscious Agent

iUniverse books may be ordered through booksellers or by contacting:

iUniverse
1663 Liberty Drive
Bloomington, IN 47403
www.iuniverse.com
1-800-Authors (1-800-288-4677)

*Because of the dynamic nature of the Internet, any web addresses or links contained in this
book may have changed since publication and may no longer be valid. The views expressed
in this work are solely those of the author and do not necessarily reflect the views of the
publisher, and the publisher hereby disclaims any responsibility for them.*

*Any people depicted in stock imagery provided by Thinkstock are models,
and such images are being used for illustrative purposes only.*

Certain stock imagery © Thinkstock.

ISBN: 978-1-4620-4138-1 (sc)
ISBN: 978-1-4620-4136-7 (e)
ISBN: 978-1-4620-4137-4 (dj)

Library of Congress Control Number: 2011913686

Printed in the United States of America

iUniverse rev. date: 10/13/2011

For Oana

once upon a time
there was a little girl
her daddy was a mime
her mommy was a squirrel

she grew into a woman
so beautiful and clever
to the craziest one
i pledge waffles forever

CONTENTS

LIST OF ILLUSTRATIONS

LIST OF TABLES

PREFACE

> Most neuroscientists feel that two orders of magnitude above and below one's central focus is 'horizon enough' and that anyone attempting four orders above and below is reckless. However, there are some who attempt such a dangerous dynamic range. They probably know that the risk of failure is the price of synthesis, without which there are only fields of dismembered parts. – Rodolfo Llinás, *I of the Vortex, 2001*

I am an experimental psychologist by training, and I have long been interested in understanding how a myriad of neurons in the brain could possibly give rise to conscious mind. I have also worked in industry as a human factors designer and systems engineer. That work taught me the value of analyzing tasks and defining the interface requirements needed to enable hardware and software systems to accomplish various functions. This book is my attempt to apply my knowledge of neuroscience and my systems engineering background to reverse engineer the functions of the human mind. My goal is to propose a synthesis of functions that reaches across what many might characterize to be a dangerous dynamic range – from the interactivity of neurons to the behavior of a conscious reasoning socially self-aware agent.

My approach follows two broad themes. The essays in the first half of the book review topics relevant to understanding the biological foundations of mind. I begin with the argument that emergence is not some strange exotic process, but rather the normal process by which physical and biological systems evolve. New organizations continually take form in nature and as they do so they introduce new properties. I call this the *organization effect*. The organization effect implies that explanations of emergent phenomena need to consider the nature of the organizations that bring those phenomena into existence and the selection processes that shape those organizations. Following this principle, this section of the book goes on to argue that many properties that we associate with mind – for example, intentionality, knowledge, and goal-directed intention – are better understood as properties of certain kinds of organizations.

A natural extension of thinking about how properties emerge from particular kinds of organizations is that if you understand the essential elements of the organization, then you understand, at least in principle, how to create systems that share those properties. The second half of the book follows this idea to describe how a system of processing streams and central coordinators creates the properties of human mind. To make this explanation more concrete, I propose a systems-level model for the design of a conscious, reasoning, socially aware robotic agent. It's a complex undertaking. However, by taking it step by step, we gradually see the properties of mind take form in our robot. Admittedly, this is an unconventional strategy. However, it is the best method I have found for explaining the emergence of conscious mind. There is no mystery once you understand how to build it. The model of mind I propose in this section is then tied back to the first theme in the last chapter, to explain why qualia, the experiential qualities of consciousness, exist, and how they contribute to the spiral of processing that makes our robot conscious.

For a long time the construction of a conscious robot has seemed to be little more than a dream. Surely many aspects of brain function remain to be resolved, and there are immense challenges in building complex networks that can keep themselves tuned, balanced, and interacting with each other in meaningful ways. However, over the last few decades visionary robotic designers have begun thinking about machine consciousness and expert neuroscientists have been seriously unravelling the structure of the human brain. As a result, I believe we now know enough about the functional organization of the brain to formulate a systems-level model of consciousness mind that can be applied to a robotic agent. As Adam Zeman notes:

> If the computational organisation of the brain is the key to consciousness, nothing, bar complexity, stands in the way of conscious machines.[1]

In the quote at the top, Rodolfo Llinás emphasizes that it is only through synthesis that we move beyond dismembered facts to see bigger pictures. However, synthesis without accessibility is of little value. No one, certainly not I, can be an expert in all these subject areas. Yet the broad themes we are exploring force us to deal with many complex and puzzling concepts. My first strategy for bringing these ideas together has been to link them to a common framework of selection-guided organization. This ties the organization effect to natural selection and its extensions in cognitive and cultural selection processes. My second strategy has been to make my model of mind as simple as possible at each level. Thus, I focus my analysis on what I believe to be the essential functions needed for a conscious reasoning socially -aware mind. That is the engineer in me. When it comes to explaining the emergent properties of mind, I have also enlisted the help of my feline companion, Tom Terrific. For some reason, thinking about Tom's mind is often easier than thinking about our own. However, the strategy that best clarified my thinking was the task of providing an essentialized account of the processes needed to build a conscious, reasoning, socially aware agent. I call him Cogley. I hope that my descriptions of Tom and Cogley prove to be as instructive for you as they have been for me.

Gary Lucas
Bloomington, Indiana
April 2011

INTRODUCTION
AS SIMPLE AS POSSIBLE

As Simple as Possible

Everything should be made as simple as possible,
but no simpler. – Albert Einstein, 1879–1955[1]

There are many functions in the brain which can operate without the support of attention, but to the extent they do so, they operate largely outside conscious awareness. In contrast, there are certain tasks that can only be accomplished when the effects of many different brain regions can be brought to bear on the problem at the same time. Attention provides a way of coordinating distributed brain regions by uniting selected feature clusters into globally connected circuits of perception, action, and feeling. This coordinated interaction is the selection advantage which attention provides. A coordinated predator has the capacity to become a better hunter. A coordinated prey animal has the potential to escape harm more readily. My thesis is simple: Attention is an evolutionary consequence of the ongoing competition for better ways to coordinate behavior. Consciousness is the feeling-centered awareness that emerges when perception, actions, and feelings are bound together in states of attention. Consciousness is not equivalent to attention, but attention determines to what a conscious mind reacts, and consciousness guides the flow of attention, so the two can never be fully separated.

Recognizing that attention and consciousness are connected makes other aspects of mind more understandable. For example, it helps us to understand why sharing and coordinating attention is so important. Consider what happens as I interact with another mind. When I talk with you, I must engage your attention. No doubt your attention shifts along the topic of discussion, first tracking my words and then exploring your own memories. I see evidence of this shift when your eyes glance upward as you reflect on my message. However, once you have processed that information, your eyes again engage mine as you react with new ideas. If you suddenly shift your attention away and cease to engage me, I may briefly continue to direct a barrage of verbal tokens toward the side of your head, but I soon come to recognize that I am no longer talking with you. To talk with you, I must continue to engage your attention. To be clear, the "I" who is attempting to engage your attention emerges from the committee of processes which guide my attention. My conscious mind is centered in my attention. Your conscious mind is centered in your attention. It is only when we coordinate our attention in shared activities that we feel our minds have met.

As I am writing this introduction, one of my favorite feline companions, Tom Terrific, has appeared on the arm of my chair. Earlier Tom had called out to me, but my attention was engaged elsewhere, and it wasn't substantially redirected by his call. Now Tom is being more assertive. Strategically positioning himself on the arm of my chair he leans close to my head and announces his presence with a soft trilled greeting. Then, having placed his paw on my shoulder, he gently tugs. When I look up, Tom begins making repeated head dips, which I have come to recognize means that he is watching intently for my reaction. As soon as I begin to move, he jumps down from the chair, raises his tail as a marker, and runs ahead. He's leading me to the location where he wants my help, and periodically he pauses and looks back to make sure I am still following. Without speech, Tom cannot direct my attention as efficiently as you might. However, aside from the more limited medium of exchange, the communication process we follow is not so different. Tom uses signs to recruit my attention and direct me to places where he wants my help, and when he does, our minds meet in moments of coordinated attention.

My mind is organized around the events that occupy my attention, and Tom's mind is organized around the events that occupy his attention. There are obvious differences in the type of communication skills available to the steering committee at the center of Tom's attention and those available to the committee at the center of my attention. There are differences in the complexity of the concepts our respective committees are able to resolve and share. Of course, there are also differences as to what events Tom's committee finds interesting compared to mine. However, the core neural processes by which the two committees engage attention are similar.

**Figure 0-1: Tom Terrific in a typical
state of focused attention.**

Our goal will be to explore the nature of the mental processes that make minds like yours and mine and Tom's possible. What kind of neural organization is necessary to make a mind conscious? How do minds develop a sense of self? How is your consciousness different from Tom's? How is it similar? And how can we tell? These questions are not new, but for a long time they have seemed insoluble. However, the wealth of new insights which have developed in cognitive science and neuroscience in recent years has brought us to a point where we can begin to frame workable – albeit still controversial – answers to these questions. We will explore the implications of these findings and discover what they tell us about minds and the states of consciousness they experience.

The range of processes that need to be explored to complete this plan is large. We'll need to dig down into the neural mechanisms supporting different adaptive abilities to understand how different kinds of knowledge are represented in the brain and how different knowledge systems interact. As we do, the goal will be to follow Einstein's maxim to make our descriptions as simple as possible but without oversimplifying them. This is, of course, a goal that we can never fully achieve. Somehow we must simplify the structures of the brain to the point that we can begin to understand how they do their work without losing a sense of the intricate interactions that make that work possible. This is necessary because it is only by acknowledging the very complexity of the interacting processes that we can begin to appreciate how conscious minds emerge from the nested interaction of simpler neural organizations.

Our initial strategy will be to anchor the emergence of mind in biological mechanism. For some, the idea that physical stuff could give rise to conscious experience seems unreasonable. However, most of us have come to accept the idea that certain physical interactions can give rise to the peculiarly unreasonable process we call life. Less than a hundred years ago, many were convinced that some *vitalistic force* was needed to explain how physical structures could give rise to the seemingly nonphysical property of life. Today the concept of life is every bit as wondrous for us to consider, but we no longer feel the need to posit the existence of a separate vital force. We have learned enough about the intricate machinery of the cell and the transcription of DNA to understand (at least in principle) how a myriad of molecular processes interacting in a cascade of signaling subsystems results in the sustained operation of cellular growth and the potential for self-organization and reproduction. Further, we accept that more complex life forms, tissues, organ systems, and complex agents like ourselves emerge from the collective interaction of living cells.

The emergence of mind took a giant leap forward in the Cambrian explosion some 545 million years ago when animal forms with cells specialized for passing signals began to proliferate. Today we call those cells neurons, and they form the basis for rapid communication in animal nervous systems. Neurons aren't simply good at producing signals; they have long, tubular axons that transport signals across vast cellular distances. This allows them to form what are effectively signal-processing circuits, multipart detection-reaction functions which can mediate complex behavioral adaptations. Massive collections of neurons in central brain areas – human infants begin with about one hundred billion neurons – have resulted in circuitry of immense complexity. A centrally coordinated architecture of neural circuits in the modern vertebrate brain has even evolved to coordinate those signals in states of attention. As we come to understand how neural signals interact to manage states of attention, the nature of the conscious mind will no longer seem quite so mysterious, although its consequences will remain as wondrous as before.

It will take some time before we are ready to describe the nature of the neural architecture that allows vertebrate minds to implement this central coordination strategy, and there are still pieces of this puzzle that remain to be resolved, but we can now assemble much of the picture. A growing body of studies implicates the thalamocortical architecture of the brain in the management of attention. Within this architecture three streams of processing interact: 1) a perceptual stream of recognition and memory, 2) a planning stream for assembling action candidates and committing to them, and 3) an interoceptive stream for monitoring ongoing status and adjusting reactivity states. The confluence of these three streams in states of attention produces a mind which can feel status changes as it perceives and acts, which can perceive its own feelings and actions, and which can react to changes in what it perceives and how it feels. This interplay between perceptions, actions, and feelings lies at the core of conscious experience. The thalamocortical architecture does not explain all the higher features we associate with consciousness, but nest this central coordination architecture in a mind that can also remember and learn, and the potential for more complex conscious experiences grows.

As Tom's entry into our discussion makes clear, an underlying goal in this work is to understand the continuity of processes that contribute to the minds of higher animals, by which I mean modern birds and mammals. Studies of brain-damaged individuals indicate that the areas of the brain that are more developed in humans, such as the prefrontal cortex, contribute to greater intelligence, but they are not what makes us conscious. Studies of conscious experience point to the reentrant connections of the thalamocortical architecture as the networks which make attention possible. It's not so hard to recognize attention as an index of conscious experience once we understand this neural structure and its link to feelings. However, this architecture is not unique to humans or even to mammals. The modern avian brain includes the same general plan of sensory projections through the thalamus and shows similar evidence for centrally coordinated, feeling-guided attention. The implications are clear. If this architecture results in consciousness in humans, then to the extent that other animal minds share a similar architecture, they must necessarily share in some of the same kinds of conscious experiences.

This talk of consciousness and animals is no doubt making some readers anxious. We have convinced ourselves that if we cannot directly experience the conscious states of another species, then we cannot ever know whether or not it is conscious. This creates a black hole of impossibility from which there is no hope of escape and no reason to try. In addition to this impossibility assumption, there are several other spirals of reasoning that keep our thinking about consciousness in the dark. We are particularly confused about the brain's ability to automate activities so that they require minimal attention, and we assume that automated behavior implies an absence of consciousness. We think of conditioning procedures as processes that produce automated behavior, which suggests that anything that can be learned through conditioning doesn't require consciousness. Finally, we tend to confuse consciousness with reasoning, and reasoning with talking to ourselves; thus we assume that if animals do not use language to reason, then they cannot be conscious. However, all of these tenuous chains of logic are based on partial information and faulty assumptions.

The tentative answers to these objections are:

- We can tell when another mind is conscious without going inside it, because conscious agents behave differently than unconscious agents. With practice, we can even recognize when other agents are in different states of consciousness. We do it all the time.

- Conditioning paradigms are training procedures, but only simple associative relationships can be trained in the absence of attention. Conscious attention expands what can be learned via conditioning procedures.

- Many common actions, like tying a shoelace, initially require our full attention to be learned and then gradually come to be largely automated. However, they don't become completely unconscious when they are automated; rather they become background tasks whose operations require minimal attention as long as the tasks run off as expected. This background allocation strategy frees up the bulk of central attention resources to process other experiences. It's a way of nesting and prioritizing attention.

- Language extends the range of features to which conscious minds can attend, but language doesn't make minds conscious. We have it backward. Tracking and decoding the meaning of words is one of those tasks that we humans can only accomplish when we consciously attend to a chain of words.

Consider the meaning of the statement, "Say that again; I wasn't paying attention."

Though these arguments may temporarily cause you to reconsider some of these objections, they probably haven't convinced you. However, our strategy will not be to fight these old battles, but to eliminate the need for them. We didn't abandon the idea that some separate vital force was needed to explain life due to some clever argument against vitalism. Ultimately, it was abandoned when we developed a new understanding of the chain of connections between DNA codes and the protein-building subsystems that make cellular operations and their potential for reproduction possible. With those connections in place, we no longer needed to postulate a separate vital force to bridge the unknowns. We understood, at least in principle, how certain physiochemical systems are capable of copying and sustaining themselves. Similarly, the best strategy for dealing with the old objections to consciousness is to make them unnecessary. We need to identify the emergent chain of organizations which enables neural signaling networks to develop dynamic states of attention, the agency to take actions that change attention, the ability to feel the value of those changes, and an ongoing sense of self-identity. With an understanding of these causal chains, we will no longer feel the need to invent special constructs to bridge the unknowns.

Unlearning old ways of thinking about consciousness will take some effort, but it can be done. In fact, the process has already begun. Simply by considering the possibility that there is a grain of truth in the argument that conscious minds are organized around states of centrally coordinated attention, you are on the way to a new way of thinking about consciousness. The ability to engage central attention states and the ability to form new concepts are features of mind that, at least in some situations, can be observed and measured. In fact, the mechanisms of attention and the concepts that occupy it are dimensions on which the minds of different species can be compared. However, we're jumping ahead of the story here.

There are many intermediate steps that we need to explore first.

We will begin by looking into the structure of the vertebrate brain to discover how minds like ours are assembled, and how varied neural organizations represent different kinds of knowledge in the brain. To keep our model as simple as possible, we'll even speculate on how the component parts might be designed and interconnected to build a conscious robotic agent. After all, *if you can't build it; you don't really understand how it works.* An emphasis on designing a conscious robot will force us to simply our thinking and focus on the essential functions needed for a conscious mind. Finally, we'll examine the nature of the developmental trajectories that shape conceptual growth and the emergence of a sense of self in our robot. Along the way we'll check in with Tom to keep in contact with the biological continuity of the phenomena we explore. Consciousness, we will find, emerges in layers of physical, biological, cognitive, and cultural coordination.

PART ONE:
THE EMERGENCE OF MIND

CHAPTER 1
THE ORGANIZATION EFFECT

According to astrophysicists, the current instantiation of the universe began about 13.7 billion years ago with an explosive expansion, known affectionately as the Big Bang. Since that moment, energy-matter states have been expanding and interacting. New kinds of aggregate organizations with emergent properties continue to take form. Particles, elements, galaxies, stars, planets, and living organisms have all emerged in branching chains of this process. Yet the closer these phenomena get to consciousness, the more uncomfortable we seem to be with the concept of emergence. We know that mind and consciousness have emerged within living systems, and yet they sometimes seem magical and unscientific. Are the emergent properties of mind connected with the properties of matter? Should we characterize them as something real? What is necessary to create a conscious agent? As we attempt to answer these questions, our strategy will be to follow Einstein's maxim to keep our answers as simple as possible, but without oversimplifying them. Our starting place will be the concept of emergence, the idea that organizations with novel interactive properties routinely come into existence. Emergence is not some strange exotic effect. It is the normal process by which nature evolves. We are agents with emergent properties. If we are going to understand ourselves, we must become more comfortable with this idea.

THE ORGANIZATION EFFECT

There is really only one science, and the various "special sciences" are just particular cases of it. This is a magnificent ideal; it is certainly much more nearly true than anyone could possibly have suspected at first sight; and investigations pursued under its guidance have certainly enabled us to discover many connexions within the external world which would otherwise have escaped our notice. But it has no trace of self-evidence; it cannot be the whole truth about the external world, since it cannot deal with the existence or the appearance of "secondary qualities" until it is supplemented by laws of the emergent type.... If we take the mechanistic ideal too seriously, we shall be in danger of ignoring or perverting awkward facts of this kind. This sort of over-simplification has certainly happened in the past in biology and physiology under the guidance of the mechanistic ideal; and it of course reaches its wildest absurdities in the attempts which have been made from time to time to treat mental phenomena mechanistically. – Charles Dunbar Broad, *The Mind and Its Place in Nature*, 1925

Experience shows that simpler stuff can sometimes come together to form something with largely different properties. For example, the element sodium is a soft, silvery metal that reacts violently with water. It will ignite spontaneously in moist air or oxygen and produces severe burns on contact with the skin. It is commonly stored under nitrogen or kerosene cover to prevent it from coming in contact with the air, and it must be handled carefully when removed from cover to avoid injury. Potentially even more dangerous is the element chlorine because, as a gas, it's harder to control. When used as a weapon, this pungent, yellowish green gas causes acute respiratory damage to anyone who inhales it, and even brief exposures can be fatal. Yet the chemical combination of sodium metal with chlorine gas produces a substance with fundamentally different properties from those of its components. Sodium chloride is a solid white crystalline compound which is routinely stored in our homes in containers with porous tops. We even sprinkle it on our food. It's commonly known as table salt.

The topic which philosopher Charles Broad introduces in the quote above is often referred to as *emergence*, the observation that component parts sometimes interact to form combinations with novel properties.[1] The term "emergence" was actually introduced by the philosopher George Lewes over fifty years earlier, when he attempted to distinguish between small, combinatorial chemical changes and large, discontinuous changes.

> Although each effect is the resultant of its components, we cannot always trace the steps of the process, so as to see in the product the mode of operation of each factor. In the latter case, I propose to call the effect emergent. It arises out of the combined agencies, but in a form that does not display the agents in action.[2]

Broad argues that the appearance of new properties in this manner effectively requires two types of natural laws: *trans-ordinal laws* and *intra-ordinal laws*. Trans-ordinal laws explain the interactions which lead to the emergence of "higher-level" substances from "lower-level" ones. They explain, for example, the principles by which sodium and chlorine combine to form a new substance compound. Intra-ordinal laws, in contrast, describe the behavior and properties found within each individual substance. There are separate intra-ordinal laws that describe the behavior of sodium and chlorine, and yet other intra-ordinal laws that describe the behavior of the compound which they form. However, the intra-ordinal laws for table salt cannot be readily predicted from the intra-ordinal laws of its components. There is a qualitative shift in the behavioral properties of table salt. That's why we consider it a new substance, or more generally, a new kind of organization.

Not every new combination of parts displays largely different properties. Combining salt with water produces a mixture that is closely related to the properties of its parts. Indeed, if we look closely, we find that the mixture has a few novel properties. It conducts electrical current better, and it freezes at a slightly lower temperature. However, many of its properties are readily traceable to its component parts. Certain combinations, in contrast, interact more strongly. When they do so, many of their interactive properties are bound up forming the new organization. The resultant interactive properties are therefore largely different from those of the parts. This is what I call the *organization effect*. The organization effect implies that the properties of a new substance often cannot be predicted fully from the properties of its parts alone, because what combinations the parts may form, which will be stable, and what subset of properties are subsequently displayed depends on interactions which occur in the context of the new organization. For example, elemental oxygen commonly combines to form O_2 oxygen gas molecules. However, under higher-energy states it may also combine to form O_3 molecules, which are commonly known as ozone. Ozone and oxygen gases have the same parts, but they have different properties because their parts are organized in different ways.

Computer scientist John Holland has explored the nature of emergent properties in some detail. Holland is probably best known as the father of genetic algorithms, computational procedures that enable modular programs to evolve new traits for problem solving much as living organisms do. Holland's thinking on emergence is summarized in two extraordinary books.[3] The first of these, *Hidden Order*, describes the underlying principles that allow genetic algorithms to discover new solutions. The second, entitled *Emergence*, describes the process more generally by showing how the interactions of component mechanisms, under certain constraints, can result in complex systems with new kinds of properties.

Although Holland's approach to emergence is based on computer simulations and mathematical analysis, he reaches conclusions which are very similar to those of Charles Broad, and a comparison between the ideas of these two men should help us think about emergence better.

Holland refers to the intra-ordinal laws which summarize the physical and behavioral properties of lower-level substances as *microlaws,* and to the intra-ordinal laws which summarize the properties of an aggregate compound as *macrolaws.* He notes that the macrolaws are always "constrained" by the properties of the microlaws below them. That is, they depend on those properties for their operation, so they cannot violate them to any great extent and remain in existence. Thus, the two sets of property laws are connected across levels, and in principle, the macrolaws can be reduced to descriptions in terms of microlaws *and* the conditions under which the microlaw parts come together and interact. Holland's microlaws and macrolaws correspond to Broad's intra-ordinal laws at each level, and the conditions under which the aggregate takes form and remains stable are part of what Broad calls trans-ordinal laws, the laws which connect the two levels.

The primary difference between these two approaches is that Holland provides a more formal treatment and emphasizes that the glass holding our view of nature is half full – nature is connected. Macrolaws emerge from the interaction of particular combinations of microlaws under certain conditions. In contrast, Broad emphasizes that the glass is half empty – nature is discontinuous. New substances have largely different interactive properties. They require their own special laws to characterize their behavior. Although Broad and Holland take different perspectives, their different emphases seem superficial once Holland adds that the sheer complexity of the task makes it infeasible to derive most macrolaws – they are only derivable *in principle* – and once Broad adds that the existence of trans-ordinal laws connecting the various special sciences has proven to be "more nearly true than anyone could possibly have suspected at first sight'.

To put it another way, all the phenomena of nature may be interconnected, but the laws of each new substance organization are rarely obvious. The laws of each new organization must be discovered. The sheer complexity of the combinatorial possibilities makes predicting the behavior of new substances infeasible. Most of the derivations of macrolaws occur in after-the-fact attempts to reverse engineer the nature of the interactions that lead to a macrolaw phenomenon. We see the regularities and describe the properties of a new substance first, without knowing how it is derived. Later, we identify the trans-ordinal laws that explain how the regularities might be connected with other phenomena, a process that sometimes requires decades or even centuries of human endeavor. It is true that with practice experts can sometimes anticipate certain emergent effects based on their similarity to effects they have already learned to recognize. The fact that they can do this at all suggests that Holland's insight is correct – nature does seem to be connected. However, Broad's insight captures the fact that new organizations tend to display properties that are often discontinuous from those of their component parts.

CONSTRUCTIONISM AND REDUCTIONISM

Contraria sunt complementa.[4]

There are two seemingly opposing ways of thinking about the organization of the world. One is a constructionist approach. The other is a reductionist one. Constructionism is the holistic position that emergent properties result from, and can only be fully explained within, the behavior of aggregate organizations. The organization effect follows from this viewpoint. Reductionism is a simplification strategy. It assumes that all the phenomena of nature are connected across levels, and it seeks to explain phenomena on higher levels by showing how they depend on the interaction of their component parts. Reductionism has proven to be a highly productive strategy in science, to the point that it is sometimes argued that everything can be explained by reductionist theories. Some even go so far as to conclude that reductionism is the more fundamental strategy for understanding the universe, and that constructionism is unnecessary, or even unscientific. This reductionist emphasis sometimes leads its proponents to conclude that complex phenomena are really nothing but the interaction of their component parts.

As we have already noted, the paradox of this *nothing buttery*[5] argument is that it seems true in the sense that each emergent organization must necessarily depend on the interaction of its parts. Yet it remains false because each emergent phenomenon can never be fully explained by rules derived only from the parts alone. The argument that a table is nothing but wooden planks and fasteners has lost contact with the organizational properties that make a table a functional, higher-order organization. A table has properties that unorganized wood and fasteners do not. The argument that mental experiences are nothing but the interaction of neurons makes a similar error. A tangle of neurons is not a mind. It takes a special organization of neural functions to bring the properties of mind into existence.

What also seems to be lost in the emphasis on reductionism is the recognition that reductionism is the mirror image of constructionism. You cannot have one without the other. As the quote above suggests, reductionism and constructionism are contrarian complements. The value of reductionism is in its ability to connect emergent phenomena with underlying mechanisms. However, reductionist investigations would be of little value if there weren't emergent properties to be explained. To be useful, reductionist theories first need to be enlightened by the discontinuous behavior of higher-level organizations. It is only after interesting macrolaw phenomena have been identified that attempts can be made to trace them back to interactions on lower levels and explain something about how they operate. After such reductionist connections are discovered, it is sometimes then possible to impose changes on the lower-level components and thereby affect the behavior of the higher-level system.

This re-constructionist turn is what makes reductionism useful. Reductively informed re-construction is routinely used in attempts to re-engineer emergent phenomena. However, it is often difficult to discover low-level modifications that transfer back to higher-level systems in an expected manner. Medical researchers routinely try to reduce health-related problems to simpler biochemical interactions, which they often attempt to treat with medicinal compounds. This is the foundation for much of modern pharmaceutical research. However, finding a compound that can alleviate a particular medical condition is notoriously challenging. Large, aggregate systems behave in marvellously complex ways. What looks promising in the test tube may have far different effects within a living cell. What works in mice may not work in humans. Indeed, the same medicine that works in one person may not work for another, or it may have unexpected side effects.

Using reductionist links to re-engineer the behavior of aggregate systems has proven to be difficult because aggregate systems often involve interactions on multiple levels, none of which can be fully predicted from the behavior of their component parts. As Nobel laureate Phillip Anderson succinctly summarized the problem, *more is different.*

> The main fallacy in this kind of thinking is that the reductionist hypothesis does not by any means imply a "constructionist" one: The ability to reduce everything to simple fundamental laws does not imply the ability to start from those laws and reconstruct the universe. In fact, the more the elementary particle physicists tell us about the nature of the fundamental laws, the less relevance they seem to have to the very real problems of the rest of science, much less to those of society.

> The constructionist hypothesis breaks down when confronted with the twin difficulties of scale and complexity. The behavior of large and complex aggregates of elementary particles, it turns out, is not to be understood in terms of a simple extrapolation of the properties of a few particles. Instead, at each level of complexity entirely new properties appear,

and the understanding of the new behaviors requires
re-search which I think is as fundamental in its nature
as any other.[6]

Some reductionist thinkers accept that emergent properties occur but seem to dismiss their importance on the assumption that the defining characteristic of emergent phenomena is their unexpectedness rather than their discontinuity. Thus, they suggest that as knowledge is gained, emergent phenomena simply go away, because they are no longer unexpected. This leads them to conclude that with "complete knowledge," only reductionist explanations will be needed. This line of reasoning is generally consistent with the argument that in principle macrolaws can be derived from microlaws and the conditions under which they come together. However, we have already noted that as macro-level phenomena grow more complex, it becomes infeasible to derive them. They must be discovered. Further, the discontinuities we discover simply don't go away as we find connections across levels. They require new laws that remain tied to the constructionist organizations in which they emerge. The assumption that, with complete knowledge, only reductionist explanations will be needed may imply faith in the connectivity of nature, but it illustrates why reductionist explanations are incomplete. Complete knowledge, to the extent it can be approximated, requires knowing about all the constructionist organizations and discontinuous properties that need to be explained. Such an approach is not reductionist.

Compounding the difficulty of forming reductionist explanations is the observation that stable phenomena at one level can subsequently serve as building blocks for yet more complex phenomena. In effect, the macrolaws for one organization may become the microlaws for the next. As this occurs, each new level introduces more properties that are discontinuous from those of the previous levels. The changes across a single shift in levels may seem small enough to be largely explained on a reductionist level. However, as Phillip Anderson noted, any attempt to appeal to reductive explanations across multiple levels soon succumbs to the sheer complexity of the task. Although the microlaws associated with the component parts determine what interactions are possible, they do not predict which of many possibilities will occur or which will be stable. Other factors – such as the proximity and density of the components, supporting environmental conditions, and chance orders of interaction – also play a role in what organizations take form and remain stable.

LEVELS, FORCES, AND BONDS

Having provided a cursory introduction for thinking about emergence, there are three related ideas that we need to discuss a little more: 1) what we mean by levels of organization, 2) the idea that forces are also emergent properties, and 3) the interconnection of forces across levels. This is where thinking about emergent properties gets really interesting.

Levels

When we think about emergent systems, we inevitably refer to different phenomena as operating on different orders or levels. However, it should be clear that there is no strict hierarchy of levels in nature. Rather there is a web of interactions of varying complexity. The concept of levels simply follows from the value of tracking macrolaw regularities separately once they have been discovered.

> When we observe regularities, we often move the description "up a level" replacing what may be difficult or even infeasible calculations from first principles. These regularities still satisfy the constraints of the microlaws, but they usually involve additional assumptions. Under these assumptions the regularities persist and a simpler, "derived" dynamics can be used. These additional assumptions are usually described by phrases like "normal" or "natural" conditions.[7]

The practical value of viewing nature in orders or levels is apparent in the branching connections among scientific disciplines. The physical properties of atoms emerge from the interaction of a subset of quantum particles and forces. Chemical properties emerge from the combinatorial flexibility inherent in the bonding affinities of atoms. Biological properties emerge in large part from the incredible functionality made possible by chemical cycles operating within systems of protein chemistry. Psychological properties emerge in biological systems with complex neural communication networks. However, as Phillip Anderson noted, the fact that the sciences can be partially connected across levels does not imply that the phenomena in any one field can be fully explained by appealing to laws from another field. Biology is not merely an applied form of chemistry. Psychology is not simply applied biology. Each field involves new organizations which display novel interactive properties that do not exist in the others. That's why separate fields of study are needed.

Forces

> All systems are selected by their ability to utilize energy; and this energy – the ability to do work – is a "force," if there is any at all, in evolution. Liberally interpreted, selection does occur in the inanimate world, often providing a formative step in the production of order.[8]

To this point we have argued that new organizations display novel "interactive" properties. However, at the simplest levels of organization the interactive properties of particulate substances are what we term the fundamental forces of nature. So when we say that new organizations display novel interactive properties we are effectively saying that they express new kinds of forces. Matter and energy, particles and their forces, or more generally substances and their interactive properties, can never be completely separated from each other. When nature reorganizes both change. In fact, it is the change in interactive properties that causes us to categories a substance as different. Exactly how substances and their forces are bound together is not always clear. The process of combining component parts to form new organizations necessarily involves work. It seems that some micro-level forces are always bound up maintaining each new macro-level organization. Those forces that are not completely bound up in this process are free to be expressed, but they appear in new combinations; thus the substance displays new properties.

This is a fundamental process in the evolution of nature. The interaction of parts modifies both the physical arrangements that take form and the forces that new structures express. Following the logic of the organization effect described earlier, this implies that although macro-level forces must always be derived from the interactions of their micro-level components, there is no reason to assume that the resultant forces can be fully predicted from the component forces alone. Some component forces are bound up maintaining the new organization. Some are expressed in varying degrees and in new combinations. Some combinations may even be influenced by the structure of the new organization, making them more or less accessible. However, it's not simply that new forces emerge in the context of new organizations; but that once new forces emerge they have the potential to influence what subsequent organizations are likely to form.

Stuart Kauffman captures the dynamics of this process with a concept he calls the *adjacent possible*.[9] As Kauffman notes, at the fringe of every ongoing network there are features which are merely one organizational link away from becoming a functional part of the network. When, for some reason, one of those new functions happens to take form, then the interactive potential in the adjacent possible necessarily changes. Some possible connections become more likely to occur, and others become less probable. As a result, the patterns in which nature evolves are threaded and self-modifying. This is what makes predicting a series of constructionist changes so difficult. Each new organization has the potential to change what subsequent organizations are likely to emerge. If that were not enough, as organizations with interesting interactive properties accumulate, they occasionally come together in combinations that promote whole new trends in nature's tendency to organize. The most significant of these trends are whole new domains of selection.

What convinces me that this line of reasoning is correct is the observation that as nature evolves, novel forces routinely emerge and set new trends in the way nature organizes. Chemical reactions continually reorganize matter on earth, and yet the interactive properties of each chemical element did not exist at the moment of the Big Bang. Atoms and their interactive properties gradually took form as the early universe evolved.[10] Thermodynamic selection processes are not adaptive; they are merely directional. However, over time thermodynamic processes gave rise to interactive molecules that were capable of supporting the adaptive processes involved in natural selection. The biological drive to survive is a primal force in nature, but it only comes into existence among living organisms. Sexual attraction is a potent force in many organisms, but it only comes into existence in certain kinds of reproductive agents. Cognition is a powerful force for guiding decisions, but it only comes into existence in agents with complex neural networks. Status, maternal care, and social attachment are all forces that emerge in cognitively enhanced social organisms. They strongly influence the structure of the social networks in which they occur, and yet they do not exist within the domains of physics or chemistry, nor can they be predicted from them. So although the evolution of new forces may not seem to be a necessary outcome of new organizations, observation suggests that it is.

Bonds

Proponents of reductionism sometimes seem to assume that all causal forces flow from the bottom up. However as John Holland notes, when the stock market begins to sell off, it becomes clear that a macro-level phenomenon, the behavior of the market, can also drive the behavior of its components, its individual investors. The logic is simple here: Whenever things are *bound* together in nature, there is the potential for downward causation. Once parts are bound together, then forces that tug on any one level must either break the bonds that hold the parts together, or the forces have the potential to influence the organizations on other levels. That's what it means to be bound together. We can never explain why the peacock displays its marvellous tail by appealing to the forces of chemistry or physics. Yet sexual selection has an effect on the physiochemical substances that give rise to the colors in the peacock's tail. There is a downward causal effect that cannot be explained on the basis of physiochemical forces alone, although it clearly influences them. The downward effect occurs because the selection of physiochemical coloring substances is bound together with the peacock's reproductive success. Change one and you change the other.

In a similar way, my biological subsystems no doubt provide upward forces that contribute to the needs which occupy many of my daily thoughts. I cannot go long without thinking about food, shelter, and the many basic needs that promote my biological welfare and survival. However, my thoughts also have effects on the lower-level substance domains of my body because there are bonds and interfaces across levels. My cognitive decisions engage the activity of neural planning circuits, which subsequently engage motor control circuits, which release transmitter substances that activate muscle fibers, which are powered by physiochemical reactions that ultimately involve a multitude of quantum interactions. Thus, when I discover that we are out of Guinness, a vast number of fundamental particles are abruptly drawn into massive displays of quantum interaction as the constituents of my physical body travel to the nearest liquor store. The movement of all those particles, in that particular trajectory, and with such vigor, simply cannot be explained without knowing about the macrolaws of human motivation by which access to good stout can gain control over motor decisions and the nested chain of interactions by which they are implemented. The movement of quarks and the taste of stout are bound together by those links.

And this is not an unusual case, as Michael Gazzaniga notes:

This sort of reality – the macroscopic level impacting the microscopic level – is all around us. In our everyday use of computers for word processing or spreadsheet analysis we use a macroscopic level of programming that in turn controls the microstructure of the computer. The emergent phenomenon is controlling the elements that generated and built it! The same applies to the brain.... We are in no way separate from the [micro-level] machine, but are only able to understand ourselves at the macro-level.[11]

CONCLUSION

The physical universe evolved as fundamental particles began interacting in the aftermath of the Big Bang. However, the universe is not static; it keeps changing. As particles combine and new structures take form, new interactive forces continue to emerge. This is why we've come this way. All the really interesting phenomena in nature – from atoms, to molecules, to complex proteins, to living cells, to neurons, to brain structures, and on to conscious minds – have properties that are largely discontinuous from those of their component parts, and yet they are all interconnected. If we want to understand emergent phenomena, we must become comfortable with the idea that the interaction of component parts can result in organizations with novel interactive properties. Further, we must become comfortable with the idea that the forces of nature can flow in both upward and downward directions. Both of these ideas are essential for understanding how consciousness emerges and why it is adaptive.

Emergence is obviously a much more complex topic than we can do justice to in one short chapter. However, I believe we have reached some important insights. Emergence is not some rare exotic outcome; it is the very process by which new organizations and forces come into existence. It's how nature evolves. Constructionism and reductionism are not conflicting viewpoints; they are inseparable and complementary ways of thinking. Though reductionism is a powerful tool for connecting phenomena across levels, evidence for connectivity across levels does not imply that macro-level phenomena are fully explainable by micro-level laws. The discontinuity between the behavior of component parts and the interactive properties expressed by more complex organizations is what causes us to categorize those phenomena as emergent, not simply the fact that their properties are unexpected. New organizations don't lose their discontinuous character once their properties are discovered. Emergence persists.

The link between emergent properties and changes in organization, what I have called the *organization effect*, has one more important implication. If changes in organization introduce new properties, then it follows that if we can reconstruct the essential elements of an organization, we should be able to duplicate its basic properties. This will be our working strategy as we explore the nature of mind and consciousness. To put this strategy in better perspective, we will begin by stepping back several billion years and exploring the sequence of events that led from the activity of living cells to the emergence of mind. Planning, cognition, aboutness, and representation are all emergent properties that we ascribe to minds. Our goal will be to discover what kinds of organizations are necessary to bring these properties into existence, because a conscious robot will need a share of all these properties.

CHAPTER 2
TUMBLING TOWARD COGNITION

Logic would seem to imply that the properties of new organizations must be a mix of the properties of their component parts. However, their properties often obey laws of a different order. This discontinuity is what gives them an emergent character. The discontinuity is also what makes new properties difficult to explain. An insight into the nature of emergent properties is what we have called the *organization effect*, the observation that when component parts interact, the organizations they form influence the dynamic properties of the whole. It follows from this observation that the properties of new organizations cannot be fully determined simply by looking at their component parts, because the organizations do not exist at the parts level. The organizations and their interactive properties only come into existence when the parts assemble.

One property that has consistently played a role in the evolution of life is signaling – that is, the tendency of some organizations to produce outputs that can be detected by others. Life as we know it could not exist without signals to coordinate activity among component parts. Indeed, messenger RNA serves as an intracellular signal for what proteins to build in the cell. It should therefore come as no surprise to discover that extracellular signaling strategies gradually evolved to coordinate the behavior of multicellular aggregates. One group of mobile multicellular organisms even developed a class of cells, neurons, which specialized in rapid directed signaling. Neuronal signaling was largely just an extension of previous signaling strategies, but its speed and directedness, coupled with the ability of neurons to form forward-looking associations, resulted in yet another kind of selection process, cognitive selection. If we want to understand mind, then we need to look more closely at how forward-looking associations contribute to cognition and planning.

TUMBLING TOWARD COGNITION

Thought grows from action ... activity is the engine of change. – Esther Thelen, "Time-Scale Dynamics in the Development of an Embodied Cognition," 1995

I t's 3:00 a.m. and you suddenly awaken to the sounds of a violent storm. Warning sirens are blaring, and you think you hear the rumbling of an approaching tornado. To reach a place of safety, you quickly need to negotiate your way down the hallway, make two turns, and take shelter in the cellar. However, the storm has already taken the power out, and the pathway is pitch-black. You anxiously feel your way along the hallway, striking the wall as you miss the first turn and banging your head on a picture frame in the next hallway. As you approach the approximate location of the cellar door, you frantically grope for the knob. Once you find it, you quickly swing open the door and stumble down the stairway, scraping your shoulder on the wall as you go. Unless you are accustomed to moving without visual feedback, you probably consider this a successful transit and are quite willing to tolerate the bumps and bruises. However, one point is clear. In the absence of good sensory feedback, planning rapid movement can be an awkward and risky business.

Adaptive systems accumulate functions that serve biological needs. This gives them a directional character. Modern plants, for example, have many directional growth strategies and even behaviors of sorts, but except for a few rapid spring-like actions, such as the closing of the Venus flytrap, their actions are slow, and they cannot quickly coordinate separate actions with each other. Many animals, in contrast, routinely coordinate their actions in combinations that enable them to react in complex patterns and sequences. The fact that animals move in this way is so obvious that it hardly seems worth mentioning. Yet, it provides us with an important insight into the nature of mind. As a group, plants have adopted a largely stationary lifestyle. They depend on widely available commodities, such as sunlight, water, carbon dioxide, and access to mineral nutrients, and they can generally procure them quite efficiently by growing extensions that increase their contact with those resources. Following this strategy they are generally not motivated to engage in more costly relocation activities. In contrast, animals as a group have adopted a more mobile lifestyle. They move to procure resources, and if they have depleted resources in one location, they may move on to other locations.

However, as noted above, movement introduces special problems. In order to move efficiently from place to place, a mobile agent first needs to decide on a direction of travel. This requires sensory receptors which can resolve the location of potential goals. In many cases, successful locomotion also requires an ability to represent the spaces and substrates in the environment in a way that enables an organism to select action plans that fit environmental constraints. To do that an agent must be able to represent information about the both the objects in the world and its own size and position with respect to them so as to determine which features it can access and what spaces it can pass through. A large animal, in particular, needs to be able to make reasonable predictions about which routes to try and which to ignore. In short, planning is essential for efficiently managing movement.

Clearly sensory receptors and motor control systems are necessary for planning movement. However, what is the nature of planning, and how are plans assembled? To answer these questions, it is best to begin with a very simple planner. One of the simplest examples of sensory-directed planning can be found in the behavior of *Escherichia coli*, a bacterium commonly found in the human gut. Like many species of bacteria, the *E. coli* bacterium has several helical, thread-like flagella with a rotary microtubule mechanism at the base of each.[1] The flagella are capable of two locomotor actions. When the flagella begin rotating in a counterclockwise direction, they assemble together in a coherent rotating bundle which propels the bacterium forward in what is commonly called a *run*. These smooth-swimming runs are terminated by *tumbles*, which occur when the flagella briefly reverse their direction of rotation. The reversal disrupts the organization of the bundle and causes the bacterium to twist and turn abruptly. After a tumble the bacterium is typically oriented in a new, and largely randomized, direction from that of its previous run.

These two actions are what *E. coli* uses for its locomotor planning. Swimming runs are longer when the bacterium senses it is approaching a chemo-attractant or avoiding a chemo-repellent. Tumbling occurs more frequently when the bacterium senses a decrease in an attractant state or an increase in a repellent state. By randomly changing direction when attractants are declining or repellents are increasing, and by running longer when positive states are encountered, a bacterium can effectively home in on favored targets and avoid noxious substances.[2] However, sensing rising and falling gradients requires a conditional decision process with a short-term memory. How does a brainless *E. coli* remember previous gradient levels, and how does it use that information to make more adaptive action decisions? The answer to this question provides a fascinating glimpse into the origin of mind.

Let's begin with the memory question. An *E. coli* bacterium has specialized chemo-receptors on its surface, which bind with particular substances. One substance it finds attractive is the amino acid *aspartate*. Aspartate receptors have an external binding site which is anchored in the cell membrane. These external sites have several internal cytoplasmic methylation sites connected to them. The external binding site becomes active immediately whenever it contacts aspartate. The methylation sites become active through a chain of enzyme reactions. These reactions take a few seconds to complete. Thus, whenever the activity of the external receptor changes, activity in the methylation sites persists at the previous level for about three seconds. As a result, the activity in the methylation sites serves as a running index for the level of aspartate binding activity which occurred three seconds earlier. This three-second lag in the activity of the methylation sites provides all the short-term memory E. coli needs to decide whether a chemical gradient is rising or falling.

Consider how this short-term memory contributes to E. coli's decision process. The locomotor behavior of E. coli depends on the combined activity of the external receptors and the methylation sites. Activation of an external receptor results in a chemical signal which simulates the flagella to rotate in the counterclockwise turns, which propel it in locomotor runs. However, the activation of a methylation site has the opposite effect. It produces a signal which simulates the flagella to rotate in the clockwise turns that cause tumbling. The net effect is that the probability of running increases when the activity of the external receptors is higher than the activity in the methylation sites. Because the methylation sites track the external receptor activity in the prior three seconds, higher activity in the external receptor occurs when the aspartate gradient is rising. The probability of tumbles increases when the activity of the methylation sites exceeds the current receptor activity. This occurs when the gradient is falling.[3] The "decision" to run or tumble simply depends on the relative activity of these two receptor systems.

The emergent nature of this decision is readily apparent. Viewed as a reductionist mechanism driven by a chain of chemical processes, the decision process used by E. coli appears to be completely mechanical. The run versus tumble decision emerges from the activity of two competing processes. In that sense, the decision might be considered "nothing but" a difference in the activity of the two receptor processes. However, viewed as an adaptive decision process, the competing chemical processes can be seen to capture an important kind of knowledge. Bacteria that swim toward food sources are more successful than those which search randomly. When a combination of mechanisms enabled E. coli to make directional decisions, those mechanisms were favored by natural selection. E. coli's plans are not highly predictive in character. They do not represent goals in any complex way.

Yet their decisions are nevertheless forward looking. They link gradient changes to actions that increase the chance of finding food and avoiding toxins. Forward-looking decision processes are the essence of planning.

E. coli bacteria have one additional behavior related to gradient following which merits special attention. When stressed the bacteria release chemicals which attract other *E. Coli,* causing the bacteria to clump together in dense collectives called quorums. The dynamics of this quorum organizing process are fascinating. Studies show that the tumble frequency of an individual bacterium strongly depends on its position within the quorum and its direction of movement. Tumbles are reduced within the quorum. However, as a cell reaches the edge of the cluster, tumbling dramatically increases until a run sends the bacterium back into the quorum. This results in a tightly packed cluster with sharp boundaries.[4] The quorum stays densely packed because individual cells move toward stronger regions of the chemical gradient which, because the gradient density is a product of signals from all the cells, is stronger near the center and grows weakest at the edges.

Now consider the implications of this quorum-forming behavior. By releasing a chemical that other bacteria can detect, the *E. coli* are essentially signaling to each other. Each signal indicates that a bacterium has encountered some sort of stressful condition. The small quantity of chemical released by one bacterium has a negligible effect on group behavior, but if many bacteria detect the stressful condition, a gradient soon develops within the group. The formation of a dense collective then acts as a barrier, enabling the bacteria in the collective to better resist conditions, such as the presence of antibiotics or viral invaders. This is yet another case in which, as Phillip Anderson[5] noted, an increase in scale and complexity leads to the emergence of novel properties. The action plan of the quorum is a multicellular phenomenon. It can only be implemented by an aggregate of cells interacting with each other. Intercellular signaling enables the quorum to coordinate a group-level action plan.

Temporary quorums and cellular colonies have special properties due to the scale of their organizational structures and the increased complexity which their signaling introduces. Multicellular aggregates can become even more complex when their organization is based on divergent cell types, cells that have been crafted to behave and signal in different ways. However, the development of different cell types within the same organisms introduces something of a challenge for genetically guided adaptation. After all, each cell in the organism has the same genetic makeup. In order for different cells to develop different structures and behaviors, the same genetic organization has to behave differently in different cells. Variations in the development of cell types probably first emerged in cellular colonies, where the location of a cell influenced its behavior. For example, cells on the surface of the colony are more likely to come in contact with foreign substances, including those produced by invading organisms. If there are specialized adaptations for reacting to such events, the cells on the surface are more likely to express them. In contrast, cells in the interior of the colony are more likely to encounter substances produced by cells within the colony. As a result, they are more likely to express adaptations that favor interactions among cells. In effect, the way the cell develops depends, in part, on what it detects in its local environment.

It is well-known that environmental factors influence the expression of genetically guided adaptations, and that external events can partly determine what features come to be expressed. However, it is the nature of adaptive systems to discover ways to tinker with their own developmental designs. Thus, it was simply a matter of time before cells found ways to manage aspects of their own developmental environment. This happened when genes began to produce intermediate products, which served as signals to promote or inhibit the transcription of other genes. If a signaling molecule binds onto a segment of DNA, so as to interfere with the ability to other molecules to bind and make RNA copies of that region, then copying of that gene is effectively turned off. If the signaling molecule attracts and promotes binding with copying enzymes, then the gene is turned on in force. Molecules of this sort are called transcription factors. Transcription factors provide a simple but highly efficient method of managing the expression of genes by selectively starting and stopping the production of certain gene products. And if some of these transcription factors are transported outside the cell, then a cell may also influence the development of its neighbors.

Unravelling the nature of these gene regulation networks has become one of the challenges for modern genetic research. The theoretical roots of this work can be traced back to the nineteenth century, when researchers such as William Bateson[6] recognized that such networks were likely. Bateson's insight into homeotic organization was triggered by the occasional appearance of errors in segmental development. For example, he noticed that an added leg segment occasionally appeared in an insect where antennae normally should have developed, or that antennae occasionally developed in a crustacean where eyes should have been, or that a stamen might develop in a flower where a petal should have been. Bateson coined the term *homeosis* to describe cases in which one structure was transformed into another. The errors in these cases were not simply deformed versions of what should have developed, but rather the development of normal structures in the wrong places. This suggested that the error somehow caused a shift in an underlying developmental plan, implying that the plan must follow some sort of segmental rules. However, the mechanisms that guided this segmental plan remained a mystery for some time more.

In the late 1940s Edward Lewis took up the search for the mechanisms underlying homeosis in Drosophila. After some thirty more years of work, Lewis[7] was able to show that there was a complex of eight "master control genes" which, acting together, directed segmental development in Drosophila. When researchers were subsequently able to analyze the structure of these eight master genes, they were able to "box in" a common pattern of 180 base-pairs in the sequences forming each homeotic gene. This common sequence of code came to be known as the *homeobox* region. Subsequent work showed that the homeobox region codes for a spiral shape which helps a homeotic protein bind onto the certain regions of a DNA strand. Once bound the remainder of the protein then acts as a transcription factor to promote or suppress the transcription of the gene. All this works because the DNA base-pair codes have evolved to include marker sequences, which indicate where the copying enzyme (RNA polymerase) should begin copying, and upstream binding areas, called promoter regions, where transcription factors can influence whether the copying enzyme is attracted or blocked from that location.

A segmental development plan emerges from this organization because each homeobox gene is sensitive to a different gradient of signals, and the cytoplasm of the egg is seeded with a beginning gradient of signals and messenger RNA concentrations, which initiate the activation of the master genes. The fertilized egg, and its initial daughter cells, are omnipotent stem cells. They are capable of developing into any type of tissue. However, as the cells divide, those in different regions of the developing cell group encounter different signal concentrations, which promote or suppress the transcription of different genes. These changes begin to program the developmental trajectory of cells in each region of the organism. These less potent stem cells can only develop into certain classes of cells. These cells then produce and encounter other signals in highly specific locations and concentrations. This results in a gradient of signals, which turn specific sets of genes on and off in each region of the developing embryo. In this way, the cells in each region of the body are gradually programmed to have just the right set of genes activated, so that they will develop the internal structures and mechanisms required for their roles in the overall body plan.

The hox genes' ability to turn on or off whole sets of genes in a segmental manner provides a mechanism for managing the differentiation of cell types using the same DNA code. The cells in different regions simply come to be programmed to behave differently. These different cells types, in turn, make the emergence of multicellular organisms possible. However, in many ways the behavior of cells in a developing multicellular organism is not that different from the behavior of an *E. coli* quorum. The bacteria in the quorum also follow signaling gradients and signal to other cells. What makes multicellular organisms so different is that the interaction of signaling gradients and transcription factors causes cells with the same DNA to develop in different ways. Although collectives of *E. coli* can form one kind of quorum, the cells in a multicellular collective are capable of forming hundreds of different interacting quorums – cell groups, tissues, and organs. Each of these quorums has its own distinct properties, due to the different features that come to be programmed into the cells of each group.

The role that genes play in guiding development is truly impressive and sometimes leaves the impression that most genes do their work during these early developmental stages. However, development is only a small part of what genes manage. Once the basic body plan of an organism has been established, most of the genes managing development are turned off. However, there are many other genes that play a role in routine cellular activities. In fact, many genes are commonly turned on and off minute by minute to support specific cellular tasks. For example, some genes must be activated so nutrients can be metabolized. Others must be activated so membranes can be maintained. Yet others are needed to ensure that enzymes are replenished or that receptor sites are refurbished. In a normal day, tens of thousands of signals are required to turn particular sets of genes on and off to support these activities in each cell in the body. Signaling is essential for life.

In modern multicellular agents some cellular quorums even produce special long range signaling chemicals known as hormones. Peptide and protein hormones are small to medium chain proteins that attach themselves to receptor sites on other cells, where they activate internal messenger systems and enzymes which regulate the metabolism of the target cell.

These regulatory effects may last from minutes to hours. Steroid hormones are signaling molecules which have a distinct, four-carbon ring structure that enables the hormones to attach to receptors which transport them into the nucleus of target cells, where they turn specific target genes on or off. The effects of steroidal hormones typically last for hours, days, or even longer if the signal is maintained. Somatic growth, physical activity, sex drives, and immune system readiness are just of few of the processes regulated by steroid hormones. Hormones, in effect, provide a high-level interface for coordinating gene activity across multiple cells and organ systems. Multicellular organisms simply could not coordinate their many complex systems without such signaling processes.

FROM SIGNALING TO PLANNING

The potential for construction and coordination made possible by genetic signaling strategies is particularly obvious when we consider the development of the nervous system. As the vertebrate embryo begins to take form, two signaling molecules, Noggin and Chordin, are released by the mesoderm notochord, the region that will become the backbone. These signaling molecules induce the overlying ectoderm to thicken to form a sheet of neural stem cells known as the *neural plate*. As the neural plate receives more signals from the underlying mesoderm, a line of cells along its center begins to draw inward toward the mesoderm. This pulls it into a U-shape, which then fuses at the top to form a tube. This quorum of cells, now called the neural tube, goes on to form the central nervous system, the spinal cord and the brain. The center of the tube forms the central canal in the spinal cord and the ventricles, the chambers in the brain which hold the cerebral spinal fluid.

Hox genes contribute to the anterior-to-posterior differentiation of structures in the nervous system, and other signaling genes establish a central-to-lateral polarity for neural development. Cells along the dorsal edge of the neural tube, known as neural crest cells, break away and migrate across the embryo in response to these and yet other signaling gradients.

Two genes with highly memorable names have been implicated in neural crest cell migration. These are known as Slug and Sonic Hedgehog. Slug produces a protein that reduces the adhesions between the cells, thereby freeing them up for migration at just the right time. Sonic Hedgehog produces a signal that encourages migration and differentiation. Depending on other target signals, the neural crest cells differentiate to form glandular tissues, cranial nerves, and peripheral nervous system structures, including the sympathetic and parasympathetic nervous systems, the epinephrine-producing region of the adrenal glands, and the enteric nervous system, which manages digestion and the flow of material through the gut. Sonic Hedgehog also appears to be essential for cranial-facial development.

Gene regulation networks and their signaling systems make the evolutionary design of the nervous system a much more tractable task. In addition to cell migration signals, some signals turn on growth cones in the neurons. These growth cones result in the growth of cellular tubes that become the axons of the neuron. However, there is no need to provide each axon with directions regarding where to go. The growth cones simply follow signaling gradients. The design of the nervous system can be varied dramatically by modifying the expression of a few migration and growth signals. Modifications that cause a specific transcription factor to be produced for longer durations can cause a particular brain region to grow larger. Changes in the concentration of particular signaling factors can cause more extensive neural connections to develop or can extend the range over which interconnections develop. Changes that cause an established targeting signal to be produced in new regions will cause some of the neural growth cones to be attracted to the new target regions, thereby establishing new pathways. Code for the right signals and neurons will migrate to new areas and send out growth cones to make new sets of connections.

A fascinating point about the development of these neural networks is that the neurons don't need to behave much differently from the *E. coli* bacteria that we considered earlier. *E. coli* detect attractive chemical gradients and propel themselves toward them. Developing neurons migrate along chemical gradients, and once they take up residence in a new location, they send out migrating growth cones which propel themselves along connection gradients to form axons. When stimulated by certain conditions, *E. coli* release chemical signals that influence the behavior of other *E. coli*. When stimulated, neurons produce chemical reactions, which propagate along their axons and release chemical signals, called neurotransmitters, that modify the behavior of the neurons they contact. Collectives of *E. coli* signaling to each other result in information-processing quorums. Collectives of neurons firing together result in information processing circuits. The roots of sensitivity and signaling are present in *E. coli*. Neurons simply have more precisely crafted mechanisms for sending and detecting signals.

Neurons, however, do have better memory mechanisms than *E. coli*. Once activated neurons may change their firing behavior for periods of minutes or longer, and more important, they can form long-term memories as a result of this short-term activity. They do this, in part, by growing more dendritic branches and synapses to connect with the neurons that signal to them. In effect, neurotransmitter signals don't simply stimulate other neurons to fire; they also promote the growth and maintenance of synaptic connections. In fact, the mechanisms for synaptic development appear to be closely related to those for growth cone development.[8] The tendency of neurons to grow more contacts with neurons that frequently signal to them is known as Hebbian learning. It's a simple learning strategy named for Donald Hebb, the Canadian neuropsychologist who first proposed it.[9]

Hebbian learning is now recognized as a basic principle of all neural organization. Research indicates that the order of firing is important for Hebbian learning. If a presynaptic neuron is activated momentarily before a postsynaptic neuron, then the synaptic connection is strengthened. However, if the order is reversed, then there is no strengthening, and there may even be a gradual weakening. The aphorism, *neurons that fire in sequence, wire in sequence*, seems to capture the nature of this order-dependent learning process. Importantly, this order-dependent learning causes neurons to form forward-looking associations among themselves. These forward-looking changes in connectivity naturally lead to predictions, because whenever a sensory input becomes active, it will tend to activate connected sensory and motor neurons in proportion to the previous degree of sequentially correlated activity between them. To the extent that similar patterns of features are repeatedly encountered in certain situations, learning about past patterns of activity provides a simple strategy for predicting what is likely to happen when a situation recurs.

REINFORCING HEBBIAN LEARNING

Although Hebbian learning is essential for forming predictive associations, unguided Hebbian learning is not a very selective process. Some cases of sequential firing may be incidental and unimportant, and other cases may be critical predictors of success in a task. A neural system that learns about all coincident events equally is not going to be a very efficient planner. Therefore as nervous systems evolved, the more successful forms found ways to tune Hebbian learning to more relevant aspects of a situation. One common tuning strategy has been to generate added signals whose timing is correlated with important changes. The value of this multisignal reinforcement strategy is obvious. If near synchronous sequential signaling stimulates the formation of stronger neural connections, then releasing signals that increase the formation of associations when critical changes are detected should increase learning for associations that predict those changes.

Almost everyone recognizes that signals for rewarding outcomes tend to reinforce Hebbian learning. This happens because cells that fire in sequences that lead to rewarding outcomes have their sequences strengthened more than those that fail to produce rewarding outcomes. However, it is gradually becoming clear that other conditions also result in signals that make learning more efficient.[10] In effect, the reinforcement of Hebbian learning is a much broader process than originally assumed. Surprisingly, although the added reinforcement signals are broadly distributed in the brain, they tend to be produced in only a few localized regions. This arrangement appears to ensure that the timing of the added signals can be tightly controlled.

Researchers are still discovering some of the more specialized reinforcement signals for tuning learning, however a number of major reinforcement signals have now been well established. The most important of them can be characterized as signals for change, saliency, reward (wanting and liking), aversion (avoiding and disliking), emotional tone, and sociality. These characterizations are, of course, woefully oversimplified. Most of these signaling systems operate on multiple levels and involve several different receptor subsystems which serve varied functions. Many effects cannot even be characterized in a simple phrase. However, even these simplified descriptions provide us with a much richer view of reinforcement. Given that the reinforcing role of many of these signals is not well appreciated, a brief overview of how these added signaling processes enhance learning seems in order.

The primary reinforcement signal for *change* is the neurotransmitter norepinephrine. Norepinephrine is released in response to sudden changes, especially when they are novel or unexpected. The primary source of this transmitter is a small nucleus in the brainstem known by the Latin term for "sky blue place," the locus coeruleus. The name suggests that this location might be a seat of peace and calm. However, the name is derived from the bluish color produced by melanin granules in the region, not from its behavior. There's nothing calming about the activity of the sky blue place. It alerts other parts of the brain, including broad areas of the cortex, when a change in input has been detected. In general, the larger the change and the more unexpected it is, the stronger the signal which the sky blue place produces. This is important because associating stimulus changes with changes in outcomes turns out to be important for causal reasoning. Novel inputs often turn out to be related to unexpected outcomes, so noticing when a change leads to an unexpected outcome provides a simple strategy for enhancing causal learning.

Saliency is a term generally used to describe the intensity of processing that a particular cue receives. The initial saliency of a cue depends on the intensity of stimulus inputs and on initial receptor sensitivity. However, cue processing is also enhanced by a number of associative mechanisms which use the transmitter acetylcholine. There are several sites for acetylcholine release, and they promote cue processing in several ways. Brainstem acetylcholine is important for managing the circuits in the thalamus that control wakefulness and attention. These ensure that cues frequently associated with the timing of acetylcholine release get more processing. Acetylcholine released from the septum promotes memory storage. Saliency for memory storage appears to depend largely on the intensity of emotional reactivity states processed in the extended amygdala. As a result, stronger memories tend to form for events associated with more intense states of emotional reactivity. When recalled, memories tend to reengage the patterns of stimulus saliency and emotional reactivity that occurred when they were formed. This is what makes memories so adaptive. They prepare you for the kinds of situations that were associated with similar cues in the past.

Acetylcholine is also broadly distributed to the sensory and motor cortex from the basal nucleus of Meynert in humans. This region signals about what might be best characterized as *task-related saliency*. Many studies have shown that simply pairing a particular sensory cue or motor action with the release of acetylcholine from this nucleus results in the reorganization of perceptual and motor networks in the cortex so as to devote more cortical processing space to the cues and actions most often encountered in those tasks.[11] The reinforcement effects of these task saliency signals are related to what we think of as practice learning. They cause more cortical resources to be assigned to frequently encountered task cues, making them more discriminable. Similarly, they cause more cortical resources to be devoted to task-related actions, making their control more flexible.

The signals most commonly thought of as reinforcement signals are those for *reward* learning. However, reward signals actually come in two varieties, *wanting* signals and *liking* signals. The wanting signal is particularly important for response learning. The neurotransmitter that signals wanting is largely dopamine. The primary brain sites for the release of dopamine are two small regions near the floor of the midbrain. Dopamine release motivates search-and-contact activities and strengthens learning for activities that lead to reward. The timing of dopamine release is quite precise. As learning progresses, it shifts to earlier stages of an activity. This promotes learning of longer response sequences.[12] The neurotransmitters thought to facilitate liking are an opiod transmitter group known as endorphins. In general, this group is thought to signal states of pleasure and well-being, and to be involved in preference learning.[13] Liking signals are produced in the presence of a rewarding event, not in anticipation of it, and when released they tend to inhibit further seeking. In effect, wanting promotes seeking that lead to states of liking, and states of liking results in feelings of pleasure that reduce further seeking.

Aversion learning signals also come in two varieties, *avoiding* signals and *disliking* signals. Avoidance is not as easily tied to any single neurotransmitter; rather, it appears to work by activating dopamine sites that motivate seeking safety by withdrawing from states of dislike. Dislike is largely signaled by circuits associated with pain, distress, or disgust. One special transmitter for signaling about pain is substance P. Strong dislike situations also activate saliency signals that promote memories about the features associated with those states. These memories in turn motivate future avoidance reactions. The onset of states of dislike also commonly triggers signals from the sky blue place which induce stress reactions. All of these signals are highly adaptive ways of promoting avoidance learning in emergency situations, but they illustrate why punishment is often not the best form of discipline. Punishment does not simply produce the opposite effects of reward. The change alerting and stress reactions triggered by punishment often have unintended consequences – in particular, the formation of unwanted fear and stress reactions.

Emotional tone is a difficult process to characterize. It appears to act less as a discrete reinforcement signal and more as a sensitivity bias, which adjusts emotional balance between approach and avoidance dispositions. The neurotransmitter which appears to play a major role in this process is serotonin. For example, increases in serotonin tend to modulate pain and aggression, to reduce depression, and to promote confidence. Decreases in serotonin tend to produce the opposite effects. Serotonin is primarily produced in a series of nuclei along the midline of the brainstem known as the raphe nuclei, a term meaning seam. Given that serotonin has an influence on so many kinds of activities, it would be incorrect to limit it to a single function, like balancing emotional tone. However, being prepared to act with the best reactive disposition for a situation appears to have the potential to promote better learning and decision making. Given that orexin transmitters from the hypothalamus, which have been implicated in approach and avoidance signaling, also project to the raphe, it seems likely that serotonin is involved, at least in part, in setting the emotional biases that favor approach versus withdrawal in cases of uncertainty.

Another commonly overlooked domain of reinforcement signaling is *sociality,* that is, signals for social stuff. Although this kind of reinforcement is still largely overlooked, it is hard to overstate the importance of sociality reinforcers. Research suggests that sociality signals enhance motivation and learning in a variety of social contexts. In fact, they are particularly important in promoting attachment, caretaking, trust, generosity, and cooperation.[14] In mammals, the neuropeptides oxytocin, vasopressin, and some endogenous opiods are thought to be the primary social reinforcement signals. We'll focus here on oxytocin, which acts both as a neurotransmitter and as a signaling hormone. Consistent with its role in attachment, oxytocin is released in human females at the time of birth and is repeatedly released by breast stimulation during nursing. In fact, additional oxytocin receptors develop in females just prior to childbirth. This enhanced level of oxytocin signaling and receptor sensitivity is thought to be a core mechanism for promoting attachments between mothers and their infants. Oxytocin and endorphins are also released in both males and females during sexual contact. These signals promote attachments which are essential for maintaining stable parental care relationships.

Less intense but important forms of social contact include non-touch "contacts" such as smiling, eye contact, sharing attention, soft melodic voice tones, and even polite conversation. These contacts promote friendship, caring, trust, generosity, and cooperation, while reducing stress reactions. Given that females tend to be more prone to stress, these sociality signals are probably especially important for countering stress and for forming extended social bonds in females. Because oxytocin pathways also project to midbrain dopamine centers and hippocampal memory regions, social contact and approval also appear to enhance learning and memory in social situations. The tendency of sociality signals to promote social bonds while enhancing learning and memory in social situations is part of what makes us such sensitive social agents. We naturally form friendships, learn social skills, and remember important social situations. It does not appear that there are separate reinforcers for negative social situations. However, it is clear that negative social cues – frowns, glares, looks of disgust, and loud and rasping voice tones – often serve as punishers. It appears that negative social cues enhance avoidance learning in social contexts.

Delta Rules and Intensity Effects

There are yet other features of this multiple reinforcement signaling strategy which make reinforcement more effective.

As researchers Wolfram Schultz and Anthony Dickinson note,[15] in many cases learning systems have evolved algorithms which vary the strength of reinforcement signals in proportion to the amount of learning left to be resolved – or as it is more often described, in proportion to the *magnitude of the error*, or the *surprisingness* of the training outcome. As a result, the greater the difference between current associations and the consequence encountered, the stronger the reinforcement effect. This strategy has come to be known as the Delta Rule in artificial neural network algorithms, and such functions have also been reported in many regions of the brain.

> Neurons in several brain structures appear to code prediction errors in relation to rewards, punishments, external stimuli, and behavioral reactions. In one form, dopamine neurons, norepinephrine neurons, and nucleus basalis [acetylcholine] neurons broadcast prediction errors as global reinforcement or teaching signals to large postsynaptic structures. In other cases, error signals are coded by selected neurons in the cerebellum, superior colliculus, frontal eye fields, parietal cortex, striatum, and visual system, where they influence specific subgroups of neurons.[16]

The Delta Rule is one strategy for weighting the strength of reinforcement signals. In other situations – cases of stress and sociality, for example – the weightings appear to depend more on the intensity of the triggering events rather than on some measure of predictive error. The stronger the triggering event, the stronger the reinforcing signal. These intensity-weighted signals provide yet another strategy for tuning Hebbian learning to make the connections more relevant and efficient. In the case of homeostatic needs, the level of need may also influence the intensity of the signal.

CONCLUSION

Here is another insight into the emergence of mind. Many

adaptations, even the seemingly mechanical reactions of *E Coli*, have evolved with a forward-looking character. However, the ability of living systems to make forward-looking predictions took a major step forward when cells capable of rapid and directed signaling evolved. Neurons aren't simply capable of signaling; they can learn *when* to signal. Hebbian leaning, the strategy by which *neurons that fire in sequence wire in sequence,* emerged from the tendency of neurons to use signaling gradients to guide their connectivity. Neurons learn to predict repeating patterns of activity by growing larger and more numerous connections, called synapses, with the neurons that often signal to them. This greatly oversimplifies the processes by which brains learn, but it helps us think about how networks capable of forming forward-looking associations, trial-and-error response learning, and declarative memories evolved. Neurons have an inherent tendency to wire in sequences, and when other circuits evolved to the point that they could produce signals that reinforced this sequential learning strategy, neural networks gained the ability to fine tune their learning and become even more predictive.

Researchers are only beginning to recognize the scope of all these reinforcement processes in the brain, but the implications are clear. If we want to build smarter learning algorithms, then using multiple signals to fine tune Hebbian learning will be an important strategy. Change signals can favor associations between novel cues and unexpected outcomes, a simple but highly adaptive "assumption" that promotes causal learning. Saliency signals can favor learning and remembering for more intense cues and outcomes. Biological processes for adjusting receptor saliency provide a way for natural selection to tune learning to ecologically relevant cues, and we can use this strategy to tune our learning algorithms to ecologically relevant cues. Reward and aversion signals can favor learning for features that lead to positive internal outcomes, while avoiding aversive ones. Emotional tone signals can adjust approach and withdrawal dispositions to fit recent situations and current mood states. Sociality signals can facilitate social bonding and learning about things that promote social approval. Following this strategy, neurons that fire in sequence will still wire in sequence, but those that fire in sequences when these added reinforcement signals are most active will be the ones that form the strongest connections.

Chapter 3
Mediating Aboutness

Signaling plays a critical role in guiding the organization of living systems on many levels. The balance of signaling activity between two opposing detector-action systems enables a brainless *E. coli* to make forward-looking decisions. A cascade of genetically guided signaling gradients manages the differential development of cell types in multicellular organisms. Yet other networks of gene-guided signaling manage the day-to-day operation of cells. Hormonal signals produced in one region of the body temporarily turn coordinated sets of genes in other parts of the body on and off. Organisms that depend on movement for their survival have extended this signaling strategy to yet another domain. Neurons, cells that specialize in rapid directed signaling, enable neural networks to plan complex movements.

In the course of managing movements, neural networks have also found ways to represent objects and actions in their plans. These interim signaling states, or representations as we often call them, in turn enable more complex forms of cognition to take form. Normally we don't think too much about all the signaling that goes on in the brain; interim signaling is just what brains do as they prepare for action. However, there's something interesting going on here. Minds, as we experience them, wouldn't be possible without the ability to represent stuff. Yet exactly what enables representations to be *about* something has proven difficult to specify. This is where an organization effect approach can help. If aboutness is an emergent property, then there must be something special about the organizations which make aboutness possible. If we want to build a robot which can represent ideas as we do, then we need to understand what enables signals to be about something.

MEDIATING ABOUTNESS

> The name "representation" does not come from scripture. Nor is there any reason to assume that the various things we daily call by that name have an essence in common, or if they do that anything people have in their heads could conceivably share it. What is needed is not to discover what mental representations *really are* but to lay down some terms that cut between interestingly different possible phenomena so we can discuss their relations. – Ruth Millikan, "On Mentalese Orthography," 1993

Animate systems need some way of representing actions, objects, and their relationships in the environment in order to form effective action plans. However, philosophers have varied opinions regarding how those mental states should best be construed. The philosopher Franz Brentano[1] introduced the concept of *intentionality* into modern philosophical thought in an attempt to distinguish mental phenomena from physical relationships. An important characteristic of intentionality is what Brentano referred to as its "relation to content." Simply put, intentional relationships are representational – that is, they are *about* something. However, the aboutness is not limited to relationships in the real world. Brentano was intrigued by the fact that objects that did not exist or actions that were impossible could nevertheless be considered in the mind. Real actions require real objects to support them. You cannot slaughter a goose for dinner unless that goose is present. In contrast, beliefs and desires can be *about* events that are errant, nonexistent, or even impossible. You can imagine being a knight and slaying a dragon, even though you have not been knighted and have never encountered a dragon. This is the puzzling relationship which *aboutness* captures. Somehow mental representations are not rigidly tied to reality, yet they are nevertheless linked to it in some important way.

Although philosophers do not agree on how to think about intentionality, they nevertheless agree that it is essential for understanding mind, and they have developed a number of strategies for explaining intentionality. However, in doing so they have often sought to characterize Brentano's "relation to content" in terms of more complex representations, in particular in terms like *beliefs* and *desires*. Some have even claimed that in order to have beliefs and desires, an agent must first have the concept of beliefs and desires. Others have argued that having a concept of beliefs and desires necessarily requires having language. This would effectively restrict mental states only to agents with language. Yet others, recognizing that beliefs and desires are concepts we use when we reason about the minds of other agents, seem to want to equate intentionality with having a theory of mind.

It seems clear that all of these approaches assume that some form of mental representations is essential for intentionality. However, given that the nature of mental representations is not well defined, none of this really helps explain how intentionality works, although it may help explain why philosophers continue to struggle with the concept. To put these issues in better perspective, we will consider two modern approaches to thinking about intentionality. One highly influential strategy for thinking about intentionality that focuses on beliefs and desires is the formulation proposed by Daniel Dennett. Another model, with a largely different perspective on how intentionality comes into play, is that proposed by Ruth Garret Millikan. After a brief description of the ideas in both approaches, we will show how Millikan's model of intentionality is compatible with the organization effect. This will not resolve all the issues regarding intentionality that philosophers worry about, but it will help us think about the organizations which will be needed to build a robot which can have mental representations.

DENNETT'S INTENTIONAL STANCE

The philosopher Daniel Dennett[2] takes a unique approach to the explanation of intentionality. His central argument is that intentionality comes from making strategic interpretations, such as those that go into deciding if someone has beliefs and desires. These interpretations, he argues, are only objective in the context of some decision criterion. Thus, making sense of intentional relationships requires taking a particular point of view, or *stance*, just as a certain point of view is required to decide whether someone has talent or style.

> My thesis will be that while belief is a perfectly objective phenomenon (that apparently makes me a realist), it can be discerned only from the point of view of one who adopts a certain *predictive strategy*, and its existence can be confirmed only by an assessment of the success of that strategy (that apparently makes me an interpretationist).[3]

Dennett's concept of intentionality as outlined in his 1987 book, *The Intentional Stance*, has been expanded in more recent works.[4] For example, he proposes evolutionary stages of mind. Simple *Darwinian* minds have adaptations shaped by natural selection. *Skinnerian* minds add behaviors shaped by trial-and-error reward learning. *Popperian* minds add inner models that allow animals to test ideas before they act on them. *Gregorian* minds add cultural artifacts, such as tools and language, to their adaptive repertoire. Dennett even argues that the intentionality of mind is built up in layered organizations.

> We are descended from robots, and composed of robots, and all the intentionality we enjoy is derived from the more fundamental intentionality of these billions of crude intentional systems.[5]

However, though Dennett argues that higher-order intentionality is derived, whenever he discusses how intentionality leads to mental events, he inevitably returns to beliefs and desires. For example, he suggests that in order to assess whether an agent is rational, one must adopt an *intentional stance* in which one assumes that an agent has higher-order beliefs and desires, and then test how well that strategy predicts their behavior. As an illustration of this method, Dennett suggests applying it to the analysis of vervet monkey alarm calls. Researchers have found that the East African vervet monkeys (*Cercopithecus aethiops*) use different alarm calls for different types of predators, and they respond to these distinct calls with different behaviors.[6] For example, when vervet monkeys hear leopard alarms, they run into the safety of the trees; when they hear eagle alarms, they look up and hide in the bushes; when they hear snake alarms, they stand tall and begin looking down in the grass around them. The question is how best to decide what these alarms mean to the vervets? Dennett suggests it can be done by adopting an intentional stance.

In the case of the vervet monkey calls, Dennett proposes describing the monkeys' behavior with intentional idioms, such as "believes" and "wants", in a way that presupposes their rationality, and then testing to see if their behavior is consistent with that level of description. Further, he argues that the level of rationality which the animals display can be scaled in terms of the degree to which the intentional idioms can be meaningfully nested. Following this logic, first-order intentionality involves simple intentional relationships, beliefs, and/or desires. Second-order intentionality involves a hierarchical relationship among these representations – beliefs about beliefs. Third-order intentionality involves beliefs about beliefs about beliefs, and so on. For example, here's how Dennett suggests this approach might be used to scale the levels of intentionality in vervet monkey alarm calls.

> Fourth Order
> Tom wants Sam to recognize that Tom wants Sam to believe that there is a leopard....
>
> Third Order
> Tom wants Sam to believe that Tom wants Sam to run into the trees....

Second Order
Tom wants Sam to believe that there is a leopard....

First Order
Tom wants to cause Sam to run into the trees....

Zero Order
Tom (like other vervet monkeys) is prone to three
flavors of anxiety: leopard anxiety, eagle anxiety,
and snake anxiety. Each has its characteristic
symptomatic vocalization. The effects on others of
these vocalizations have a happy trend, but it is all
just tropism, in both utterer and audience.

We have reached the killjoy bottom of the barrel: an
account that attributes no mentality, no intelligence,
no communication, no intentionality to the vervet.[7]

As this example illustrates, Dennett's approach equates
mentality and intelligence with a representational hierarchy
of beliefs and desires. However, as researchers Thomas
Suddendorf and Andrew Whiten note, much of the intelligent
interpretations which Dennett ascribes to this nested hierarchy
can actually be done with *secondary representations*, as
proposed by Josef Perner.[8] The secondary representations
simply need to pose alternative interpretations of the world
which can be compared.

Perner's term secondary representation does
not refer to a level of recursion but to a new kind
of representation that is entertained concurrently
with another.... The advantage of simultaneously
entertaining a second model is, of course, the
ability to collate the two – that is, to bring them into
propositional relation.[9]

For example, an observer monkey may perceive that Sam
is making a leopard call, and the observer may be motivated
to run into the trees. However, at the same time the observer

may notice that rather than running into the trees, Sam is trying to steal food. These two views of the world could then be used to resolve the best course of action. Because these two interpretations motivate different reactions, a more complex resolution may result. For example, the observer might be motivated to threaten Sam and discourage him from stealing his food, while staying near a tree in case the leopard call is accurate. However, if the observer has been fooled by such tactics in the past, he may simply ignore Sam's call entirely and confront him at the food site. Note, however, that there is no need for the observer to conclude that Sam wants him to believe there is a leopard before he can counter Sam's deception. Competing concurrent representations of Sam's behavior are enough to motivate a resolution. Indeed, when we are being deceived, it is often such competing representations that motivate our reactions. We are told one thing but see evidence of another. Most of the "I think he wanted me to believe" descriptions only come into play when we retell the story. We usually don't think that way in the heat of the moment. It's typically not needed, and given that it takes more cognitive resources to reason that way, it would usually be a waste of time to do so.

Dennett's writings on intentionality have been widely cited, and many have come to accept the idea that a nested logic of beliefs and desires is how we should think about intentionality. However, the intentional stance has not proven to be a very productive research strategy. The nested idioms are difficult to apply, even to our own experience, and they just get in the way when we attempt to think about simpler kinds of systems. Indeed, in trying to follow this logic, some are left claiming that simpler "intentional" systems, like thermostatic controls, must also have beliefs and desires. In effect, they claim that the thermostat *believes* that the temperature has dropped and *wants* the furnace to increase the temperature. However, for most people this strategy only seems to make thinking about intentionality more confusing. It is not that beliefs and desires are not sometimes useful categories for thinking about the representations and motives of other agents, but they don't appear to be at the core of many representations, and applying intentional idioms to simple mechanisms like thermostats doesn't seem to help. Something is missing in this analysis. Before we can make more sense of the concept of intentionality, we need to explain how aboutness works in simpler systems, and how we get from simple aboutness to beliefs and desires.

MILLIKAN'S INTENTIONAL ICONS

A more basic approach to intentionality can be found in Ruth Millikan's thinking about how communication systems work. As Millikan[10] explains, intentionality does not require representations which are as complex as beliefs and desires; it merely requires a functional signaling system. She notes, for example, that even the calls and innate signals used in animal communication are "intentional" in the sense that Brentano proposed because they are *about* something and they have an "intentional content" that is separable from the physical world. For example, such signs can refer to things that are errant or nonexistent. A juvenile vervet monkey may produce eagle calls when the object sighted is merely a falling leaf, and yet others may react as there was a real eagle.[11] As Millikan explains, even the waggle dance that a honey bee scout uses to signal food locations has an aboutness to it which is not directly tied to the physical world. As research has shown, if the bee is confused, and the orientation of its dance is errant, then new foragers will be led to fly in the wrong direction.

Millikan uses the terms *producer* and *consumer* to describe the sender and receiver roles in a signaling exchange. These terms emphasize the active role played by both parties. In what turns out to be an important insight, she notes out that in order for a system of signal usage to evolve and to be maintained, it must serve an adaptive function for *both* the producer and the consumer. If there is not a benefit for both parties, a *proper function* in Millikan's terms, then the party which fails to benefit will have no reason to continue to use the signal. While these arguments are often phrased in terms of natural selection benefits, they are not limited to evolved signaling systems. The same logic applies to signaling maintained by learning or by agreed-upon conventions. For signal usage to be maintained, it must provide sufficient benefits to both the producers and consumers so that they will continue to generate and react to the signal for that end. It is only under those conditions that the signal has an intentional content.

There are three supporting concepts which Millikan uses to explain how intentional content emerges in signaling systems.[12] The first of these is what she calls a *mapping.* Simply put, a mapping is some connection between a detection process and some adaptive reaction. In animal communication, this is typically a neurological connection between a receptor-recognition system and a response system which results in a signal-related trait. Take the example of the European rabbit's danger thumps. The danger thump includes a producer mapping, which results in signaling behavior. That is, whenever a potentially dangerous condition is recognized, neural signaling between danger recognition processes and the rabbit's motor system motivate the rabbit to stomp its hind feet. The consumer mapping involves a complementary signal-related trait such that whenever a rabbit hears a foot stomp, it becomes alert, looks around, and, if the consumer rabbit is in the open, it seeks cover.

Millikan's second insight, and a key to understanding aboutness, is that the communication signal serves as a *mediating event* between the traits supported by the two mappings. That is, the signal serves to coordinate the detection and signaling behavior of the producer mapping with the reaction to the signal managed by the consumer mapping. In the case of the rabbit, the thumping sound is the mediating event. The producer trait, stomping when a dangerous condition is recognized, produces the sound. The consumer trait, running for cover, is triggered by the thumping sound. Thumping coordinates the producer's detection with the consumer's reactions and, in doing so, mediates a two-part adaptive function. The mediating event is separate from the function of the two traits, but it plays an essential role in coordinating the traits in an adaptive combination. From the point of view of the producer and consumer systems, the signal is *about* the functions it mediates.

The third idea which Millikan emphasizes is what I call the *adaptive alignment* of traits. It is based on Millikan's recognition that in order for a signaling strategy to be effective, the producer and consumer traits must be aligned so that they play complementary roles in a function which provides mutually beneficial outcomes. In the case of innate communication systems, this means the two traits must co-evolve so as to work together. For example, the danger thumps produced by the European rabbit are highly adaptive, because the rabbits have evolved a thumping action pattern, which is released by signs of danger, and because they have co-evolved alerting and cover-seeking dispositions, which are released when they hear thumping. The producer and consumer traits work together, and the combined function is beneficial to both roles. No single rabbit takes much risk in thumping, and yet all benefit from the observations of each other. However, if the two traits weren't aligned in this way, there would be no adaptive outcome.

When we put Millikan's ideas within an organization-effect framework, it becomes clear that the organization which gives a signal its aboutness is not just the producer or the consumer traits; rather it is the co-aligned organization formed by the combination of the two. As Millikan notes, it is the interaction of traits in co-aligned function, not the organism's "thoughts," which provide a signal with its sense of meaning. The rabbit's thump is *about* avoiding dangerous things. The rabbit shows that it knows this meaning on an adaptive level by producing thumps when dangerous events appear and by reacting to the thumps it hears by seeking cover. If the rabbit stops and reflects on the function of the signal, when the thump occurs, the meaning of the signal is not changed. The thump still mediates the same function. However, thinking about the function of a signal – that is, re-representing it in the context of other activities – sometimes enables the signal to be linked with yet other functions.

There is one additional feature of a social communication system that requires special consideration. When we examine a signaling process from the perspective of the producer and consumer agents, it becomes obvious that there are two aspects to the aboutness of a signal. The producer detects an event and produces a signal. The consumer detects the signal and reacts with some complementary action. This means that the signal has two subcomponents of meaning. To put it another way, the aboutness of the signal points in two directions. The referential meaning of a signal depends on the mappings that cause the producer to signal. The consequential meaning depends on how the consumer reacts to a valid signal. However, the consumer is always less certain about the referential conditions and the producer's motives, while the producer is less certain about what consequential actions a consumer may take. This uncertainty adds to the complexity of coordinating actions between two independent agents. In many cases, there may even be a need to evaluate the honesty of the signaler and the cooperativeness of the consumer, if those traits are not well-known, before signaling.

These producer and consumer aspects of meaning correspond to two common categories of word usage: *descriptive functions* and *directive functions*. Descriptive functions emphasize the referential meaning of a word – that is, what the signal producer has apparently detected. Directive functions emphasize the consequential meaning of a word – that is, what the consumer is expected to do when he hears the signal. Most signs, like the vervet monkey's leopard call, or a human's warning of fire, take on both descriptive and directive functions. Signal users often consider both meanings in their reaction to them. If Tom is far away and calls "leopard," then Sam may focus on the referential meaning of the call and look to see where the leopard is located, before he takes directive action. However, if Tom is nearby, then Sam is more likely to focus on the consequential meaning of the call first and climb the nearest tree before he considers the referential meaning. In the end, though, both meanings usually play a role in how an agent interprets a signal.

Millikan's analysis of signaling systems simplifies our thinking about them by breaking their "intentional content" into more fundamental components. A signal functions as a mediating event to coordinate the actions of producers and consumers. The mediation works because the producer and consumer traits have been aligned, by evolution or learning, so as to play a mutually beneficial function. These complementary roles are the key to understanding intentionality. The meaning of the signal is not rigidly tied to the physical world, because the signal merely mediates a function and may be subject to error or bias. Just as Brentano predicted, the "added something" which gives a signal its subjective character is not a physical thing. Aboutness results from the fact that the producers and consumers of a signal are adaptively aligned to use it to mediate a functional outcome. The meaning of a signal is a property of that adaptively aligned organization.

BEYOND SIMPLE INTENTIONALITY

By linking intentionality to adaptive signaling processes, Millikan has provided us with a simple yet flexible concept of intentionality. The aboutness of a signal has a biological character, as Searle suggested, because signals mediate adaptive outcomes.[13] However, intentionality is not limited to mental processes. Even simple bacterial cells like *E. coli* have internal signaling processes that perform adaptive functions. What we characterize as mentality emerges in exchanges among rapidly acting, highly directed neural networks which are capable of establishing interim signaling states that can be evaluated and recombined to guide decisions. The term *representation* has been used in many ways. However, Millikan suggests that it is best reserved for these combinatorial states of aboutness, because interim states and recombinations add new levels of complexity to aboutness.[14] Under such conditions, a combination of signals can represent more than the sum of its parts.

Think of it this way: Each producer-consumer exchange has a local meaning, but the meanings also interact so as to enable more complex combinatorial states to be represented. Consider what happens when our feline companion Tom steps on a thorn. A pain receptor in Tom's paw detects the thorn piercing through the paw tissue and fires, creating a mediating event which releases a neurotransmitter signal at synapses in his spinal cord. That signal is referentially about the injury. Some spinal interneurons then send inhibitory signals to the extensor motor neurons. Those signals are consequentially about suppressing the stepping action that led to pain signals. Other spinal interneurons send signals to the flexor motor neurons. The flexor motor neurons in turn release a transmitter signal at the muscle sites they innervate. The muscles recognize the transmitter as a signal to contract, and contracting those muscles causes Tom to lift his paw.

This sequence of signals is adaptive, because each producer-consumer pair has been selected and aligned to use the signals it receives to mediate specific functions, and the collective of functions has been aligned to produce a well-coordinated sequence of actions. Meanwhile, other signals from the spinal cord have informed somatic status networks in Tom's brainstem about all this peripheral signaling activity. Those status networks send signals that impinge on central attention pathways, and when they gain attention, Tom becomes consciously aware of them. Tom interprets those signals as signs of an injury. That interpretation motivates him to pause and lick his paw. All the complex behaviors which you and I and Tom display emerge from the combinatorial interaction of coordinated signaling processes. Some, like the reflexive withdrawal of Tom's paw, have been arranged by natural selection. Others, like Tom's conscious decision to pause and lick his paw, have been arranged in part by learning and by high-order cognitive processes.

The same kind of coordinated functions emerges when

signals are exchanged between social agents. For example, the word "purple" has come to be *about* the color purple, because you have learned to use that label to direct your attention to a specific color range. The word "dog" has come to be *about* canine animals because associations with "dog" cause you to recall memories of that type of animal. I can even combine those words to activate associations for experiences that are nonexistent. All I need to do is draw your attention to the two associations at the same time, and you will think of purple dogs. However, those words only serve to trigger that mental representation if we have both acquired similarly aligned associations for the words. If we lack a common set of associations – for example, if you only understand Chinese words and I speak American English – then my saying "purple dogs" will not have any aboutness for you. In effect, signals only have meaning if they can coordinate agent systems in some related function. If the signals lack shared associations, then they are meaningless.

One obvious advantage of combining signals in interim states of aboutness is that what can be represented is more flexible. Some combinations may even be used to represent the states of other representations. This is the function of the representations we call beliefs and desires. However, such representations only tend to be used under certain conditions. For example, we might conclude that after Tom stepped on the thorn, his representations changed so that he believed he was in pain, and he desired to lick his paw. If Tom used language to describe his experiences, then it is likely that he would have reported about them in just those terms. However, it is unlikely that Tom would have taken the time to tell himself that he believed he was in pain before he licked his paw, just as you would be unlikely to tell yourself that you believed you were in pain before you attended to your injuries. Metarepresentations aren't usually needed for direct reactions.

Beliefs and desires, as we generally use the terms, represent states of confidence in perceptual and motivational dispositions. To declare I believe something is to say I consider my past perceptions reliable indices of the state of the world. To say I desire something is to say that I am aware of being motivated to take actions to gain that item. Occasionally I may even call some of these measures of confidence to attention as I attempt to reason about my own behavior or the behavior of another agent. Dennett suggests that rationality always depends on using such metarepresentations. However as we have noted, we don't need metarepresentations to explain many cases of rational thinking. The resolution of conflicting secondary representations is often enough. Thus we can usually reason without considering how confident we are about our representations.

A focus on beliefs and desires leaves out these simpler kinds of reasoning. In contrast, Millikan's explanation of how representations can be coordinated in adaptive functions makes it easier to understand how rational behavior emerges. Thinking in terms of coordinated, signal-mediated functions also enables us to make sense of simpler kinds of intentional systems. For instance, it allows us to think about thermostats without having to attribute beliefs to them. The functions which bring the thermostat-furnace organization together serve human needs, but there is no adaptive outcome for the thermostat or the furnace. They don't benefit from warming. They are simply tools that produce and react to signals. Humans arrange for signals from the thermostat to reach the furnace, and for the furnace to treat those signals as temperature related. Within that co-aligned framework, the signal has a kind of aboutness, which is meaningful *for the human designers*. However, the signal has no meaning to the thermostat, and the thermostat has no need and no internal processes to entertain beliefs about how confident it is in its representations.

Thinking in terms of coordinated function even helps us make better sense of Dennett's intentional stance strategy. Taking an intentional stance is to assume a functional alignment with respect to some signal. It is this alignment in function which enables producer and consumer systems to use signals to mediate tasks. Looking for a similar alignment can sometimes be a useful strategy to help test the function of a signal. However, a focus on beliefs and desires is not usually necessary. It's not that it's wrong, but it only accounts for a subset of rational decisions. It doesn't put mentality in context. Adaptations emerge from the interaction of signaling functions on many levels. Messenger RNA molecules serve as mediating signals. Neurons use coordinated signaling networks to represent events and actions. Mental representations in turn enable higher-order planning processes, a few of which may even include beliefs and desires, to take form. To paraphrase Dennett: We are descended from signaling systems, and composed of signaling systems, and all the intentionality we enjoy is derived from the more fundamental intentionality of those billions of crude signaling systems coordinated in the service of more complex functions.

CONCLUSION

Admittedly we have ridden fast and hard through these ideas, and we have left out the contributions of many thinkers on intentionality. However, our goal has not been to provide a historical overview of the thinking on intentionality, but rather to frame the concept in ways that cut between interestingly phenomena, so we can think about them more clearly. To that end, I think we've gotten most of what we came for here. In Ruth Millikan's thinking about biological signaling, we've found an account that connects intentionality with natural selection and learning processes. This account helps us understand the nature of the organizations which bring aboutness into existence. Aboutness emerges when co-aligned systems use a signal to coordinate separate activities in a combined function. The meaning is not some internal property of the signal, but rather a property that emerges from the way co-aligned systems use the signal. Millikan's model also helps us recognize why signals have two flavors of aboutness. The meaning of a signal depends both on the referential traits that the producer of a signal brings to the communication process, the events that cause her to signal, and the consequential traits that the consumer brings to the process, as well as how he reacts when he encounters the signal. The implications are clear. If we want the signals we use to have meaning to our robots, then the robots must have recognition and action systems that are aligned to use the signals to mediate similar adaptive functions. Only then can our signals have similar meanings to them.

Meanwhile, our feline friend Tom has stepped on another thorn, and signals between receptors in his paw and somatic status networks in his brain have mediated a shift in his attention. As this happens, Tom pauses as if to ponder on all this signaling. Could it be that simple? Does the meaning of a signal simply emerge from its ability to mediate an outcome by connecting separate functions? Does rationality emerge from the ability of layers of co-aligned neural organizations to combine signaling functions to achieve yet more complex, adaptive ends? Can signals enable you to communicate aspects of your internal representations to similarly coordinated agents? We can't be sure what Tom thinks about all this signaling, but there is obviously something adaptive going on here. At least, Tom seems to feel better now that some of those attention-getting pain signals have him licking his paw again.

CHAPTER 4
A LATTICE OF KNOWERS

Signs, words, and neural transmitters are all kinds of signals. When we assume that *aboutness* is a property of such signals, then it seems altogether mysterious. How can a physical event come to contain such properties? Where are they stored? Something seems to be missing in this characterization of aboutness. However, once we recognize that aboutness results from the role that a signal plays in coordinating traits between two co-aligned parties, then representational qualities become more understandable. The meaning of the signal is not contained within the signal. The meaning emerges from complementary producer and consumer traits. It is only when the producer and consumer use a signal to mediate co-aligned functions that a signal conveys a meaning.

This chapter explores how co-aligned networks in the brain interact to create layered knowledge systems. Thinking about the organization of the brain is a considerable challenge, and it is easy to be intimidated by the task. In order to keep our discussion as simple as possible, we'll start with a functional overview and dig down into structural details in later chapters. The functional overview we begin with is a way of thinking about how different layers of knowledge work together. It's a network organization called a lattice hierarchy. Thinking about the interaction of knowledge managing adaptations within the lattice will help us explain a few more emergent properties of mind, including what it means to act with intention.

A LATTICE OF KNOWERS

That the middle motor centers represent over again what all the lowest motor centers have represented, will be disputed by few. I go further, and say that the highest motor centers (frontal lobes) represent over again, in more complex combinations, what the middle motor centers represent. In recapitulation, there is increasing complexity, or greater intricacy of representation, so that ultimately the highest motor centers represent, or, in other words, coordinate, movements of all parts of the body in the most special and complex combinations. – Hughlings Jackson, "Evolution and Dissolution of the Nervous System," 1884

One summer afternoon, shortly after I moved into a house on the outskirts of Bloomington, I was walking through my backyard when I noticed an unusual blob of something on the ground directly in front of me. Almost immediately I felt strangely alerted and froze in place. Then the blob moved ever so slightly, and my perception coalesced. The blob was a loosely coiled snake. I quickly took a few steps back for safety and then began looking more closely at the markings on the snake. In short order, I concluded that this was an Eastern garter snake. I then retrieved a stick and escorted the snake back into the woods behind the yard. All in all, this would have been a fairly uneventful incident were it not for my subsequent recognition that cognitive processes on at least three separate levels of my mind had detected the snake, that each process had taken a different amount of time to resolve the situation, and that each had resulted in a different action. Although the processing stages are not always quite so separated and obvious, this sort of layered decision processing goes on constantly in the brain.

Our aim in this chapter is to provide a framework for thinking about various kinds of knowledge and how they are managed in the brain. There are two features of brain organization which we need to consider in this process. The first is the fact that knowledge in the brain tends to be modular – that is, localized. The second is that control functions within the brain tend to be hierarchical. In order for both of these to be true, it follows that knowledge has to be re-represented across levels of control, as Hughlings Jackson suggests in the quote above. Jackson was describing the hierarchy for motor control. However, it turns out that a similar hierarchy occurs within perceptual and motivational domains. This control hierarchy has implication both for how knowledge is represented and for the time it takes to make different kinds of decisions. It also has implication for what we think of as intention. So we have a lot to cover. We'll begin with a description of the modular and distributed nature of knowledge. We'll end with a hierarchical description of intention.

A COGNITIVE LATTICE

As Hughlings Jackson suggests, brains don't simply evolve by adding new knower modules; they add modules in different control layers. The basic segmented structure of the vertebrate brain – *hindbrain-midbrain-basal forebrain-forebrain* – is an ancient design. It is apparent even in the brains of primitive jawless fish.[1] This layered organization results, in part, because many somatic and sensory inputs are routed into the brain via connections in the brainstem. As a result the lower brain regions provide the first and most basic processing, and higher brain regions have the benefit of receiving preprocessed information from the lower regions. Because these higher levels depend on this preprocessing, brains tend to evolve by adding low-level knower modules first and then adding middle- and higher-levels modules to refine the lower-level inputs. To simplify the reprocessing of information, the higher-level modules tend to employ more flexible algorithms, like learning, memory, and reasoning processes, which reorganize the low-level inputs.

The lower-level modules, in contrast, tend to be more feature-specific and less flexible.

Randy Gallistel characterized this layered architecture as a *lattice hierarchy, a* structure in which control depends on both bottom-up activation and top-down modulation.[2] This latticed organization is found on many levels in the brain, including within the networks of the prefrontal cortex.[3] However, the lattice structure, which will be our focus in this chapter, is a hierarchy of cognitive decision processes, which range from reflexes, through pre-organized behavior systems,[4] habit learning, memory, and logical reasoning (see Figure 4-1). Not all the brain structures that make this cognitive lattice work are physically arranged in a lattice, but it will help to think about how they work by recognizing that functionally they form a latticed control structure. In general, the knower modules near the bottom of the lattice tend to be faster and more biologically prepared, whereas those at higher levels tend to be slower and more adaptively flexible.

The hierarchical structure of information processing in the brain has led some to suggest that the upper layers and their top-down control effects are the key to higher intelligence. Some even assume that these higher-level structures have largely taken over control from the lower-level processes, making the lower-level processes less important or even obsolete. In particular, it is common to hear the claim that many innate behavioral dispositions have been replaced by learning and reasoning in higher animals. However, as William James notes, the lower-level modules remain important precisely because it is the interplay among layers of knowledge that results in higher intelligence. Humans do not lose their biologically prepared dispositions just because they can also think logically. Rather, the lower-level dispositions serve as uniquely tuned traits that determine what features are most likely to be processed and passed on to higher-level faculties.

> Man has a far greater variety of *impulses* than any lower animal; and any one of these impulses, taken in itself, is as 'blind" as the lowest instinct can be;

but, owing to man's memory, power of reflection, and power of inference, they come each one to be felt by him, after he has once yielded to them and experienced their results, in connection with a *foresight* of those results.[5]

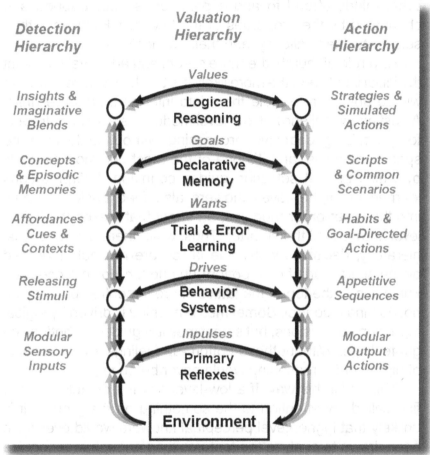

Figure 4-1: A lattice hierarchy of cognitive traits.

William James' description highlights why low-level modules remain so important in a lattice structure. The higher levels of the lattice receive most of their input from lower levels, with the lower levels serving as preprocessors for selected modules of perception and output processors for particular actions. Each higher level assimilates the lower-level sensory analyses into their representations, and similarly, each higher level assumes the functions of the low-level output modules and delegates its outputs through them. It follows from this structure that the quality of the decisions made within the lattice will necessarily be biased by the variety and quality of the inputs with which they begin, as well as the kinds of outputs they can execute. What minds attend to and expand on is bootstrapped and channeled by the properties of the low-level knowers – their selectivity, their saliency, and their connectivity.

As a rule of thumb, the more salient and adaptively relevant the initial inputs are, the more likely it is that higher-level knowers will be able to assemble them into intelligent adaptive units. Adjusting the saliency of low-level modules is a simple strategy for prioritizing cognitive processing without building more specific modules, and it's one of the most common methods by which natural selection guides cognition. Interpretations learned by higher-level knowers also feed back as inputs into the lower-order modules, an effect that makes the lower-level knowers all the smarter. Thus, while there is a functional hierarchy, decisions within the lattice are routinely triggered by the interaction of both bottom-up and top-down processes. Functions at the top of the lattice share in control, but they do not dominate control. Sometimes reasoning is driven by logical top-down conclusions, but sometimes it is guided by bottom-up gut feelings. Without those bottom-up feelings, many aspects of higher-order reasoning would never be engaged.

Think of it this way. If a low-level alerting module had not first called my attention to the something on the ground, it is unlikely that higher-level perceptual knowers would ever have been drawn to analyze the features long enough to recognize them. Initially, the snake was recognized by a biologically prepared module that triggered a freezing response. This strategy probably evolved to help early mammals avoid capture by reptilian predators, because moving cues are more likely to be noticed. With the activation of this knower module, I was then conscious of the presence of something. However, it was a crude low-level awareness which promoted caution but lacked memory recognition. It wasn't until the object moved that I – or more correctly, a part of my brain with more complex perceptual skills – recognized this something as the member of a specific threat category, a snake. With that categorical recognition in place, I engaged in well-learned avoidance behaviors. Then, with additional inspection and memory searches, yet other knower modules were able to refine the category of snake, reduce my fear with top-down motivational changes, and engage more reasoned decisions. However, the reasoning didn't start with the high-level analysis. Until lower-level knowers recognized there was something to be known, the higher-level knowers had no cause to be engaged.

This layered processing occurs even when we are engaged in what we think of as high-level cognitive activities like reading. For example, research has found that when a common word is presented visually for a brief duration, about 20 milliseconds or less, a person will not be able to recognize and say what word was presented. However, when they measure the physiological responses of the reader, they find that the reader reacts to the emotional meanings of many short words. This implies that more than one kind of word recognition occurs during reading. The parts of the brain that recognize and react to the emotional meaning of a word are not dependent on the parts that recognize more cognitive associations. In effect, just as I recognized the garter snake in several stages, we routinely recognize and react to the meaning of words in sequential stages, because knower modules in different parts of the lattice engage different aspects of meaning at different rates.

LATTICED PRIORITIES

An important function of the valuation hierarchy in the lattice is to motivate and prioritize decisions so as to reach resolution in a timely manner. In general, strong activations at lower levels of the lattice (impulses and drives) signify a greater sense of urgency and promote faster decision making. Because decisions are made faster at these levels, an agent's behavior tends to be more controlled by knowers on these levels when valuation states are stronger. When motive states are strong, we don't stop and think. We react impulsively with highly prepared actions and well-practiced habits. In contrast, we reason best when drive states are low to moderate, and when we have more time to explore memories and plan before committing to a decision. In effect, there is a constant interplay between the tendency of the top-down processes to suppress impulsive decisions, providing more time for reasoning, and bottom-up systems pushing to resolve need states quickly. This interplay adds to the dynamic nature of decision processing in the lattice.

A primary reason for the difference in decision speeds is the complexity of processing that goes into the decisions at each level. Lower-level systems of reflexes and drives make decisions based largely on threshold levels for particular inputs. Once a sufficient threshold is reached, they are drawn into reaction. The middle-level systems of biologically prepared actions and automated habits compete in the motor selection circuits of the basal ganglia. This competition takes more time to reach resolution than a simple threshold measure, but the algorithms here are still generally fast. In contrast, decisions involving memory searches are substantially slower. It takes about a half second to establish the network connections that support declarative memory, and memory searches often take several seconds more. As a result, under strong incentive states, drive-primed action patterns and habits tend to be favored over decisions that require memory searches and reasoning. Even when a decision is not resolved quickly, high drives may latch attention onto immediate cues, making it difficult to shift attention and explore memories with much efficiency.

Perhaps the account of a clever fox terrier called Peter will help to illustrate these relationships.[6] Peter had learned to open a wooden box by putting his nose under the latch and lifting it up. It was a game for him, and he was rewarded with praise and social attention. In a short time Peter became so skilled at doing this that as soon as the box was presented, he lifted the latch and opened the box. One evening, after this box-opening behavior had been well learned, Peter was sent to bed without his usual supper. Because he was usually fed but once a day, he was hungry the next day. That morning a freshly cooked bone was placed in the box. Peter was then allowed access to the yard where the box had been placed. Peter smelled the bone and soon approached the box. "When he saw the latch he ducked his head as if intending to lift it, but desisted. He then sniffed excitedly at the box and pushed it with his nose for some time. Eventually, however, he gave up and returned to the boot-room. After a few minutes, he came out again into the yard and sniffed in the same way at the box. Twice he pushed the latch from behind, but did not put his head beneath it. After a while he returned to the boot-room and showed no signs of revisiting the box." Later in the day, after the odor of the bone was diminished, Peter was again given access to the area. "He went straight to the box, lifted the latch in the most business like way, and took out the bone."

As this account illustrates, even though lifting the latch was a well-learned habit, Peter did not associate it with acquiring food, so when the food incentive was strong, he did not "think" to lift the latch. Instead, he persisted in smelling, rooting, and typical canine food-getting strategies. It was only when the saliency of food-related cues was weaker that he remembered his box opening skills. As the report goes on to explain, this same procedure was repeated two weeks later with similar results. When the bone presented a strong scent, Peter was distracted. He did not reason that the way to gain the meat was to lift the latch and open the box. Yet once the scent of the bone was reduced, Peter was able to open the latch immediately. After this second trial, however, Peter apparently connected latch opening with access to food, because on subsequent trials he was no longer so distracted that he failed to open the box. Once the new association was linked into the lattice, the smell of the food motivated latch-lifting behavior.

LATTICED DECISIONS

NASA has just landed a semi-autonomous roving geologist robot (acronym SARG), on the moon. SARG's mission is to explore the lunar terrain and look for potential sources of water and useful minerals. You have been assigned to manage this task. The work is similar to that of the older Martian rovers, but SARG is far more sophisticated. Like his Martian rover predecessors, he has a manipulation arm which can make contact with interesting geological features. SARG also has a wide array of tools and test probes at the end of his arm which he can employ in a number of established test protocols. However, there's a small control problem that needs to be resolved if you are to manage the rover's work efficiently. The moon orbits the earth at an average distance of around 239,000 miles. The speed of the radio signals that communicate with the rover is about 186,000 miles per second. This introduces a delay of nearly 1.3 seconds between when you send a command and when SARG receives and executes it. Further, the video information from the rover is similarly delayed, so what you see is what SARG observed 1.3 seconds earlier.

Given that you are unlikely to react to the input in less than half a second, the total delay between the video input you see and the time when your reaction to it is implemented by the rover is over three seconds. One strategy you might employ would be to move in short steps, wait three to four seconds to get feedback on your new location, and then move again. This strategy might be necessary if the visual input from the rover were so limited that you could not get a good view of the distant terrain. However, you have several high-resolution cameras to provide you with adequate visual feedback. Further, you have a large area to survey, so a step-and-wait strategy would be cumbersome and slow. Fortunately, with practice you can learn to think ahead and start turns before the rover gets too close to an obstacle. This strategy is not so different from driving a large boat: the faster the boat travels, the farther ahead you need to look to steer effectively, but with practice you can become quite proficient.

There are, however, problems with this strategy when there are both nearby and more distant obstacles to steer around. As the SARG acronym suggests, the solution has been to make many systems on the rover semi-autonomous and even to allow some of SARG's algorithms to learn based on their experience. For example, the rover has a neural network guidance system, which is trained before launching to operate in the kinds of terrains it is expected to encounter. In this training, the guidance network learns how to recognize usable paths and how to avoid common obstacles. It also learns to recognize and avoid dangerous conditions, such as holes or cliff edges. Similarly, it has a semi-autonomous traction network that detects wheel slippage and other navigational problems, and it compensates by automatically adjusting the grip angle or amount of weight on a particular wheel. It even has the ability to shift from wheel-powered driving to a slow, insect-like stepping pattern in difficult terrain. There is also a tilt and balance network that enables SARG's torso to remain upright and stable even when operating on inclines. Sitting atop the torso are two largely autonomous antenna modules, which automatically adjust their orientation so as to maintain the communication link with you as the rover moves.

All these semi-autonomous systems make your job as the rover manager simpler. You view a video screen of the terrain in the direction of travel and, using your mouse, you mark locations where you want the rover to investigate. This provides SARG with high-level guidance about the general trajectory of travel. However, the rover's actual path will be a blend of your guidance and the many low-level, path-seeking, and obstacle-avoidance algorithms that the various subprograms implement along the way. The rover can speed up in open areas and slow down when the terrain is rough, and along the way it will update the video information so you can monitor its progress and perhaps make adjustments to the route as you get a better view of the upcoming terrain. The decision process for path selection is not localized in any single module. It is distributed across multiple systems with different specializations and different operational time frames. When the rover reaches a new location, it will begin its standard survey automatically. Based on the features it detects, it will then choose some targets for more detailed analysis. However, you can also review the survey data and give priority to specific targets to optimize the search based on your insights.

If we look into the history of the semi-autonomous modules on SARG, we discover that a number of them – for example, the reaching algorithms, the basic object detection algorithms, the path calculation algorithms, and the protective avoidance strategies – have largely been passed down from earlier rover projects. They provide the low-level modules supporting the current design, and they represent proven functions contributed by the rover's evolutionary heritage. Of course, some of these functions may be supplemented by learning algorithms to better tune them to work in new situations. For example, the rover may need to adjust for differences between the conditions in which it was trained on earth and the actual conditions it encounters on the moon. Powdery surfaces behave slightly differently under the lower gravity of the moon than they do in training on earth, and because they weigh less, the rover's wheels may not dig into the surface well. As a result, the rover may need to adjust its traction algorithms as it encounters new surfaces, much as we humans adjust our traction algorithms whenever we encounter sandy, muddy, snowy, or icy surfaces.

SARG may even learn to anticipate some of the instructions that you, the executive manager, provide him. For example, if you begin avoiding certain features that look suspicious, simply to be more cautious, the rover's learning algorithms may detect this strategy and adjust to avoid similar features. You might, for example, even mark certain features on the screen as hazardous to inform the rover where *not* to travel. Learning to recognize new kinds of hazards would enable the rover to plan safer routes. However, there may also be cases in which the rover recognizes a potential hazard condition before you do. Perhaps the surface around your selected target turns out to contain a deep layer of powder that impedes traction. Rather than risk getting stuck, the rover may pause and sound an alert designed to prompt you to supply added guidance before it proceeds. In this case, your top-down guidance would actually be initiated by low-level alerting systems on the rover.

Within a few weeks of working in tandem with the rover, you come to rely on many of its algorithms. After all, some of them have even been learned from you. Over time as the rover learns to anticipate your decisions and as you grow familiar with its behaviors, you begin to give SARG more latitude in making routine navigational decisions. Some routines may have even become so well learned by SARG that you simply select targets of interest and wait for the rover to report back when it has finished its tests. After all, if it encounters a problem, its low-level systems will take defensive action and alert you as needed. This strategy frees up more time for you, as the executive module in the team, to plan the next target for analysis or to report to your associates about previous findings. Putting more of the decision load on the rover allows each of you to work in parallel and enables the project to move ahead much more efficiently.[7]

Now consider the parallels with the lattice control hierarchy we introduced earlier. The control structure we are describing includes semi-autonomous knower modules connected in a hierarchy that provides both top-down and bottom-up control dynamics. This is the sort of control structure we find in modern vertebrate brains. Even the communication delay has a counterpart. The declarative memory networks in the human brain take about five hundred milliseconds to link up sensory inputs with declarative memories, and decisions based on memory searches or verbal self-discussions can take several seconds more. As a result, most motor decisions are routinely delegated to dedicated robot-like control modules that can resolve decisions more quickly. There are even parallels with how you interact with the lower-level systems. You, the executive, are part of a team. You make many of the high-level targeting decisions, but lower-level modules execute the commands. In fact, you learn to rely on those systems to do what they do best, while you, the conscious agent, do what you do best. If the lower-level systems need more support, they will alert you for input. When I first froze in response to seeing something on the ground in front of me, it was a low-level brain module that first recognized a potential danger, took defensive action, and alerted the higher layers of the lattice to encourage additional processing.

LATTICED INTENTIONS

There is one more topic where the latticed organization of the brain provides us with added insight into the properties of mind. That topic is intention. It should be clear that the holder of a trait need not be aware of the trait's adaptive function for the trait to serve its end. This is true both for traits acquired by natural selection and even for those acquired by some kinds of learning. As long as traits are called into service at appropriate times, they tend to increase the likelihood of advancing an agent toward adaptive ends. Thus, collections of highly prepared actions, such as those prioritized by drive states and guided by sign stimuli, are often sufficient to allow simple minds to forage for food, find mates, and care for their young without logically reasoning about the function of their actions. I describe such actions as *endirected*, and I propose the term *endirection* to describe their purposive character. Agents that act in this manner can be said to behave *endirectionally*. Their actions can be described as *endirectional,* or purposive. The outcomes of their actions can be characterized as end states, and their persistence in seeking those outcomes, despite some failures, can be said to display drive.

There is, however, something special about the directedness of traits that emerge at higher levels of the lattice. In contrast to purely endirected traits, systems capable of trial-and-error learning, declarative memory, and logical reasoning often involve processing circuits which become consciously aware of the consequences that endirected traits produce. For now, let's assume we can operationalize this as the process of attending to the consequence of an action in the course of making the decision to act. When an agent attends to the consequence of an action, the consequence may even come to serve as a goal, or as an added source of incentive that favors the action, assuming the consequence is positive. Given this added insight into the function of their actions, agents that act with some conscious consideration of the consequences of their actions are described differently. When they consider outcomes before acting, they are said to act intentionally, and their actions are said to be *intentional* or purposeful. The outcomes of their actions are characterized as intentions or goals, and their persistence in seeking goal-directed outcomes, even in the case of adversity, is said to display will.

There are several reasons for wanting this extended family of terms. All the actions of an agent contribute to its endirected character, but only some of those traits are processed in decision systems that are capable of acting with intention – in the psychological sense of being consciously goal aware. Further, actions must have the potential to serve some end, before being aware of their function leads to them being selected intentionally for that purpose. Endirection is therefore a broader concept, and we can talk about the endirection of an agent's action without knowing whether he considered its consequence on any specific occasion. The distinction between endirection and intention also allows us to consider the teleonomic character of adaptive functions in other contexts. For example, despite all attempts to deny it, in a broad sense living systems continue to find new ways to adapt. In effect, evolution makes a kind of progress. It's not intentional progress in the sense of knowingly seeking a goal. It is endirectional in character. However, when a large number of endirectional traits accumulate in a system, the character of a system changes. It begins to persist in end-driven activities.

Thus, even without a well-represented goal, endirected knower modules tend to assemble adaptive, drive-related action plans. The modules at the bottom on the lattice decision hierarchy (see Figure 4-1) are those that we typically characterize as being endirectional – that is, as being traits for which the agent is unaware of their consequences. Those at the top of the hierarchy are ones which we generally characterize as being more intentional – that is, involving decision processes that are usually aware of the consequences of the actions they promote. Having an ability to consider the consequence of an action does not guarantee that skill is always employed before taking action. However, with increasing experience, agents with this ability tend to form associations between actions and their outcomes which influence their decision to act. In effect, their actions become more intentional.

Actions that are controlled at the trial-and-error learning level of the lattice hierarchy are especially interesting in this regard. They are often guided by conscious attention in the acquisition phase, but once the actions are automated as habits, their consequences may no longer be taken into consideration *in any detailed way* before they are executed. However, despite this shift in attention, we still generally consider such actions to be largely intentional, because they were trained with intentional support and their consequences have been learned. Even if the consequence of a learned action is not always considered before it is initiated, the very act of initiating it will tend to prime representations that are consequence aware. If the consequence is not acceptable, then that awareness introduces the opportunity, and sometimes even a "moral duty," to recognize the error and inhibit the action before it goes too far. In fact, failure to inhibit an action is routinely treated as tacit acceptance of its consequence.

Complications of this sort are why lattice structures give rise to such varied degrees of intentional decisions. Sometimes the consequence of an action is considered prior to the decision to take action. Sometimes the same action may be engaged by bottom-up processes with little consideration of the consequence. Most of the time, however, actions are guided by a combination of interacting bottom-up and top-down processes each contributing some weight to the decision. The weight given to the consequence of an action may be difficult to ascertain, even for the agent making the decision. If someone asks you to account for your decision, you can usually supply a reasonable, goal-directed explanation. However, in many cases such reasoned explanations come after the fact. They are not formally reasoned through at the time of the action, although associations with an action's consequence may have influenced its choice to some extent.

The graded nature of intention is particularly obvious when we examine the rules that the legal system uses to attribute responsibility for an action – for example, when someone is suspected of having caused the death of another. If the evidence suggests that the suspect planned to cause the death, then the subject will be charged with premeditated murder. Premeditated murder is an act involving the highest measure of intention – that is, an action guided by conscious planning. If there is evidence that the suspect was aware of the possibility of serious injury but acted more on impulse and without extended planning, then he or she is likely to be charged with second-degree murder, an act in which the outcome was less clearly considered. If the suspect was not aware that he or she had put another at risk but acted in careless disregard of the possible consequence of the action, then the subject will usually be charged with manslaughter, an action for which the outcome should have been considered ahead of time, but was not. However, if the suspect was acting with due care and was unaware that his or her actions were putting another at risk, then the death will be ruled accidental, not intentional, and the suspect will not be charged at all.

In short, we view intention as a graded property, and we attribute intentional responsibility for an action to the degree that its consequences were consciously considered ahead of time. If we are to develop robotic agents with the ability to make intentional decisions as humans do, then they will need a similar multilevel control hierarchy. Consider SARG, for example. He has all the low-level automated control modules he needs to perform routine tasks, but you provide him with high-level guidance. However, if we could find a way to enable him, or a more advanced robot, to attend to the consequences of his actions before making decisions, then he could begin to make the higher-level decisions too, and we would consider his actions to be intentional. To make this happen, we will need to add some executive modules to the control lattice of our robot.

We must also ensure that the consequences and goals, which are represented in those modules, can gain attention before the robot takes action.

CONCLUSION

Developing an appreciation for the flow of control in a lattice hierarchy is a key step in understanding the nature of mind. Many characteristics of mind, such as the tendency to engage emotional, habitual, and logical reactions in successive stages, are more readily understood in the context of this latticed control structure. A lattice of knower structures and connections among them also helps us understand how different kinds of knowledge interact so fluently in the brain. The lattice architecture takes advantage of skills and knowledge on different levels to produce decisions. It doesn't impose all the control from above. In fact, the lower-level modules often have a critical effect on what the higher-level knowers notice and learn. That is what makes the lattice so adaptive. Knowledge can be distributed across multiple levels of control, and yet reasoning is not limited to processing on any single level. Further, as we have noted, a lattice architecture can also make the control of robotic systems smarter and more efficient.

A lattice hierarchy also helps us understand how intentions come into play and why it is often difficult to evaluate the intentions of an agent. We attribute different properties to different emergent levels of mind, and we judge the behavior that is generated at different levels by different rules. Actions guided solely by low-level decision modules tend to be endirected, not consciously considered. In contrast, the consequences of actions guided by higher-level planning layers are often consciously evaluated before committing to them. Those actions are considered intentional. However, many decisions are a joint product of both endirectional and intentional processing, and it's often difficult to determine the extent to which the consequences of actions are consciously considered prior to acting, even for the agent who is making the decisions. Once we can develop a robot with a similarly latticed control hierarchy, he too will be unsure about his intentions from time to time.

CHAPTER 5
A THREE STREAMS ARCHITECTURE

The lattice hierarchy of cognitive traits is an idealized model of distributed mind. It provides an overview of how different levels of function interact in the brain, but it doesn't tell us much about the physical organizations which make those functions possible in the brain. Our goal in the next few chapters is to begin to develop a model of mind which ties particular neural structures and pathways to particular functions in the brain. We'll use what we know about the brain as a guide to define the network architecture that a robot will need to become a conscious, rational agent. Obviously, we will have to simplify many functions to make that model feasible. The aim, therefore, will not be to provide a complete explanation of how the human brain works, but rather to essentialize the functions sufficiently well so that we can begin to think about how to build a robot with similar properties. After all, if you can't build it, then you really don't understand how it works. We begin with an essentialized model of the architecture for managing three broad flows of information in the brain. I call it the three streams architecture.

A THREE STREAMS ARCHITECTURE

Gallia est omnis divisa in partes tres. – Julius Caesar,
De Bello Gallico, ~50 BC

The usual approach to describing how the brain works is to divide it into a number of anatomical structures, attribute functions to each, and then try to show how those structures interact to produce more complex behavior. This is something like trying to explain how a computer works by labelling each of the microchips inside it with a function and then noting all the interconnecting paths etched into the circuit board. It is basically the approach that many neuroscientists use, although they must first obtain some estimates of the function of different structures by removing modules or cutting connections to see what functions are lost. The main problem with this approach is that there are a multitude of different modules in the brain, and the details quickly become overwhelming. Indeed, it would take thousands of interconnected personal computers, each performing different tasks, to approach the processing complexity and memory resources of the human brain. We need a simpler strategy for describing how the brain works.

To make sense of all this complexity, we will start with a high-level model of the flow of information in the brain. As in all models we will have to make some simplifying assumptions. This will mean we won't model the workings of the human brain perfectly; rather we will try to extract the essential architecture and functions. We'll then use these functions to design a biologically inspired robot; I call him Cogley,[1] with cognitive functions similar to those of humans. Although it won't be a perfect model of the human brain, if the model does its job, it will provide us with insights that place the essential functions in a more meaningful framework. To that end the first feature of our model will be to characterize the brain as being composed of three vertically integrated information-processing streams. These are the pathways involved in action planning and decision making, perceptual analysis and memory processing, and interoceptive evaluation and management.

WHY STREAMS?

The idea that there might be different processing streams in the cortex of the modern vertebrate brain first gained attention following studies of the visual system by neuroscientists Leslie Ungerleider and Mortimer Mishkin.[2] While working with macaque monkeys, these researchers found that, although visual processing began in the occipital cortex at the back of the brain, the processing circuits soon diverged in two broad paths as they left the occipital lobe. One pathway progressed along the lower half of the temporal lobe. This was labelled the ventral stream. A second pathway followed an upward course into the parietal lobe. This pathway came to be known as the dorsal stream. When Ungerleider and Mishkin interrupted processing in the ventral stream, their monkeys lost the ability to recognize which of two objects was novel and which they had seen on the previous trial. When they interrupted processing in the dorsal stream, the monkeys were unable to decide which food cup was closest to an object. However, the monkeys with lesions in one stream could still perform the activities supported by the other stream. Based on these findings, Ungerleider and Mishkin suggested that the ventral stream was the pathway used for object identification, and the dorsal stream was used to identify the position of objects. They characterized the difference between the two streams as a distinction between *what* and *where* processing.

In subsequent studies of the visual pathways, researchers Melvyn Goodale and David Milner noted that the analytical functions of these two streams were more complex than a simple what-where distinction implied.[3] For example, *how* motor functions are also located in the dorsal stream. The performance of a female patient, whose name is abbreviated as DF to protect her identity, illustrates some of these complications. DF has bilateral occipitotemporal lesions that effectively disable her ventral stream processing for vision. In effect, her *what* processing circuits are blind. She lacks the connections needed to see objects in a way that would allow her to recognize and name them, and she is incapable of describing their size or shape. However, the visual-motor skills supported by the dorsal stream are not blind at all. When asked to grip small objects of varying sizes, she reaches directly toward them and accurately adjusts the spread of her fingers and thumb to grasp each object.

An interesting property of these visual-motor perceptual skills is that they only seem to be usable when DF actually tries to interact with an object. If she is asked to demonstrate the size of the object by showing a distance between her fingers and thumb, she guesses wildly. This suggests that the size of a grasp is under dorsal stream control when she is reaching for an object, but not when she is adjusting her fingers to demonstrate the size of the object from memory. That requires ventral stream control, which in DF's case is defective. Similarly, when DF is asked to post a card in a slot which can be rotated at various angles, she rotates her wrist accurately to insert each card. However, when she is asked to describe the angle of the slot she just saw by tilting her hand, she is largely inaccurate. Again, the dorsal stream appears to be used to negotiate the angle during the motor task; however, ventral stream processing is apparently required to demonstrate the angle.

In contrast to DF's deficit, Marc Jeannerod and colleagues reported on a patient, known as AT, who has bilateral occipitoparietal damage.[4] This damage disables her dorsal stream processing, effectively making her *how* and *where* processing skills blind. As a result, AT's ability to reach out and grasp small objects between her fingers and her thumb is grossly impaired, although she can accurately describe the size of the object verbally, and she is near normal in her ability to report sizes by scaling the distance between her index finger and thumb. Further, this scaling ability is not impaired by imposing a delay of several seconds between when AT sees the object and when she is asked to describe its size. This is significant because patient DF's ability to negotiate features using the dorsal stream is lost if she is not allowed to respond within two seconds of seeing an object. In effect, ventral stream processing results in perceptual details that can be held in memory for extended periods of time, whereas the dorsal stream detects programming features that support ongoing actions but that are not remembered beyond the brief time needed for motor execution.

Based on a number of such studies, Goodale and Milner eventually concluded that the two processing streams respond to the world differently because they are optimized for different kinds of tasks. The ventral stream is optimized for recognition and familiarity decisions, a detailed *what* function supported by connections to declarative memory networks. In contrast, the dorsal stream is optimized for spatial-motor planning and action decisions, which includes both *where* and *how* information, such as location, angle orientation, and grip size. These features are evaluated as an action is executed, but they are not remembered for more than a few seconds in that pathway. The finding that the cortex involves largely separate processing regions for visual action planning and for perceptual recognition suggested that other sensory modalities used for motor planning and object recognition might also diverge in separate pathways. Indeed, this has now been confirmed. Separate perceptual and action pathways have been reported for both touch and audition.

The idea that there were two largely separate sensory processing streams in the cortex changed the way that neuroscientists began to think about how external senses like vision and hearing were processed. However, there is another kind of sensory processing system in the brain that also needs to be considered: the system of inputs from the internally oriented receptors of the body, the interoceptive system. This processing stream detects aspects of interoceptive status and how you feel, and it adjusts the reactivity states of emotional and motivational systems to compensate for status needs. Sensory inputs to the interoceptive stream include a broad assortment of somatic status measures, such as those that detect pain, temperature, itch, gastric fullness or distress, and centrally located homeostatic monitors like those that regulate hunger and thirst. Reactivity, as I will use the term, refers to systems that manage output dispositions in the interoceptive stream. These outputs include the activity of the autonomic nervous system, motivational states guided by the homeostatic monitors, and emotional dispositions guided by other status evaluation modules.

The reason this third stream wasn't recognized as a distinct processing stream earlier was that it wasn't apparent that these inputs were well connected. Many interoceptive receptors are distributed throughout the body, and others are located deep within the brain. In addition, these senses project to lobes of the cortex which sit below the parietal and temporal lobes and are more difficult to study, so the connections were not immediately obvious. However, recent work has shown that the interoceptive pathways are well integrated beginning early in the brainstem, and that this stream has a very important function.[5] It gives rise to feelings about internal status (distress and need) and states of motivational and emotional reactivity (the readiness to act). These feelings are thought to be critical for higher-order representations of self, and for connecting feelings of self-status and reactivity with decision-making processes.

If the ventral stream can be characterized as serving *what* functions, and the dorsal stream as serving *how* and *where* functions, then the interoceptive stream is probably best characterized as serving *why* and *how much* functions. The inputs into this stream carry the basic information that an organism needs to keep itself within acceptable status ranges. When conditions venture beyond acceptable bounds – for example, when food reserves begin to run short – the reactivity modules activate physiological and behavioral reactions to correct the situation. It is these states of reactivity which set the tone of the autonomic nervous system in support of taking action. They also motivate behavioral choices and trigger reward learning in the extended circuits of the dorsal stream. Similarly, it is these states of reactivity that determine what cues most need to be recognized, remembered, and recalled by the extended circuits of the ventral stream. In short, the interoceptive stream plays a critical role in guiding the activity of the other two streams, and thereby promoting the survival and well-being of an organism.

CONCLUSION

Thinking about three largely separate processing streams in the brain provides us with insights into how cognitive functions are distributed in the brain and why different modules of knowledge are not equally accessible. The networks that support motor planning and decisions are largely segregated from those that involve perceptual recognition and memory. Similarly, the networks that process feelings of self-status and those that set reactivity states are largely separate as well. Thus there are three largely segregated processing channels in the brain. Still, it should be clear that the three streams are never completely independent and that some of the most interesting cognitive functions depend on the ability of modules in one stream to influence the activity and decisions of those in another. However, before we consider those interactions, it will help to look more closely at the functional organization within each stream to see how the organization in each stream contributes to its effects.

As usual, knowing about the organization of the networks in the brain will help us understand how each stream comes to have special properties. This will be important because Cogley will need to have the functionality of all three streams if he is to come close to displaying properties typical of a human mind. The three streams architecture will be the foundation for Cogley's design. While we are reverse engineering the brain to gain insights into how to design Cogley, we should also recognize that determining what properties are important for Cogley may also help us understand ourselves. After all, we are designing Cogley to behave as we do.

CHAPTER 6
THE PLANNING STREAM

We are now ready to look more closely at the brain systems that manage the three streams. We'll begin this task by looking into the organization of the extended pathways of the dorsal stream, a system which I describe functionally as the planning stream. The cortical circuits in this stream include the classic dorsal stream paths, which are responsible for spatial mapping (where processing), recognizing affordances (how processing), and the paths which connecting that information with motor-program areas. Activity in these networks continually assembles potential action plans. The subcortical part of this stream extends through the basal ganglia. This circuitry further prioritizes action candidates based on motivational states and past stimulus-response associations. It then selects the best-fit candidate for output. Taken together, the cortical and subcortical components of the planning stream provide an impressive architecture for recognizing action possibilities and selecting the best action plan for the moment.

THE PLANNING STREAM

> A *selection problem* arises whenever two or more competing systems seek simultaneous access to a restricted resource. Consideration of several selection architectures suggests there are significant advantages for systems which incorporate a central switching mechanism. We propose that the vertebrate basal ganglia have evolved as a centralised selection device, specialised to resolve conflicts over access to limited motor and cognitive resources. – Redgrave, Prescott & Gurney, "The Basal Ganglia: A Vertebrate Solution to the Selection Problem?," 1999

As cognitive agents we often have the experience of consciously planning our actions. However, it doesn't take much introspection to recognize that we are rarely aware of many of the components that go into our plans. For example, when I decide to pick up a cup of coffee, I usually don't reason about how I should turn my torso to put my arm in position for the reach, or how I should rotate my wrist to get the best grasp on the cup's handle. Such actions just seem to happen. Indeed, as the behavior of patient DF suggests, I can apparently make such decisions even when I cannot describe the size of the movements or the angle of rotation I must negotiate. Somehow a collective of knower modules in the dorsal stream manages to assemble my actions plans largely on its own. It's all rather convenient, but it makes it difficult to determine who is in charge. How do modules in the dorsal stream manage to plan actions so automatically? And why is it that I think of myself as the agent in charge of the planning? To begin to answer these questions, we need to look more closely at how action candidates are assembled and how the best candidate is selected for execution by the networks in the planning stream.

Perhaps the best studied collection of motor-planning circuits in the cortex is the complex of spatial memory fields, affordance extractors, and motor-programming subsystems which guided reaching and grasping in the macaque monkey. After decades of study, a high-level theoretical and computational model of these networks has been assembled. It is known as the FARS model, an acronym for Fagg, Arbib, Rizzolatti, and Sakata. Andrew Fagg coined the acronym to acknowledge that his model was built on the collective work of several groups of researchers, in particular the work of Michael Arbib and colleagues in the United States, Giacomo Rizzolatti and colleagues in Italy, and Hideo Sakata and colleagues in Japan.[1] We'll use the FARS model as a guide for thinking about how action plans are formed.

An overview of the FARS model is shown in Figure 6-1. The labels adjacent to each box are common abbreviations for the brain regions proposed to support each function. Visual inputs project to the modules on the left. The box on the top left labelled "Object Recognition" represents ventral stream perceptual contributions from the networks in the inferior temporal cortex (IT). The box labelled "Object Features" represents the dorsal stream feature detectors in the caudal intraparietal sulcus (cIPS). The box on the left labelled "Hand & Motion Detection" represents the motion detecting networks situated in the superior temporal sulcus (STS). The box on the bottom left labelled "Object Locations" represents the contribution of several spatial mapping regions (MIP, LIP, and VIP). With the exception of the IT and the STS, the analysis regions on the left side are in the parietal cortex. The motor-programming regions are situated in the frontal cortex. Cue-specific learning also influences motor decisions via additional connections in the frontal cortex. However, to keep the presentation simple, the cue-specific inputs have been omitted from this diagram.

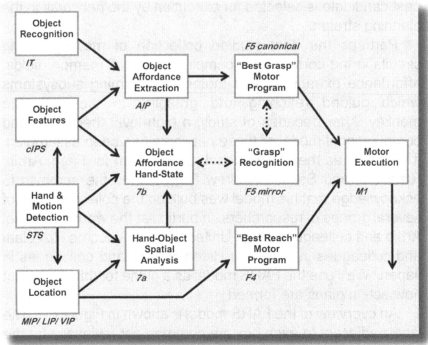

Figure 6-1: Diagram of the FARS model.

The FARS model is complicated, but the basic process by which it operates can be summarized fairly simply. Each box in the diagram corresponds to a region of cortical processing in which network subcomponents compete to represent various aspects of information relevant to reaching and grasping decisions. The idea is that each region is a specialized knower module which recognizes classes of best-fit representations needed for constructing an action plan. Some networks are specialized to detect affordances, features that support certain kinds of actions. Some regions monitor hand position and movement. Some regions integrate information from other regions to determine best-fit combinations. The spatial maps track the location of objects in various motor coordinates. For example, the maps labelled MIP, LIP, and VIP represents inputs from the medial, lateral, and ventral intraparietal cortex respectively. These regions track objects in three kinds of body spaces.[2] LIP tracks objects in visual space. This region enables a monkey to plan approaches to objects. MIP tracks objects within reaching space. This region is particularly important for planning arm movement and grasping. VIP tracks objects in the space around the face. This region is important for coordinating arm movements with the mouth during hand-to-mouth feeding.

In normal operation each module processes several different possibilities at the same time. For example, the module representing grasp affordances is likely to detect several potential affordances on a cup: its handle, its cylindrical shape, and the edge along the rim at the top – all possible points of grasp. These different affordances provide several ways of grasping the cup. Other states of information – such as the location and proximity of features, the starting position of the hand, and whether or not the cup is full of liquid – would then favor some plans over others. Because the networks operate in parallel, they don't need to wait for inputs from other modules to start processing. However, cross-connections among these modules enable them to influence each other, so parallel activity soon converges on compatible combinations. This happens because as dominant features in each region become active, they inhibit competing features while supporting compatible features.

As a result, action programs that best fit the task features and starting positions take form as the networks reach a consensus about which component parts have the most mutual support. However, it wasn't always that automatic. A lifetime of experience has gone into training the planners. In fact, your first six months of life were largely spent learning to identify affordances for simple actions and connecting them with the appropriate motor programs. The planning networks had to learn what shapes were good for grasping, how to recognize when those affordances were within reach, what combination of affordances and actions were compatible, and what combinations were not. The combinations that actually succeeded in making contact and then bringing objects closer for inspection were the ones that got their connections reinforced the most. Although most of the system components are now well trained, the process continues as new affordances and new action combinations are learned.

The push and pull of compatible and incompatible inputs in these networks also provide some insight into how other brain systems can influence an action plan. For example, if object recognition modules in the ventral stream are strongly focused on a cup's handle, because other network systems have fixed the eyes on that feature, then the reach planning modules would encounter more mutual excitation for plans directed toward the handle, and more inhibition for those that were not. In fact, if this bias was strong enough, and the handle was not accessible, activity in preparation for reaching might even spill over to motivate supporting actions, such as a change in position to make the handle more accessible. The fact that perceptual analysis and gaze direction may also contribute to an action plan suggests how flexible the process is. Anything that favors one action candidate over another has the potential to influence what plans are assembled. Given that slight changes may favor different plans, it appears that several potential action candidates may stay active at this stage, leaving the cortical plan highly flexible.

EFFECTIVITIES AND INTEROCEPTIVE EVALUATIONS

There is yet one more module in the FARS model that needs to be addressed. In addition to motor-program networks in the F5 motor region, labelled as "F5 canonical" in the diagram, researchers have identified a second type of motor-planning network in this region. These networks are active not only when a monkey produces an action, but also when a monkey observes another monkey, or the human experimenter for that matter, performing a particular action. Given that the neurons in this network respond both to actions that are made by a monkey and to actions that the monkey observes, they came to be called *mirror neurons*.[3] This network, labelled "F5 Mirror" in the diagram, is shown adjacent to the F5 canonical network. Observations indicate that mirror neurons respond specifically to the functional effects of actions, like grasping an object, placing an object on a table or manipulating an object, and not merely to feature combinations such as seeing an object near a hand. Thus, these networks really seem to recognize actions as movements that have particular effects.

Subsequent work has shown that neurons in these networks generalize to similar kinds of actions (grasping by humans or monkeys, grasping in varying styles) and may even respond if the completion of the action is occluded. Some neurons in this region have also been shown to react to mouth movements and to action-produced sounds. Similar action-recognition activity has been demonstrated using neuroimaging procedures. In one study, motor regions in humans which responded to biting and chewing actions also responded to the observation of monkeys and dogs biting and chewing, although the response to the chimp and dog actions was weaker than to those for actions by humans. While the F5 mirror neuron networks respond to fine-motor hand and mouth movements, mirror neurons for other actions have been found in posterior parietal motor areas.[4] Taken together, these studies suggest that mirror neuron networks are widespread in the motor system, that they recognize the effects of actions, and that they generalize to similar actions even across species.

The "mirror neuron" label for these networks calls attention to the fact that recognizing actions in others has the potential to facilitate imitation and socially guided learning. A major theoretical emphasis in early studies of these networks was the assumption that these neurons might promote imitation and social awareness.[5] However, the circuitry needed for visual imitation is much more complex than simply mirror neurons. Mirror neurons are part of the motor system. They are not perceptual networks. In order for them to react to observed actions, they need inputs from areas that do recognize actions. Research suggests that the action-recognition inputs to the mirror-neuron networks originate in the biological movement-detection areas of the superior temporal sulcus (STS) and adjacent regions of the temporal lobe. These movement-detection regions are known to project to affordance and early movement planning areas in the posterior parietal cortex. Those regions, in turn, project on to the ventral premotor area where the mirror neurons for grasping are located. It appears that movements are recognized visually in the superior temporal lobe and are then associated with affordances and motor-program areas associated with those affordances.[6]

Also lost in the early thinking about the role of mirror neurons for imitation is the basic finding that mirror neurons react to functional units of action. The fact that the mirror neurons are collocated with the networks that manage action programs suggests that mirror neurons may not only cause an action plan to be "motivated" by seeing another agent perform it, but more important, they may cause an action plan to be "motivated" based on the effect that it produces. This possibility led Patricia Zukow-Goldring and Michael Arbib to suggest that mirror neuron networks could be characterized as *effectivity* networks – that is, networks that learn about the effects of actions.[7] The question remains, however, how is it that a network in the motor region learns about the functions of an action? I believe the key to understanding this is to note that the region in which these mirror neurons reside receives inputs about affordances in the course of motor planning, and once an action is executed, it receives a new set of affordances. Thus the effects of an action can be learned as the feed-forward change in affordances which an action produces. Adding the effects of actions to the decision process would result in plans that aren't simply favored because they are feasible, but because they also lead to more favorable outcomes.

Now here is an interesting point about this possibility. Because there are reciprocal connections between the affordance recognizing areas in the parietal cortex and the motor planning areas, if the effect of an action is recognized as a change in affordances, then the mirror neuron networks would tend to activate feed-forward changes in affordance representations in the parietal cortex as an action is considered. In addition, because the parietal cortex is linked to movement-detection areas, changes in affordances would also activate feed-forward associations with the biological movement areas in the superior temporal sulcus (STS). It turns out that feed-forward sequences are exactly the kind of inputs that the movement-detection areas use to recognize movements.[8] Movements are, after all, sequential events, and they come to be recognized based on sequential shifts in motor information. A link between motion-detecting areas in the STS and changes in affordances in the mirror-neuron networks would therefore connect action representations with their affordance-changing effects.

One more link is needed, however, to evaluate the effects of actions more fully. That would be to link actions and their affordance-changing effects to internal evaluation networks for status and emotional reactivity. There are several paths by which this seems to work. First, the movement-detection areas for facial features send projections on the fusiform facial area, where the emotional expressions can be detected. Movements related to the torso and limbs send projections to the lateral occipital area, where emotional body language signals can be detected. All these regions also connect with the amygdala, the anterior cingulate cortex, and the anterior insular cortex, where feelings of status and reactivity are managed.[9] Thus, the connections among mirror neurons, affordances, and biological movements also link to feeling states.[10] This circuitry is thought to enable action planning to be influenced by affordance changes and their interoceptive consequences.

THE DECISION MAKERS

While the networks in the cortex provide both feasible and functional action candidates, the cortical networks are largely plan promoters. They are not the primary decision makers. It's the subcortical networks in the basal ganglia which resolve plans and send them on for execution. This makes the basal ganglia one of the most important network complexes in the brain. It's where action decisions are made. Just as the cortical planning networks involve many interconnected modules, so do the basal ganglia decision networks. Some eight modules with strange-sounding Latin names are usually cited as members of this decision system. However, logically the basal ganglia can be divided into two major territories.[11] The input functions are localized in a region just below the cortex known for its striped appearance. This is the territory of the "striped gang," the *striatum*. The output functions are localized in a lower region which is noted for its pale color. This is the territory of the "pallid gang," the *pallidum*.

In our simplified model, the primary function of the members of the striped gang, the striatum, is to assign priority weightings to the components of each action candidate. The striatum receives sensory and motor activity inputs from the planning regions in the cortex, as well as from many subcortical regions. Thus, it usually begins with well-organized plans. It combines these planning inputs with motivational biases from the interoceptive stream. These inputs serve to adjust the activation levels of different action candidates based on current drives and need states. Learning in the striatum also results in stronger connections for those stimulus-response combinations which have led to rewarding outcomes in the past. When these learned associations bias decisions on a regular basis, we call the actions they promote habits.

The priorities assigned by activity in the striatum are the inputs which project on to the pallid gang, or pallidum. However, these inputs follow two paths. One path, called the direct path, relays activation levels that energize the network regions supporting various action candidates. The other path, called the indirect path, acts like a brake to inhibit response activation. Exactly how these two paths interact remains an important topic of research. However, the circuitry in the pallidum is generally arranged so that actions which compete for the same motor resources inhibit each other. That is, they activate the braking circuitry for their competition. This mutual inhibition enables the circuitry to select between alternative action candidates. The idea is that whenever an action plan is inhibited, its ability to inhibit other action plans is inhibited in proportion.[12] As a result, as one candidate grows more active than its competitors, it begins to inhibit them more, and they inhibit it less.

Assuming that at least one action candidate meets some minimum activation level, the result is that the output path for the action candidate with the highest activation state, the best-fit candidate by this algorithm, is no longer inhibited. As a result, the inhibitory brake is released. Ordinarily the neurons in the pallidum inhibit the activity of the brainstem motor-program generators. However, once actions are selected, their brainstem connections are disinhibited. This allows the motor-program generators for those actions to run to execution. Because the changing activation levels in the pallidum also feed back to influence the decisions in the striatum and the cortical planning circuits, all three regions remain synchronized and can contribute to ongoing decisions.[13] Cortical action plans, striatal activation patterns, and competition among competing candidates in the pallidum all shape the execution plan. The first-level motor-program generators integrate movement combinations, such as those needed for various gaits or actions patterns. Lower-level pattern generators manage opposing muscle activity within each movement. The multiple layers of control result in an incredibly flexible control system.

The presence of both excitatory and inhibitory pathways in the pallidum introduces yet other possibilities for controlling actions. For example, by enhancing the inhibitory pathway the onset of an action can be better controlled or it can be suppressed entirely. Partial inhibition might also be used to control the timing or force of an action. However, there are some potential weaknesses in this design. Degradation of dopamine neurons in the excitatory pathway can introduce signaling imbalances which result in spurious start and stop decisions. This imbalance is thought to be the cause of the movement tremors that occur in Parkinsonism. This is why drugs that enhance the effects of dopamine are commonly used to treat Parkinsonism. Increasing excitatory dopamine activity makes it a little easier to keep the control signals in balance.

Dopamine signaling in the planning stream also has learning and motivational properties. There are two small regions in the midbrain, one named for its dark color, the substantia nigra (SN), and the other named for its location in the floor of the midbrain, the ventral tegmental area (VTA; "tegmentum" means floor). These two tiny midbrain regions are the primary sources of the neurotransmitter dopamine for the entire brain. This is important because dopamine plays a critical role in regulating motivation and reward. Tonic levels of dopamine provided by the SN appear to set the balance of excitatory and inhibitory motivation in the basal ganglia. In fact, the SN is considered to be a key member of the pallid gang. Though the VTA is not formally considered a part of the basal ganglia, it plays a critical role in training the networks of the planning stream to learn new behavioral sequences.

The VTA sends dopamine neurons to the striatum and broadly to the executive, motor-programming, and memory areas of the cortex. The pulses of dopamine act as reinforcement signals. They cause the neurons in those target areas to form stronger connections among whatever planning circuits are active at the time of dopamine release. As a result, whenever interactions between sensory features, motor-planning circuits, and drive states result in actions that produce dopamine signals, the active neural connections are strengthened. This makes it more likely that they will subsequently win the competition for selection under similar circumstances, an effect commonly known as reward learning. In the case of punishment, the VTA releases dopamine when the aversive state is reduced or avoided. This enables the planning stream to learn about actions that reduce and/or avoid punishing outcomes. Not surprisingly, it seems that reward learning has stronger effects on the direct/excitatory pathway in the pallidum, while aversive consequences have stronger effects on the indirect/inhibitory pathway.[14]

The description above is a simplified account of how reward learning is implemented in the brain. Several other neurotransmitters are also involved, including glutamate, which drives striatal neurons into a pre-activation upstate. Upstates are early stages in action selection. Once an upstate is activated, a response can be selected and sustained more easily. Another transmitter, acetylcholine, plays a role in releasing dopamine. Endorphins, neuromodulators associated with feelings of pleasure and states of well-being, are also released when positive end-states are detected. The endorphins appear to determine preferences for events associated with rewards, a reinforcement effect that has been termed *liking*.[15] Dopamine, in turn, appears to be the signal associated with the motivational states that result in wanting and seeking. Research suggests that as cues and actions come to predict rewarding outcomes, they too begin to trigger some dopamine release.[16] As a result, actions that lead to those cues also come to be rewarded. When a rewarding outcome is reached, the liking circuits are activated and further seeking is deactivated. It's a simple but elegant design. Plans that lead to rewarding outcomes have their connections strengthened, and plans with stronger connections tend to be selected more often.

A PLANNING COGLEY

The model of planning and decision making presented above is a much simplified account of the processing that actually occurs in the human brain. However, digging into that level of detail all at once would simply overwhelm us. Our best strategy is to give Cogley a simplified version of the algorithms in this stream for now, while keeping an eye on how we might make his decision algorithms more sophisticated as we move ahead. So what will Cogley need to be able to engage this simplified version of planning and decision-making skills? First, he will need competitive motor-planning modules, like those proposed in the FARS model, to enable him to disambiguate the most accessible action components and assemble them into a small set of viable action candidates. Given the detail covered in the FARS model, we'll not say much more about the design of these planning algorithms except to note that Cogley will need similar planning circuits for all his actions. There must also be planning modules that support locomotor activity, and yet other planning networks to support the vocal-motor skills he will need to learn speech.

To provide Cogley with human-like planning and decision algorithms, he will need high-level planning networks that assemble feasible and functional action candidates. His mid-level decision complex should work in two stages. The input component of the mid-level network should be designed to prioritize action candidates. The output component should be designed to select best-fitting candidates. The prioritizing algorithms should adjust priority weights for each action candidate based on the initial activation levels provided by the planning networks, learned stimulus-response associations, and motivational effects due to current need states. This will enable Cogley to decide which action candidate best fits not only the action possibilities generated by the high-level planning networks, but his current needs. It will also cause him to favor stimulus-response combinations that have proven to be effective in the past. That is, it will cause him to develop habits which, on average, result in more successful decisions.

Cogley's decision networks should include both excitatory and inhibitory paths. The inhibitory path should act as a brake that suppresses outputs until a decision is reached. The winning action candidate should come to inhibit others more than they can inhibit it, thus releasing the inhibitory brake and connecting the outputs with the motor-patterns generators needed to execute the actions. The execution path will need master motor-pattern generators to manage high-level movement patterns and subordinate motor-pattern generators to manage the movements of each output appendage. Part of the planning inputs should also include an ability to activate the brake voluntarily. Strong braking activity should be able to inhibit an action, while milder braking should be able to modify the timing and/or the intensity of a response. This will provide Cogley with a finer level of action control.

Cogley will also need a reward signal generator so that whenever he experiences a more preferred interoceptive state, the sensorimotor components that were active at the time of the transition will have their connection weights strengthened. A transition to a less preferred state should generate a suppressing signal, and avoiding less-preferred states should have a rewarding effect. These reinforcement and avoidance signals should enable Cogley to favor actions that lead to more rewarding states and to avoid actions that lead to less preferred outcomes. The reward signals should project both to the high-level planning networks and to the mid-level prioritization networks. These adjustments will tune Cogley's planning and decision algorithms so that he learns to make better plans and better decisions.

Like all of us, Cogley will not always be in a situation where the first action he can take will bring him reward. Usually, he will have to take a series of actions that put him in a more positive situation. So how do we ensure that Cogley can plan accordingly? I believe the key is to arrange his planning networks so that they favor activities that produce positive effectivities – that is, positive changes in affordances. This means that Cogley will need effectivity networks, which associate the effects of actions with particular motor programs. These networks will be Cogley's parallel to the effectivity functions we proposed for the mirror neuron networks. Cogley's effectivity networks should learn what changes in affordances are associated with an action. Reciprocal connections between the effectivity networks and affordance recognizing areas in the planning networks will be needed to produce feed-forward associations between actions and the changes in affordances they enable. Cogley will also need biological movement detection networks so that actions can be recognized. Links between the movement-detecting networks and the affordance networks should enable the feed-forward effects in the affordance networks to trigger feed-forward associations in the movement networks. These in turn should be connected to networks that learn to anticipate interoceptive changes, so that the interoceptive effects of actions can be evaluated.

Such a feed-forward, affordance-effectivity planning algorithm will naturally lead to sequences that bring Cogley closer to discovering rewarding situations, providing that his effectivities and affordances are appropriately weighted. In order for Cogley to plan well, much of his early development must be devoted to discovering affordance-effectivity links in his world. This means that Cogley's early experience should involve a lot of exploration and play. Many theorists have questioned why animals bother to play. Play seems to serve no immediate function. However, to the extent that many of the affordance and effectivities a sentient agent may use are not innately defined, they must be learned. Play is an obvious way to learn what actions an affordance supports and what effects taking an action can produce, as Eleanor Gibson notes.

> After a decade of research and thought on this problem, I have come to some (to me) rather obvious conclusions. First, nature has not endowed the infant with the ability to perceive these things immediately; babies spend nearly all of their first year finding out a lot about the affordances of the world around them. (Of course, we keep on finding out ever after, though not quite so assiduously.) Second, learning about affordances entails exploratory activities.[17]

If we apply Eleanor Gibson's ideas to Cogley, this means that from early on he should be motivated to investigate objects and play with accessible features so that he can discover what kinds of actions they support and what kinds of effectivities those actions produce. A playful robot will require constant supervision in his youth, but playing means that we won't have to teach Cogley everything. In many cases, all we will need to do is provide him with interesting situations and objects, and he will train himself. Of course, we'll need to balance Cogley's curiosity with some fear of novelty to keep his explorations cautious. However, if we want Cogley to be truly creative, we'll want this kind of exploration and curiosity to continue for his entire lifetime.

CONCLUSION

The extended dorsal stream, which we have described here as the planning stream, is obviously a highly complex processing pathway. However, by taking a functional approach and focusing on essential components, we can begin to understand how it works. The neural pathways in this stream support action planning, prioritization, and selection. Candidate action plans are assembled largely in the cortex based on the interaction of a number of distributed knower modules. Some modules detect affordances, stimulus features that support particular actions. Some modules recognize effectivities, which I have proposed can be evaluated as the changes in affordances that an action produces, and the associated interoceptive states with which those affordances are associated. Some modules track feature locations in body-centered, spatial coordinates. Other modules track the position of the body and limbs as they prepare for action. Candidate action plans emerge as compatible fits between component features enhance each other, whereas incompatible fits inhibit each other. In this manner, the most compatible action candidates and the most desired effectivities for a situation are constantly being assembled and promoted by the networks in the cortex.

Decisions in this stream are managed by a mixed gang of subcortical networks known as the basal ganglia. The input half of this collective – the striped gang, or striatum – prioritizes action candidates based on their activation levels in cortical and subcortical planning networks, based on their probability of success as determined by prior learning, and based on their associations with ongoing interoceptive needs. The prioritized inputs from the striatum are then passed on to the selection algorithms of the pallid gang, the pallidum. This is where the best-fit action candidate is selected. A special feature of this circuitry is the presence of an inhibitory brake. Decisions are made as incompatible action candidates inhibit each other while competitive algorithms release the brake for the best-fit candidate. The brake is a wonderful addition to the decision algorithm. As we shall see later, it also introduces some unexpected and very important properties, because the longer an agent holds the brake on, the more time he has to plan.

Another feature of this stream is that connections are constantly being refined by reinforcement signals. This means that the skills in the planning stream keep getting better as an agent gains experience. The learning never stops. New affordances are always being discovered. Effectivities are constantly being revalued. New action patterns are continually being learned. Creating a similar planning stream for Cogley will no doubt present many challenges. However, it will be critical to his success. Feasibility planning, functional planning, candidate prioritizations, best-fit selection algorithms, and reinforcement signals will help ensure that Cogley recognizes opportunities, weighs their importance, selects the best action for the moment, and becomes more skilled as he gains experience. These are important planning skills for any agent.

Obviously, this is still a rough sketch of the control architecture needed to help Cogley make intelligent decisions. Many details still need to be worked out to shape these strategies into an effective candidate generation and selection system. However, I believe the key elements of the process are here. A robot with these planning skills will be able to learn to select actions that best fit environmental constraints, external opportunities, and his needs. He will also be motivated to change his priorities as his needs change. Further, because he can learn to control the brake, he will be able to delay or inhibit actions at times. It will no doubt take many tries before a highly workable design takes form. However, we are already beginning to design robots that use such bio-inspired algorithms to solve the response selection problem, as Redgrave and colleagues characterized it in the opening quote. And once we get well-designed motor selection algorithms working, we should also be able to apply them to more cognitive decision-making tasks.

CHAPTER 7
THE PERCEPTUAL STREAM

The planning stream is the path in which action planning and decision making are accomplished. The modules in that stream have percepts of sorts, but they are dedicated to detecting affordances, feature details that support actions. They don't perceive features in larger pattern combinations, and they don't connect the features they detect with memories of past experiences, although they do learn what features best support particular actions. The stream we consider in this chapter is the perceptual stream. This pathway includes the ventral stream perceptual processing pathways in the cortex and their extensions into the declarative memory networks in the medial temporal lobe. These extensions support the formation and recall of semantic and episodic memories. Semantic associations enable perceptual items to be recognized by connecting them with meaningful tasks, feelings, and contexts. Episodic associations enable past events to be reconstructed and re-experienced. Detecting complex perceptual patterns and connecting them with co-occurrence associations requires a much different architecture than action planning.

THE PERCEPTUAL STREAM

> Within the spatial maps of the hippocampus, the "well-grounded" navigational thoughts of groundhogs may be organized around similar neural principles as the soaring spatial thoughts of falcons. – Jaak Panksepp, *Affective Neuroscience*, 1998

A few years ago Sandra Lê and colleagues examined a thirty-year-old man, identified by his initials as SB.[1] SB had experienced a bout of encephalitis at the age of three which left him without any ventral stream functionality for vision. However, apparently because his illness occurred at such an early age, he had learned how to compensate for his limited visual recognition to some extent. He had even learned to report on some of what he saw using the visual information from his dorsal stream visual pathways. The remarkable thing about SB was the specificity of his visual deficits. His visual-motor skills were well developed. He played casual sports, including some moderately fast-action games like ping pong. He could ride a motorcycle. He could juggle a little. He could even catch two tennis balls tossed to him at the same time. Obviously, when it came to actions, he could see.

When SB went through a cafeteria line, however, his visual deficits became clear. He had no trouble reaching for the plates holding items on display in the cafeteria line. However, he couldn't choose a reasonable meal this way because he couldn't recognize what foods were being offered based on visual inspection. He had similar problems with people. He could see their location and avoid running into them, but he couldn't recognize different individuals by looking at their faces. In contrast, his hearing was normal and he had no difficulty identifying people by their voices. Further, he was quite good at recognizing objects tactilely and had learned to read using Braille. In short, his recognition skills were normal when he used auditory or tactile cues, but he could not visually recognize most objects so as to name them, although he could see well enough to play ping pong or ride a motorcycle.

When SB was shown standardized pictures of common objects – a cup, a foot, a rabbit, a comb, a car – he could not identify most of the objects so as to name them, although he could obviously see features in the pictures. For example, he identified the picture of the car by its wheels. In addition, he was surprisingly good at drawing copies of complex figures. His success on the drawing task, however, revealed an odd strategy. He never drew an object as a whole, but rather he assembled his pictures in a systematic fashion by drawing the individual features of the object one at a time. Similarly, when shown pictures of faces, he reacted as if he saw only individual features. He could not tell which pictures showed normal faces and which had the features scrambled, although he could compare any two pictures on a feature-by-feature basis and tell if they were the same. When trained on the scrambled face task, he learned to label facial features, and with practice he learned to tell when some features were out of place by visually tracking certain spatial relationships, like nose above mouth and eyes above nose. However, even after this training he still could not differentiate between happy versus sad faces. He simply seemed incapable of integrating the features into more complex visual percepts. It was as if when he saw a face, he saw all its features separately (an eye, a brow, a nose, a mouth, a chin, etc.), but he never saw the features bound together as a single face. Thus he couldn't recognize people by their faces or recognize emotional expressions based on subtle changes in features.

SB's skills and deficits highlight the information-processing differences between the dorsal and ventral streams. The dorsal stream learns to parse visual inputs into features that serve as affordances – percepts that support various kinds of motor actions. However, although the affordances are refined and expanded to guide the planning of new actions, they are not grouped together as larger units. When reaching for a cup, it wouldn't be helpful for the motor-planning networks to link all the affordances of the cup together. The planning networks connect features with specific locations and motor programs, but they don't connect different features together beyond what's needed to generate an action plan. Connecting different features together would make the task of selecting a specific action plan more difficult. Instead, features like the cup handle compete with features like the cup rim for control of different plans, and plans using one affordance inhibit those that use another.

In order to integrate features into larger perceptual units, like objects or faces, a different kind of network organization is needed. Rather than mapping discrete features to actions, the ventral stream binds perceptual features into more complex hierarchically connected combinations. The nature of this hierarchical organization follows a pattern first detected by the work of visual researchers David Hubel and Torsten Weisel.[2] Hubel and Weisel found that individual cells in the primary visual cortex responded not only to the presence of light but to features such as brightness edges. Further, the orientation of the features was important. Some cells would fire only in the presence of a vertical edge at a particular location in the visual field, and others might respond to a different orientation in the same region of the visual field. These orientation-sensitive cells were called *simple cells*. In addition to simple cells, Hubel and Wiesel identified *complex cells,* which responded to a specific orientation in a larger region of the visual field. The complex cells appeared to react to inputs from a range of simple cells. Hubel and Wiesel also found yet higher-order cells, called *hypercomplex cells,* which responded to combinations of inputs from simple and complex cells.

Early studies in the lab of Charles Gross expanded on this hierarchical organization. This work found that successive regions of visual pattern recognition extended into the inferior temporal cortex, and that there were both feed-forward and feedback connections among these networks.[3] Inspired by this architecture, neuroscientist and programmer Stephen Grossberg later developed a hierarchical, two-level network program which used both feed-forward and feedback links. This network was able to learn to recognize patterns quite well.[4] In this hierarchy, programming modules in the upper-level networks competed for the best-fit recognition of patterns using the features detected by the lower-level networks. When the upper-level networks began to recognize a familiar pattern, feedback links to the lower-level network modules enhanced the expected feature representations, even for incomplete features, making those features more likely to be detected. By establishing resonant states between what features were expected in the high-level pattern and what features were represented on the input level, the program was able to recognize patterns based on feature-sets that usually occur together, even if some features were missing or only weakly represented.

Grossberg called the algorithm for recognizing perceptual patterns using resonating feed-forward and feedback links *adaptive resonance.* When a novel pattern is repeatedly detected by the upper-level networks in an adaptive resonance system, the upper-level gradually begins to recognize it as a new pattern; thus the recognition skills grow with experience. Further, once a pattern is learned, it can be treated as a functional unit in subsequent decision processes. For example, it can be recognized as the same kind of object and connected to the same outputs, even when the recognition is based on a different subset of features each time. This is essential for a good pattern-recognition system, because in the real world input samples are often noisy and incomplete. You only see part of a face or one view of an object. A system that is capable of quickly recognizing best-fit patterns based on partial data is much more likely to be adaptive than one which requires an exact match on every occasion.

While simple adaptive resonance models tend to use two-level hierarchies, research suggests that the perceptual regions of the visual cortex involve adaptive resonance hierarchies across multiple levels. Patterns recognized on early levels become the input features for later levels. For example, visual analysis begins in a region of the occipital cortex known as V1, the primary cortical site for visual input. Processing then extends in a hierarchical fashion through new analysis regions, beginning with region V2. These initial regions provide low-level feature analysis for all vision. It is at the next stages, supported by regions V3 and V4, that the dorsal and ventral pathways begin to diverge. The dorsal stream recombines inputs from these early feature analysis regions to detect affordances and determine their locations. These networks extend into the parietal cortex. The ventral stream passes these features to a continuing series of processing hierarchies that stretch into the inferior temporal cortex.

Different features detected in the ventral stream – color patterns, shapes, textures, three-dimensional objects, and higher-order feature-sets – are detected in distributed regions of the inferior temporal cortex. Some of these combinations have been estimated to involve as many as twelve levels of pattern analysis. Detailed relationships among features are processed in an adjacent region of the occipital cortex known as the anterior fusiform gyrus. These networks are capable of analyzing the relationships among features needed to recognize faces, or to recognize letters and words. Patient SB's visual inputs never reached networks capable of this level of resolution, so he could never resolve the relationships involved in faces or written words. He only saw the individual features.

There is another interesting feature of this hierarchical analysis process. As the patterns detected in each higher perceptual region grow progressively more complex, the network's ability to specify their position in the visual field becomes less specific. This trade-off occurs because the higher-level networks would need to be much larger if they were also to represent the locations of each high-level pattern. It appears that object locations are anchored to lower-level visual features, which are also accessible to the dorsal stream. More complex patterns and objects are represented in later processing regions, which link back to the localized features. Interestingly, Wolf Singer, Charles Gray, and colleagues found that although the different features of an object are processed in separate regions of the visual cortex, they tend to result in synchronous firing rhythms.[5] The features of the object share a common firing rhythm. This finding suggests that synchronous firing is somehow part of the process that enables separate features to be bound together in larger units.

DECLARATIVE MEMORY

Outputs from the perceptual networks project on to the second half of the perceptual stream, the declarative memory complex. The memory networks are rather complex, and we'll have to go slowly. However, the associations formed in this stream are an important part of conscious experience, so we need to understand them. In mammalian brains, the networks for declarative memory are situated just below the temporal auditory and visual perceptual areas. This region is known as the medial temporal lobe. The initial networks in this region make up what is known as the parahippocampal region. This is thought to be the region where semantic (i.e., meaningful) memory associations are formed. These networks in turn send projections on to a mid-level semantic network and then on to the hippocampal region, where episodic memory associations are formed. Episodic associations enable past sequences of experience to be reconstructed. Semantic and episodic memories complement each other and work together in many ways. However, it seems best to begin our description by considering how they operate separately.

Semantic Associations

We don't know all the details, but semantic memory networks appear to form co-occurrence associations across multimodal domains. These co-occurrence links result in more complex kinds of perceptual and conceptual associations. These extended associations are meaningful in the sense that perceptual units are linked to familiar spatial, behavioral, emotional, and social contexts. These contexts are in turn linked back to other contextually related features, connecting them in a contextual web of related items. Because the interoceptive cortices also contribute heavily to these memory networks, the memory items are also linked with interoceptive feelings. It is these interconnected associations which result in the experience of recognition; that is, the ability to connect items with familiar tasks, situations, feelings, and other contextually related items. Because words are also perceptual units in humans, percepts and concepts are interconnected with words by these co-occurrence associations. This is why naming an item provides good evidence of recognition. It shows that the item has been correctly linked to functional contexts and categories.

A high-level overview of the organization of the declarative memory areas in the macaque monkey is diagrammed in Figure 7-1. As shown, there is a web of interconnections which extends across several discrete brain regions. In fact, there are actually two first-level input regions. These are the perirhinal cortex and the parahippocampal cortex. Studies with humans have found that the perirhinal cortex is more active during item or object recognition, whereas the parahippocampal cortex shows more activity when subjects are asked to recall scenes or context-related information.[6] This seems to occur because the perirhinal cortex receives all its major projections from perceptual regions in the ventral stream, while the parahippocampal cortex receives a broader range of inputs from both ventral stream areas and from the affordance and spatial processing areas in the dorsal stream.[7]

This dual-input architecture has led to a model of semantic memory which is characterized as *item-in-context* memory.[8] The basic idea is that the perceptual inputs to the perirhinal cortex are bound together as multimodal "items," such as objects, persons, actions, and, in some agents, words. In contrast, the perceptual inputs to the parahippocampal cortex appear to be bound into broader contextual and categorical associations. Cross-connections between the perirhinal and the parahippocampal input regions are thought to link items with their contexts, and contexts back to other related items. It is not exactly clear how to categorize the range of associated contexts. However, it seems that they often involve tasks-related situations, emotional states, social relationships, and/or spatial associations.[9]

The categorical connections for item-in-context memory can be quite specific. For example, human patients with selective damage to memory areas often show categorical losses.[10] For example, selective damage due to strokes or injuries can cause a patient to lose the ability to recognize a certain class of objects by name, or to relate a class of objects to their function, although perceptual processing in terms of the patient's ability to find matching pictures or to draw the objects remains unimpaired. Sometimes the errors reveal evidence of semantic associations, even when recognition is inaccurate, as one study made clear in its title, "Calling a squirrel a squirrel but a canoe a wigwam."[11] In that case, the contextual links between American Indians, canoes, and wigwams were intact, although the associations between specific items and their names were not always correct.

The two initial input regions of the declarative memory system send outputs to a middle layer, the entorhinal cortex (see Figure 7-1). This region recombines its own set of multimodal inputs with items and contexts processed by the two initial regions. It is not clear how well abstract concepts and categories take form in the initial input regions, but it seems likely that they are well represented in this second tier of semantic associations. There is also evidence that some spatial relationships are mapped into relational networks here, in part based on pre-processing prior to hippocampal input, and, in part due to the post-processing of associations formed in the hippocampus.[12] Given that the networks here are capable of forming relational mappings, and that not all the inputs to the entorhinal cortex are spatial, it seems likely that other kinds of relationships (e.g., social status hierarchies) may also be represented in relational maps in this region.

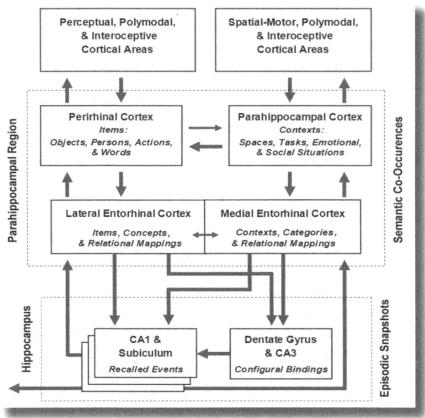

Figure 7-1: Medial temporal lobe memory regions and proposed functions.

Episodic Associations

The mix of items, contexts, and relational associations processed in the entorhinal cortex serve as inputs into the hippocampal region. The spiral shape of the structures forming the hippocampus is what gives this region its odd name. The seahorse, genus *Hippocampus*, has a curled tail, and so by analogy, the spiral shape of this region has come to be identified with this same label. However, there is another label used for this region which is also related to the spiral. Some early researchers thought that Ammon, the Egyptian god with a ram's head and horns, provided a more classic label, and this terminology also became popular. As noted in Figure 7-1, several regions in the hippocampus have come to be labelled as numbered "CA" regions. The CA designation is actually an abbreviation for the Latin reference to the horn of Ammon, *cornu Ammonis*. If we were naming this region today, we would probably simply call it the memory spiral. Whatever label we use, it should be clear that the networks in this spiral have special properties.

It is within this spiral that clusters of features come to be bound together in episodic memories. Episodic memories are *situationally related, co-occurring associations.* They capture things that happen to occur together. There is currently much research into how these networks encode sequential episodes in memory. However, several years ago Howard Eichenbaum and colleagues proposed an intuitive model that helps to explain the process at a systems level.[13] The central idea is that input features that happen to occur at the same time are constantly being cross-connected by configuration-capturing cells on the input side of the hippocampus. Originally these cells were thought to capture only spatial associations and were called "place" cells. However, subsequent work has shown these cells also encode other features, even behaviors. Eichenbaum and colleagues therefore chose to characterize them as "event" cells, cells that connect a small number of co-occurring features together as an event. When the features are relevant to spatially connected events, they can be used to reconstruct spatial associations and path decisions. However, when they are relevant to situational or behavioral events, they can also be used to reconstruct episodes of past experience and guide situational decisions.

Perhaps the simplest model for thinking about these event cells is to think of them as capturing snapshots of features that happen to gain attention together. Of course, we must understand that the snapshots in the hippocampus include all kinds of perceptual, conceptual, and contextual elements, not just visual features. The idea is that the event cells are constantly forming configural associations among active items and contexts. Some of these snapshots may capture features localized in time or place (such as the objects on a desk), while others may encompass broader contextual features (the spatial arrangement of furniture in a room, or the sequence of rooms encountered along a hallway). Some features may be encountered for brief time periods and occur in a few snapshots – say, objects seen as you pass by a particular doorway. Other features may be encountered for longer periods and may be included in many of the snapshots, such as the hallway you walked through and the package you were carrying as you began your walk.

The net effect of capturing all these configural snapshots is that the more often a feature is noticed, the more likely it is that it will be included in a number of different configural snapshots. As a result, the recall of a feature can activate associations with several other snapshots. The connectivity between features in overlapping snapshots suggests how episodic memory sequences can be reconstructed on recall. A cue triggering the recall of any one snapshot will tend to activate features connected to others and trigger the recall of yet other snapshots. As a result, simply by shifting attention to particular items in a series of configurations, a string of associated snapshot events can be recalled. And because the configurations contain many kinds of items and contexts, the connectivity is highly flexible. For example, a different episodic sequence can be reconstructed simply by focusing on a different branch of cues. Spatial sequences can be used to reconstruct map-like relationships, whereas other features can be used to reconstruct the temporal flow of events. If you paid enough attention to generate repeated snapshots, then you can probably even track back to recall in what room you left that package you were carrying.

Maintaining configurations as separate snapshots requires a neural architecture that uses a sparse storage strategy.[14] This is an architecture in which information is stored in largely separate places. Overlap in the storage process results in blended associations. Once associations are blended, they cannot be recalled as discrete events. This highlights a key difference between what are typically thought of as *learning networks* and what are considered *memory networks.* Learning networks involve a lot of overlap and form blended input-output associations which average successes together across trials. They generally form associations slowly, often taking many trials to learn best-fit input-output rules. Configural memory networks, in contrast, must form associations quickly to capture elements that occur together in one situation. Sparse storage enables the separate events to be stored and recalled as discrete events. This is where a recent finding changes how we think about the episodic memory networks. For a long time, researchers thought that all the neurons in the brain were formed by the time of birth, or soon thereafter. However, in the dentate gyrus of the hippocampus, there is a layer of late-stage neural stem cells which produce as many as five thousand new configural memory neurons each day.[15] Conditions that promote neurogenesis, such as exercise or eating lots of blueberries, can apparently double that number.

Given the need for sparse storage to support configural memories, this daily supply of new neurons makes perfect sense. It ensures that the input networks of the hippocampus can always form unique configurations for different events, because they always have new neurons waiting to be connected in new feature configurations. However, research also indicates that many of these newly formed configuration cells die within a few weeks. This explains why most episodic snapshots are short-lived. We don't maintain highly detailed configural memories for most experiences for very long. There is only a short window of time, ranging from hours to days, in which most of the details of a recent situation can be recalled. However, it appears that those episodic configurations which are frequently accessed tend to be maintained for months or longer. New configural associations may even form when a previous configuration is recalled.[16] That would have the effect of refreshing the memory, recruiting more configural cells, and extending the time over which the configuration would be maintained.

Consistent with the snapshot model of episodic memory

formation, recent findings have confirmed that episodic associations do in fact depend on overlapping configurations formed in the input regions of the hippocampus.[17] As mentioned earlier, there is also evidence that the recall of feature configurations facilitates the consolidation of relational co-occurrence mappings in grid-like regions of the entorhinal cortex, making some episodic associations more permanent.[18] Indeed, there is evidence that the cells forming new episodic configurations may even fire spontaneously from time to time, reinstating the configural associations in a manner that appears to help to consolidate them in longer-term associations.[19] In some cases, these configural flashbacks even manage to reach conscious awareness. If you hear a song on the morning radio, and an hour later you find yourself humming the same tune, it could be due to one of these consolidation flashbacks breaking through to conscious awareness. Similar flashback associations are known to occur during REM sleep and to influence the content of dreams at times.

Re-Membering

Understanding the network organizations that enable perceptions and memories to form also helps us understand how such experiences are reconstructed from memory. Memory recall begins when retrieval cues, features from ongoing perceptions, trigger co-occurrence associations with contexts and related items. Those co-occurrence associations trigger backward links to the perceptual inputs with which they were first associated. A point to note here is that the perceptual networks don't pass on the perceptual patterns they have detected to the memory networks. They simply pass a signal that they have recognized something. As a result, the memory networks don't have to store perceptual experiences in a new place. Rather, they store backward-connecting pointers to the original perceptual networks. These links enable the memory networks to bind perceptual features in a web of semantic and episodic associations. When a memory is recalled, the perceptual associations are reactivated via the backward links to the perceptual networks and the co-occurring percepts are re-experienced.

The potential for interactions between semantic and episodic processing suggests that the two memory regions complement and support each other in a web of connectivity. Perceptual inputs and interoceptive feeling states, together with some early sensorimotor affordances and spatial cues, are routinely mapped into functionally related item and context co-occurrence associations in the semantic memory networks. These semantic associations even persist as short-term activation states for as long as a minute or more after the perceptual states end. This sustained activity has the effect of holding recent items in short-term memory. It seems likely that this sustained activity also contributes to the formation of more overlapping feature configurations in the hippocampus. This presumably increases the degree of connectivity among semantic memories and episodic configurations. Similarly, when episodic associations are recalled, they reactivate items and contexts, thereby aligning semantic associations with features bound in episodic memory.

Apparently due to the depth of connectivity in these networks, this web of semantic and episodic memory associations takes form rather slowly. Studies suggest that it commonly takes about a half second for memories to be link up with new perceptual events. However, once memory associations stabilize, perception is significantly augmented because memories supplement the experience of the present with meaningful associations from the past. The advantage of forming all these interlinking semantic associations and episodic configurations is that a few perceptual cues in a particular context may be sufficient to activate a whole group of functionally related associations from past experiences. As a result, an odor in the air can be perceived as more than merely a sensory experience. If it comes to be recognized as the smell of cookies baking, then it is likely to remind us of other cookie-related experiences. It may even reactivate associations related to childhood experiences, like helping mom bake cookies.

Sometimes the memories are so attention capturing that we don't even notice what perceptual inputs triggered their recall; we are simply walking near a bakery, and suddenly we find ourselves recalling childhood memories.

Memories with strong emotional linkages, like those for the day you forgot to set the timer and the burning cookies set off a household panic, are particularly likely to be remembered for long periods of time. This is because memory storage is influenced by the intensity of interoceptive reactions. The strategy of controlling memory storage based on interoceptive activation levels is known as *memory modulation.*[20] Memory modulation is managed by saliency signals originating from the medial septum and the diagonal bands of Broca in the septum. These signals seem to be the origin of a rhythmic slow wave activity pattern known as hippocampal theta. The mechanisms are not well understood, but stronger theta activity appears to promote stronger memory associations and perhaps the formation of more frequent episodic snapshots. The subjective experience of time slowing down under strong states of interoceptive arousal is probably related to such increases in attention and episodic memory activity. The idea is that the more configural snapshots that form in a particular time period, the slower time seems to flow. When such detailed memories are recalled, they are similarly experienced in slow motion.

An interesting feature of memory modulation is that it is something of a two-way street. Because the declarative memory networks receive a broad array of inputs from the interoceptive cortices, interoceptive associations are always significant parts of any memory. As a result, whenever a memory is recalled, the reconstructed experience does not simply include perceptual associations but also part of the interoceptive states that were active when they were stored. The late Russian neuroscientist Olga Vinogradova argued that the interoceptive activity serves as a sort of selection and latch mechanism.[21] The idea is that memories compete for recall based on the strength of the interoceptive states that they activate. As a result, the memories with the strongest interoceptive associations in a situation are the ones most likely to win the competition and gain attention. Once a memory is recalled, the interoceptive associations serve as something of a latch. Only memories with stronger interoceptive associations are likely to break through and displace them, at least until attention shifts for some other reason.

Latching onto memories based on their importance, as weighted by interoceptive associations, has obvious adaptive potential. It means that the most important memories related to a situation are likely to dominate attention. However, it could be risky to latch onto past associations for long time periods, especially when other things are happening. To prevent this latching process from locking onto memories too long, Vinogradova notes that there is a region in the lateral septum which seems to provide a way of interrupting the memory latch.[22] Conditions known to interrupt the memory latch are stimulus change, especially novel or surprising changes which trigger change alerting signals from the sky blue place, the locus coeruleus. Thus, whenever events change suddenly, memory processing tends to be interrupted so that the new situation can be promptly evaluated by recalling co-occurrences linked to the new situation. It's a simple but very functional latch-and-interrupt process. The medial septum and the diagonal bands of Broca in the septum send latching signals whenever strong interoceptive associations are recalled. The lateral septum sends an interrupt signal whenever perceptual conditions change.

A Perceptual Stream for Cogley

Our emphasis has been on visual perception, and we'll mostly stay with that here. However, it seems worth noting that each sensory domain requires its own analysis strategies. Visual patterns have a spatial character, whereas auditory patterns are sequential in nature. A sequence of sounds, like the hum of an engine followed by the screech of tires, the crunch of metal, and the tinkle of falling debris, can come to be recognized as a distinct perceptual event: a car crash. Similarly, a series of phonemes can come to be perceived as a word rather than just separate sounds. This means that the networks that detect sound patterns must differ from visual pattern recognizers. In particular, short-term memory buffers will be needed within the auditory networks to help recognize sound segments and to connect them into useful sequences. However, once a sequential auditory pattern is recognized, it can be linked with memories in much the same way as a complex visual pattern. All that is required is a link between the perceptual analyzer network and the memory networks. This means that the memory networks can represent multimodal items without concern for what kind of perceptual system detected the input patterns. All the memory networks need to do is reactivate the relevant perceptual network. Once reactivated, the pattern will be reconstructed in the appropriate perceptual framework.

Using visual perception as our model, it seems likely that pattern recognition can be implemented using a webbed hierarchy of adaptive resonance networks. The upper-level networks in each stage should learn to detect patterns in the lower-level inputs, and when recognized, the expected features should be primed by backward links to the previous level networks. In effect, the representations should interconnect in both bottom-up and top-down directions. A capacity to use as many as twelve processing stages will be needed to match the complexity of analysis that occurs in human visual perception. A special detail area for the analysis of subtle visual pattern relationships, like those involved in facial recognition, will also be needed. Outputs from higher-level patterns recognizers should then be fed into multimodal pattern recognition networks. In the brain these multimodal processing areas are often called association areas. A rich, multilevel mix of associative, perceptual, and interoceptive inputs should then serve as inputs into Cogley's declarative memory system.

Deciding how Cogley's declarative memory networks should be organized is a major challenge, because although we have some basic models for how the declarative memory system works, there are still many aspects of the design which remain to be resolved. Perhaps we should start with a requirements list.

Cogley will need a declarative memory system which is capable of item-in-context memory mappings.

- The item-in-context mappings must be capable of forming both spatial and non-spatial associations.
- The input networks must be capable of forming semantic co-occurrence associations – that is, thematically meaningful cross-connections – for both spatial and non-spatial contexts.
- The later stage networks must be capable of forming episodic co-occurrence associations – that is, situational cross-connections with items and both spatial and non-spatial contexts.
- Once activated semantic associations should be temporarily sustained so as to provide short-term memory pointers to recent items of activation.
- A small set of cues must be sufficient to retrieve both contextually relevant semantic associations and situationally connected episodic memories.
- Episodic associations should trigger the recall of connected sets of experiences. Although these episodic events may be triggered by current perceptual cues, because they are bound together, they may often compete for attention with ongoing perception.
- A memory modulation process will be needed to ensure that stronger memory connections are formed in times of stronger emotional reactivity.
- Memories should generally compete for recall based on their importance as indexed by the strength of the interoceptive associations they activate.
- It should also be possible to guide memory recall selectively based on attention to retrieval cues and contexts.
- A latch-and-release mechanism must be available to lock relevant memories in place on recall while still

enabling memory processing to be reset when the situation changes.

- The memory system needs to be capable of making some sort of goodness-of-fit assessment for the memories retrieved by a cue, or for the cue's novelty and the absence of retrieved associations. Further, it must be capable of signaling this assessment to non-memory systems.

Addressing the Requirements

Of course, not all of these tasks can be accomplished solely by the networks of the declarative memory system. However, Cogley's memory networks must be capable of supporting these functions. Given the number of requirements, perhaps we should address them one at a time, although it should be clear from the discussion that many of these processes interact. We'll try to follow a biologically inspired design strategy. However, given that we don't fully understand how all these memory processes work, in some cases we'll have to invent plausible mechanisms to support them. There appears to be a lot of variability in human memory associations, so there is no reason to assume that we cannot establish similar memory associations in Cogley. Just as in humans, Cogley's co-occurrence associations will be highly individualistic. Our aim should simply be to include enough of the essential properties of human memory to provide Cogley with memories that have similar operational characteristics.

Item-In-Context Memory. As described earlier, there are two first-level input paths into the semantic memory region. These input paths carry a different mix of information. One is more involved in item and object recognition, and the other is more involved in spatial and contextual recognition. Following our bio-inspired design strategy, it seems appropriate to give Cogley a similar dual-input memory architecture. Let's call these two input paths the item path and the context paths, respectively. There should be a broad mix of perceptual inputs into both these paths. However, the item path should be proportionately more visually dominated and produce more concise representations.

The context path, in contrast, should receive a more even mix of perceptual inputs and should include inputs from early spatial-motor planning networks as well as interoceptive inputs, so that Cogley's item memories can be linked to both spatial and non-spatial context domains.

Spatial and Non-Spatial Associations. The fact that both spatial and non-spatial contexts are found in memories suggests a division in the flow of information within the memory system which we can use in Cogley's design. Much of the connectivity among spatial cues appears to depend on associations made in the hippocampus. Therefore one path of information flow through Cogley's mid-level semantic network should be optimized to make sure that spatially relevant context information reaches Cogley's episodic memory networks so that he will form spatial memories. Item information should flow through a separate path to the episodic memory networks, where it can be linked with the spatial configurations. Given that it is less clear how non-spatial contexts like tasks and emotional states are managed in human brains, let's assume that they also follow a separate path for Cogley. This should provide Cogley with the opportunity to form situational memories that connect items with both spatial and non-spatial contexts.

Meaningful Semantic Associations. Cogley's perceptual networks should bind features together based on spatial-temporal coincidence. That is, sets of features that occur together in time, and which are spatially collocated, should be treated as connected perceptual units. Temporal contiguity should also be important for connecting semantic co-occurrence associations between items and contexts. However, to be meaningful the associations need to be connected along some other dimension. Given that we know that the strength of a memory is correlated with the strength of interoceptive changes with which it is associated, it seems reasonable to use co-occurring interoceptive state categories as an added grouping factor. This would mean that non-spatial contexts should tend to form around interoceptively valued tasks, including language-centered tasks in humans, emotional situations, and interoceptively charged social scenarios. It also means that spatial contexts may include interoceptive associations that make them more meaningful.

Short-Term-Memory Activations. Semantic networks appear to be capable of holding recent experiences active in short-term memory. In fact, once co-occurrence associations are activated in the parahippocampal cortex, they remain partially activated for about a minute, and for longer if they are occasionally revisited. We will need to ensure that Cogley's semantic memory networks can sustain recognition units for similar short-term periods. For starting values, let's assume that these associations lose about half their activation in the first fifteen seconds and most of the remainder in the next sixty seconds. This sustained activation should serve to facilitate the formation of co-occurrence links among items and contexts in semantic memory and provide more overlapping inputs to the episodic memory networks. Re-attending to items should partially refresh them and cause them to be sustained longer.

Situational Episodic Associations. By sustaining the activity of semantic associations, short-term memories will provide a rich context of associative links to supplement Cogley's interpretation of his ongoing perception. This mix of item and contextual associations will also provide a steady stream of inputs into Cogley's episodic memory networks. The episodic networks should capture feature snapshots – that is, situational co-occurrence associations for active perceptual events. These events will be configural connections among items and both spatial and non-spatial contexts. A high rate of capture, say about two per second, seems about right when interoceptive states change rapidly. It may even be useful to capture several configural snapshots at the same time when interoceptive states of strong. Lower capture rates, perhaps as few as one every few minutes, may be sufficient when changes in interoceptive activity are slow. When a configuration is activated on recall, all its associated item and context links should become activated together and flow back to activate the semantic networks, which should flow back to activate the perceptual networks. This mix of events should enable Cogley to reconstruct episodic sequences and re-experience them on recall.

The beginning strength of episodic configurations should be a function of the interoceptive associations which were active when the snapshot was formed. The strength should then degrade exponentially until it reaches some cut-off level, at which time the node should be freed up for reuse. Let's say a strong snapshot should take about seven days to be fully degraded if never refreshed. A weak snapshot may be lost in a few hours. This should limit detailed memory recall to a medium-term time range and free up episodic memory space for other medium-term associations. Snapshots should have their strengths renewed on each access, so that they will be remembered longer, and new snapshots should form when memories are recalled in new contexts, introducing new items to the mix of configural associations. This should enable frequently recalled snapshots to be refreshed repeatedly.

A background reactivation process, a flashback process, will also be needed to strengthen semantic associations. Flashbacks should refresh episodic snapshots and ensure that important episodic memories are remapped into more permanent co-occurrence networks before the configural snapshots decay. Evidence suggests that there are map-like regions in the entorhinal cortex where spatial information is represented in relational ways. Cogley will need similar mapping spaces in his mid-level memory network for connecting spatial relationships. This will provide longer-term storage for spatial information. Given that part of this network also carries non-spatial information, it seems reasonable to extend this mapping strategy to non-spatial relationships. The goal should be to provide a system for mapping any kind of transitive semantic relationships.

Memory Retrieval Cues. The ultimate value of having all these interoceptively weighted semantic and episodic associations will be their ability to help Cogley make better decisions. To make this work, whenever perceptual analyzers activate memory links, the items and contexts activated in semantic memory should trigger the recall of associated memory items and events. These should feed back to reactivate associated perceptual and interoceptive states. In effect, Cogley's perception will constantly be supplemented by associations from past experiences, and those added associations will enhance his ongoing perceptions and feelings. This will cause him to re-experience aspects of past co-occurrence associations whenever current perceptions trigger their recall.

Because memory associations will propagate back to activate some of the affordance regions shared with the dorsal stream and to interoceptive networks related to motivation, Cogley's declarative memories will be indirectly linked to action planning. Recalled motive states and affordances will not be as intense as real perceptions, and they won't be linked with active spatial locations, so they are unlikely to dominate motor planning except during imagined activity. However, activating memories that share compatible features should influence Cogley's planning by biasing his attention to relevant objects and affordances and by enhancing interoceptive dispositions that motivate particular actions. This will enable memories of previous situations to influence the action candidates Cogley considers when he encounters similar situations.

Episodic Events Should Compete for Attention. When items and contexts reach the episodic memory networks, the configuration that best matches those inputs should be activated together as outputs. This will ensure that Cogley recalls past episodes of experience as discrete events. The complexity of the episodic events should often be rich enough that they cannot be fully established unless they gain attention. This means that attention to episodic memories will often compete with attention for ongoing perceptions. It will take us several more chapters to describe the neural organizations required for Cogley to establish and manage attention. However, it should be obvious that attention will be an essential process for managing episodic memories. First, because features that gain attention are more likely to be stored in memory. Secondly, because attention to features recalled in memories can serve as retrieval cues for searching subsequent memories. By shifting attention among related items and contexts, Cogley should be able to search through his episodic associations and prompt the recall of related events. In this way, he should be able to reconstruct selected episodes of attention from past experiences.

Memory Modulation. To ensure that Cogley's memories will be adaptive, he will need a memory modulation process which reinforces associations for items and contexts that co-occur with strong interoceptive experiences. In humans, the interoceptive saliency signal is thought to be generated in the septum, a part of the extended amygdala. Cogley will need a similar interoceptive signaling module. This module should be sensitive to the level of interoceptive activity, and its signals should increase the likelihood that Cogley stores memories for items which are associated with stronger interoceptive states. Because interoceptive inputs will be broadly represented in Cogley's memory networks, similar interoceptive states should come to be re-activated whenever Cogley recalls those memories.

Memories Should Compete for Recall. In order to ensure the recall of best-fit associations, memory recall should compete for attention based on the strength of the interoceptive states that are co-activated with them. We have already noted that interoceptive states will be part of Cogley's memory recall. Following Olga Vinogradova's model, the level of interoceptive arousal that is activated as memories begin to be recalled should provide a competitive force which enables those memories with the strongest interoceptive support to win the competition for recall and compete for attention. In effect, interoceptive states will not only be important for memory storage, they will also determine what memories are likely to be recalled. This will assure that the most meaningful memory associations, in terms of interoceptive value, are the ones most likely to be stored and the ones most likely to gain attention on recall.

Attention to Retrieval Cues Should Guide Memory Recall. We have already argued that attention to items and contexts should guide the retrieval of memories in a top-down way.

Interoceptive activations, in contrast, provide a bottom-up mechanism for guiding memory recall. Obviously these two retrieval strategies will need to be balanced, such that when interoceptive activations states are stronger, they should dominate the retrieval process. However, when interoceptive states are low to moderate, then attention to retrieval cues should have more influence. This should enable Cogley to manage his memory recall better while still biasing his recall toward more interoceptively valued memories. Learning to manage memory recall will take some practice, but if we arrange for the recall of targeted memories to be rewarding, then over time Cogley should learn to manage his memory recall.

Latch and Release Mechanisms. If memories compete for attention based on their associated interoceptive activity, then other things being equal, this should cause those memories with strong interoceptive associations to gain attention, at least until shifts in attention occur for some other reason. This would mean that Cogley should sometimes explore various aspects of his memory for extended periods of time without noticing what is going on in his immediate environment. Therefore, it seems reasonable to provide him with a mechanism which can interrupt ongoing attention to memories, in effect releasing the memory latch, when unexpected changes occur.[23] Olga Vinogradova argued that the lateral septum provides a signal which interrupts memory processing in humans. Change alerting signals from the sky blue place, the locus coeruleus, are thought to be one of the primary processes that activate the lateral septum and interrupt ongoing memory processing. We will need to give Cogley a similar change-alerting module and ensure that it has connections with the memory interrupt circuitry.

Goodness of Fit and Attention Value. Based on the strategies and algorithms described above, whenever perceptual inputs match memory patterns with strong interoceptive associations, those patterns should be recalled. When no pattern match is found, the memory networks should react with a novelty response. This novelty signal should cause Cogley's memory networks to begin to learn co-occurrence associations with the new pattern. Given this behavior, it is also important to provide the cognitive networks outside the memory system with information about the strength of the interoceptive associations activated by memory recall, or the novelty of perceptions that fail to link with past experiences.

We don't know much about this process, but the primary pathway projecting outside the memory system, the fimbria-fornix path, originates largely in the subicular complex. Recent research suggests that this complex receives inputs from both semantic and episodic memory regions; it therefore seems to be a likely location for signaling about these evaluations. Some of the outputs from this complex are directed to the ventral striatum and the prefrontal cortex, where they help coordinate action planning with memories.[24] Cogley will need a similar support structure to signal about his memory evaluations. When a pattern is matched and co-occurrences are activated, this module should provide a signal weighted by the strength of the co-activated interoceptive associations. This signal should encourage networks in the planning stream to align with the recalled memory associations. When the pattern is novel, attention to the perceptual features should be favored by the learning algorithms.

A Visual Memory Bias

Although we are considering the organization of the perceptual stream, there are a few additional organizational issues that deserve comment. The declarative memory networks operate with what appear to be rather general purpose algorithms. They form co-occurrence associations based on whatever kinds of inputs happen to occur together. Evolutionarily, this makes them surprisingly flexible. As Jaak Panksepp suggests in the quote that opened this chapter, the hippocampus of the groundhog and the falcon appear to operate in a similar manner, although they receive largely different kinds of perceptual inputs and end up supporting very different kinds of tasks.[25] A more detailed comparison has been made between the well-studied memory systems of the rat and those of the macaque monkey.[26] This comparison shows that the rat's inputs to the perirhinal cortex item/concept region are dominated by olfactory inputs, and the monkey's inputs to this region, like those of humans, are dominated by vision.

This is one of the fascinating things about different kinds of minds. They conceive of the world in different ways, in large part because the co-occurrence associations they form are weighted differently for different sensory modalities. For example, Tom's item recognition memories appear to be dominated more by auditory and olfactory inputs, with vision being a lesser weighted input for recognition. Although Tom uses his vision superbly for orientation and dorsal stream action planning, much like patient SB, he doesn't use vision nearly as much for object recognition. This does not imply that Tom cannot recognize anything visually. His ventral stream for vision isn't completely disconnected, like it is for patient SB. However, Tom places less weight on vision for recognition, just as we humans place less weight on olfaction for recognition. This means that Tom's item and context categories are necessarily organized much differently from mine. For example, his hippocampus takes many more snapshots, which include auditory and olfactory items, than my hippocampus does.

We primates are largely visual creatures. Vision dominates our perception, how we recognize things, and more generally how we represent and think about the world. So, if we want Cogley to develop cognitive concepts and skills similar to those of his human partners, then we will need to ensure that his visual sensors produce salient input signals and are the predominate inputs to his item memory networks. His balance of inputs should be weighted, like those of humans, with vision first, audition second, touch third, and olfaction last. This will ensure that Cogley is more likely to assemble human-like percepts and use vision as a guide to forming most of his higher-order concepts. However, auditory inputs should be represented well enough to ensure that Cogley will also be able to recognize words as items in semantic memory and associate them with his visually dominated concepts. This will happen in large parts because words will be part of the social context in which Cogley learns about other items. Important feeling states should also be represented as items, although proportionally, they should probably not be represented in any more detail than olfactory information. It seems likely that most feelings states will be recognized as associated contexts.

CONCLUSION

It should now be obvious why the planning stream is largely separate from the perceptual stream. These two pathways require very different kinds of network organizations to optimize the work they do. The planning stream does simpler perceptual analyses, which make it faster; it links simple spatial-motor percepts with compatible motor programs to form action plans. Processing in the perceptual stream is slower because it resolves patterns on multiple levels. However, it is capable of detecting much more complex patterns and of linking those patterns with others. Semantic memory networks link items with co-occurring contexts – spaces, tasks, emotional, and social situations – and contexts with co-occurring items. These co-occurring associations are the essence of what we think of as recognition, the ability to connect perceptual units with meaningful tasks and labels and to frame them in useful ways.

Episodic networks, in contrast, capture episodes of conscious awareness in what are basically configural snapshots. These snapshots make it possible to place items in spatially connected relationships, and to reconstruct perceptual experiences along sequential storylines. The networks required for episodic associations require a different architecture than those for semantic associations, but associations in each of these memory regions serve to supplement ongoing perceptions in a situation. With a functional perceptual stream in place, Cogley will experience the world on two levels. One grounded in the present, and one grounded in the past. In this way, he will always experience more than is present in the immediate situation. Further, by shifting his attention to relevant episodic associations, he will be able to reconstruct, and partially re-experience, past episodes of attention.

Admittedly, the perceptual architecture proposed here is rather complicated. However, we need this level of detail to provide Cogley with a functional perceptual-processing stream. We have yet to explain how Cogley will manage his attention and why he will experience consciousness. However, it should now be obvious that declarative memory is largely a tool for remembering and supplementing conscious experience. It is linked with feelings and is guided by attention. It connects the items of perception with contextual associations that make them meaningful. With a functional perceptual stream in place, Cogley will become a more perceptive and more conscious agent. However, we have some work to do before Cogley reaches consciousness. First, we'll need to provide him with networks that enable him to represent feelings. That will be the goal of the next chapter. Subsequently he will need networks that enable him to focus his attention. Meanwhile, Tom has begun exercising more to promote the level of neurogenesis he needs to be able to capture more episodic snapshots of his experience. However, he refuses to eat blueberries, even though they have a similar effect.

CHAPTER 8
THE INTEROCEPTIVE STREAM

The perceptual stream extracts perceptions and expands them in the context of memories for related experiences and feelings. However, though memories may activate feelings, memory networks are not where feelings are perceived. That happens in a third processing stream, the interoceptive stream. This stream includes the input pathways for the internal perception of somatic status and the reactivity pathways which adjust motive and emotive states in reaction to status evaluations. The reactivity pathways include most of the networks which border the underside of the exteroceptive cortex, a loose collection of structures commonly referred to as the limbic system. Although the motivational role of the limbic system has long been recognized, until recently the homeostatic relationship among status and reactivity networks has not generally been appreciated. However, the connectivity of pathways in this stream has now been established, and its importance is hard to overstate. Activity in the interoceptive stream guides and prioritizes processing in the other two streams. In fact, interoceptive input is essential for logical decision making, because feelings of status and reactivity underlie the values which networks in the other streams must take into account to make appropriately valued decisions.

THE INTEROCEPTIVE STREAM

> Feelings let us catch a glimpse of the organism in full biological swing, a reflection of the mechanisms of life itself as they go about their business. Were it not for the possibility of sensing body states that are inherently ordained to be painful or pleasurable, there would be no suffering or bliss, no longing or mercy, no tragedy or glory in the human condition. – Antonio Damasio, *Descartes' Error*, 1994

It's late in the afternoon on Wednesday September 13, 1848, and a remarkable accident is about to occur. Phineas P. Gage, a twenty-five-year-old construction foreman, is managing a crew which sets explosive charges to clear paths for the railroad. Today, the crew is clearing rock for tracks being laid near Cavendish, Vermont. As usual, Gage oversees the drilling and packs the charges himself. His tool for the task is a personally designed, three foot seven inch-long iron bar, tapered at one end. The bar is one and a quarter inches in diameter and weighs over thirteen pounds. A new hole has been drilled in the rock, and Gage has just placed a measure of powder and a fuse inside the hole. However, he's momentarily distracted by a call behind him, and when he looks back he fails to notice that his crewman has not yet placed the packing sand in the hole to cover the powder. When Gage begins ramming his tamping iron in the hole to pack the charge, he's striking against raw powder. Suddenly, the powder explodes; shooting the iron bar outward like a massive bullet, and Phineas Gage is directly in its path.

As neurologist Antonio Damasio later reconstructs the incident, Gage's head is thrown back by the force.[1] The tapered end of the bar enters his head through his left cheek, pierces the base of his skull, and exits through the top just behind the forehead, resulting in extensive damage to the ventromedial prefrontal cortex. Gage is dazed and is carried to a nearby oxcart by his crew, but amazingly he's awake. Within a few minutes he's talking, and he sits upright in the cart for the three-quarter-mile ride into town. When the cart stops Gage gets out with minimal help and is escorted into the local hotel. Gage converses with the hotel owner while the local physician is summoned. The physician is astonished. There's a funnel-like opening on the top of Gage's head. Inside the wound he can see brain tissue throbbing with each heartbeat. Yet the patient is lucid and recounting his accident. Subsequent newspaper reports describe Gage's apparent lack of deficit as something of a miracle. However, things aren't quite what they seem. Gage can still walk and talk normally, but he is a changed man. The injury has disconnected the logical planning areas in his prefrontal cortex from his social feeling networks. His personality and his judgment will never be the same again.

THE INTEROCEPTIVE LATTICE

Feelings depend critically on a system of interoception and reactivity.[2] We'll refer to this pathway as the interoceptive stream. Our goal in this section is to learn enough about the functions of this stream to provide Cogley with a similar architecture. Broadly defined, the signals in this stream represent internal perceptions regarding somatic and emotional status, and they engage somatic and emotional output dispositions which help to manage them. The status inputs include those from central brain regions and peripheral inputs from the sympathetic and parasympathetic nervous systems. The sympathetic status inputs are bundled together in the first layer of the ascending spinal cord. The neurons in this layer provide a broad array of somatic information, including sharp pain, dull pain, tissue temperature, mechanical stress, itch, and irritation related to immune system activity. Most of these inputs project directly to the parabrachial nucleus (PBN) near the top of the brainstem, where somatic status is first summarized.

Parasympathetic status inputs to this stream arrive by way of the facial, trigeminal, glossopharyngeal, and vagus nerves. These inputs converge on the nucleus of the solitary tract in the medulla and are then relayed to the PBN. They provide ongoing feedback related to heart rate, blood pressure, respiratory functions, taste, swallowing, and the general condition of the stomach and gut. Another source of input to the PBN located in this region is the area postrema, which contains a chemotoxic trigger zone sensitive to toxic substances in the blood. Signals about abdominal distress from the vagus nerve and signals of systemic distress from the area postrema project to a processing module known as the vomit center. When highly activated the vomit center initiates the stomach purging reflex for which it is named. When less activated it produces signals of distress we call nausea. These signals project to the PBN. Thus, the PBN has access to a broad array of measures about the status of the body.

The PBN is not only the first integration site for somatic status information; it also influences some output reactions, such as respiratory control and taste aversion. In addition, there are dense projections from the PBN to the periaqueductal gray (PAG), a gray colored output region lying along the central aqueduct in the midbrain. The PAG manages motor expressions for many core emotional behaviors such as rage, fear, and sexual behavior, and it has direct connections with the nuclei that manage autonomic nervous system activity in support of these reactions. The reactivity states and autonomic arousal levels also influence the tone of vocalizations. In fact, stimulation of the PAG can produce some emotionally related vocal outputs. Although the reactions triggered at this level appear emotional, they are largely reflexive and don't really result in directed emotional feelings. For example, if the PAG is isolated from higher-level networks and then stimulated, emotional-like reactions can be produced. Rage is one example. However, the reaction is called shame rage, because it isn't directed at anything. The networks at this level simply produce the motor components of the reaction and set autonomic arousal levels in support of them. They don't feel or direct emotions, but they normally pass information about their status evaluations and reactivity states on to higher-order interoceptive networks, where they can be felt and directed.

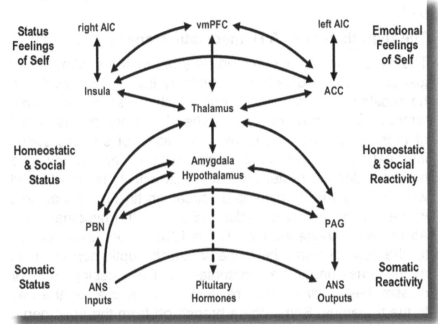

Figure 8-1: Major paths of interoceptive connectivity and processing.

Abbreviations: ANS, autonomic nervous system; ACC, anterior cingulate cortex; AIC, anterior insular cortex; vmPFC, ventromedial prefrontal cortex.

As illustrated in Figure 8-1, the interoceptive networks are generally arranged in a lattice hierarchy with connections to the cortex channeled through the thalamus. The brainstem interoceptive networks have reciprocal projections with two mid-level interoceptive sites in the base of the upper brain, the hypothalamus and the amygdala. These are some of the most critical and complex interoceptive control networks in the brain, and they are considered major components of the limbic system. Many of the mechanisms in these regions are still not well understood. However, given that they contribute to what an agent feels and how they are motivated, we need to gain a basic understanding of what these networks do and how they are interconnected, because Cogley will need similar interoceptive networks if he is to feel and behave like his human partners.

The Hypothalamus: A Homeostatic Manager

The hypothalamus lies immediately below the thalamus and just above the midbrain. It is centrally involved in regulating homeostatic states such as hunger, thirst, stress reduction, temperature regulation, and general activity cycles. Most of these functions require the integration of somatic status information with motivational dispositions, and many also trigger endocrine system activity. This activity is coordinated with daily light-dark rhythms by circadian timing mechanisms in the suprachiasmatic nucleus (SCN) of the hypothalamus. As its name suggests, the SCN is located just above (supra to) the optic chiasm, the place where the optic nerves cross as they pass under the hypothalamus. The crossing point, or chiasm, lies a few inches directly behind the eyes on the way to the thalamus. Some fibers branch off from the optic nerve at the chiasm and innervate the SCN. These fibers provide information about external light levels which the SCN uses to entrain its internal rhythm to the twenty-four-hour light-dark cycle of the outside world.

Projections from the SCN go to a sleep-wake center in the hypothalamus which is situated just in front of the optic chiasm. This center coordinates sleep cycles and daily activity levels with the circadian rhythms maintained by the SCN.[3] Other projections from the SCN go to the brainstem where they influence daily cycles of body temperature and autonomic activity. Yet other projections go to the paraventricular nucleus of the hypothalamus, which has neural connections with the pituitary gland. The pituitary is located just below the optic chiasm and releases a number of regulatory hormones, which target various glands and organs in the body, most notably for their behavioral effects – the thyroids, ovaries, testes, and adrenals. In most cases, the pituitary hormones trigger a cascade of other hormones. These regulate activity in target regions of the body by turning specific genes on or off in those regions. In effect, the hypothalamic-pituitary connection provides a downward control interface between the neural systems regulating interoceptive status and endocrine systems managing gene activity in the body.

The Amygdala: An Emotional Manager

Although the hypothalamus is largely dedicated to homeostatic regulation, the other mid-level interoceptive system, the almond-shaped amygdala, is dedicated to managing states of emotional reactivity.[4] Perhaps the best way to explain the function of the amygdala is to note that it has two main processing divisions. There is an affective output division which manages the magnitude and balance of emotional states of reactivity. These set the balance for many reactivity states. There is also an associative input division, known as the basolateral complex, which receives sensory inputs from the thalamus and the cortex and learns to use those cues to anticipate and guide reactivity changes. We'll begin with the affective division of the amygdala, because the outputs in that division are what the associative division learns to manage.

The first thing to know about the output division is that it is not confined to the almond-shaped structure traditionally known as the amygdala. The affective control functions actually extend across several adjacent brain regions, forming a larger control network that is commonly referred to as the extended amygdala. This extended version of the affective amygdala is interconnected by a bundle of nerve fibers known as the stria terminalis. The stria terminalis also connects the amygdala with the hypothalamus. Prominent structures along this path that are generally considered part of the extended amygdala are the septal area, the bed nucleus of the stria terminalis, and the nucleus accumbens shell area, a motivational area in the striatum. These regions are involved in a number of affective dispositions which have broad implications for social behavior, including aggression, fear, stress, caretaking, social bonding, and sexual interest. Many of these vary substantially between males and females. It should therefore come as no surprise to discover that many brain regions in the extended amygdala and hypothalamus develop differently in males and females.[5]

The nuclei in the associative division of the amygdala are more cortex-like and appear to be phylogenetically newer than other parts of the amygdala. These nuclei receive an array of multimodal sensory inputs from thalamic nuclei and some directly from the cortex. They in turn connect broadly with affective output regions across the extended amygdala. Some of the connections are biologically prepared. For example, loud noises, the barring of teeth, and low-pitched growls trigger fear and defensive reactions, and the sight of fleeing prey can trigger pursuit. In addition to such innately prepared reactions, the associative division of the amygdala learns to connect external cues with subsequent affective reactions. This learning enables exteroceptive events to influence interoceptive dispositions[6], an effect I'll call *valuation learning*.[7] Valuation learning is a particularly adaptive process, because it enables an organism to anticipate interoceptive changes and be better prepared for them. As we noted a few chapters ago, the biological movement areas in the temporal lobe have connections with the amygdala. As a result, the interoceptive effects of actions, both those produced by self and others, can readily be learned.

The extended amygdala also releases reinforcement signals which facilitate associative learning in other parts of the brain. This enables both innately prepared connections and learned associations to influence learning throughout the brain. Positive associations in the amygdala activate networks in the lateral hypothalamus. These networks in turn send special neuropeptide signals, called orexins, to the ventral striatum and to ventral tegmental dopamine areas. These signals engage motivational states that activate approach and investigatory behaviors. Negative associations activate regions in the medial hypothalamus which release orexins in response to fear and stress. These pathways project to the ventral striatum and ventral tegmental areas and motivate withdrawal and avoidance.[8] As a result, interoceptive associations formed in the amygdala trigger reactions in the hypothalamus which can promote approach or withdrawal decisions in the planning stream. Further, because the target systems in the ventral tegmental area release yet another reinforcement signal, dopamine, this link also facilities reward and avoidance learning.

Extended Reinforcement

In an earlier chapter, we noted that several kinds of reinforcement signals have evolved to tune Hebbian learning. It turns out that most of these signals are managed by regions in or connected with the extended amygdala and hypothalamus. In addition to the orexin-mediated signals from the hypothalamus, which manage approach-withdrawal dispositions and dopamine release, there are cholinergic reinforcement signals that originate from the basal nucleus of Meynert, the medial septum, and the diagonal bands of Broca in the septum. The basal nucleus of Meynert informs perceptual and motor areas in the cortex about affective activity using the neurotransmitter acetylcholine. This signal promotes cortical reorganization for cues and actions that are active when decisions are made in the basal ganglia.[9] In effect, the signal tunes the cortical sensory and motor systems to be more sensitive to cues involved in common learning tasks. The medial septum and the diagonal bands of Broca produce reinforcement signals which promote memory associations.

Social contact releases sociality reinforcement signals, such as oxytocin, vasopressin, and some endogenous opiods. Oxytocin, which is sometimes characterized as the "tend and befriend" hormone, is produced in the supra optic and paraventricular nuclei of the hypothalamus. Interestingly, oxytocin acts both as a hormone and neurotransmitter. Oxytocin is produced at high levels at the time of childbirth. Indeed it is one of the main hormones involved in triggering the birthing process. It is also released in both mothers and infants during nursing and the contact associated with holding and physical care. When detected by networks in the bed nucleus of the stria teminalis, oxytocin results in feelings of trust and caring, which promote interpersonal bonding. The mutual activation of oxytocin in mother and child is thought to account for the strong bonds that form between mothers and their infants. In fact, oxytocin appears to be critical for normal development: If infants do not receive a sufficient level of comforting contact in their first year, they do not develop normally. Instead, they grow listless and may even die. Those who survive tend to be socially defective.[10] Similar effects have been found to occur in monkeys. The changes in social behavior are associated with persistent decreases in cerebrospinal oxytocin.[11]

In humans the release of social reinforcement signals like oxytocin has been extended to non-touch forms of social interaction. For example, smiling, eye contact, attention sharing, and soft and melodic voice tones appear to have socially reinforcing effects. Eyebrow lifts as a signal of interest, positive head nods, and polite conversation are also thought to accentuate these signals. In fact, it seems likely that any social interaction you would describe as a close, touching, or friendly experience releases oxytocin. As noted earlier, in addition to the distribution of oxytocin to areas involved in maternal care and social bonding, there are also projections of oxytocin to the ventral tegmental area and to parts of the hippocampus. These distribution patterns appear ideally situated to enhance learning about social events in the planning stream and to promote the storage of socially relevant memories in the perceptual stream. The ability of oxytocin to enhance attention, bonding, learning, and memory in social situations is why it is considered one of the primary reinforcement signals for social development.

Summarizing the systems we have described so far, the brainstem components of the interoceptive lattice monitor a broad array of somatic status conditions and generate emotion-like behavior patterns and autonomic states of reactivity. The hypothalamus is largely sensitive to homeostatic needs, hunger, thirst, and temperature regulation. It also manages motivational states associated with approach and withdrawal and coordinates neural activity with endocrine system activity. The extended amygdala is more sensitive to social and emotional needs – fear, sex, caretaking, and social bonding. Regions in the extended amygdala are also responsible for the release of several reinforcement signals that influence learning, memory, and sociality. It seems clear that many of these status and reactivity states sometimes gain consciousness. However, at this level of the lattice they are pre-conscious sensitivities and moods. Antonio Damasio characterizes these pre-conscious states as parts of the *proto-self*. He suggests that they are precursors to feelings of self.[12]

The Interoceptive Cortices

We are not yet ready to explain how the moods of the proto-self become conscious feelings. That will have to wait until we describe the architecture of the thalamus. However, we can now say something about where complex feelings of self are represented. Ascending paths in the interoceptive lattice project through the thalamus and connect the signals from the proto-self with two primary interoceptive regions in the cortex, the anterior cingulate cortex and the insular cortex. The insular cortex is the primary cortical integration center for status feelings, and the anterior cingulate cortex manages feelings of emotional reactivity. Feelings of both status and reactivity are then re-represented in association areas of the anterior insular cortex. These multiple layers of representation are thought to contribute to the complexity of subjective feelings. Downward projections from the cingulate, augmented by the amygdala and the PAG, modify the activity of the autonomic nervous system in readiness for action. In this manner, status and reactivity states are integrated in the PBN and PAG, expanded and re-represented in the hypothalamus and the extended amygdala, and then re-represented again in the interoceptive cortices. The high-level re-representations, in turn, have downward effects on these same systems.

Summary feelings of self-status appear to be re-represented more strongly in the left anterior insular cortex; summary feelings of motivation and emotional reactivity appear to be re-represented more strongly in the right anterior insula. As Bud Craig documents, the right and left anterior insula cortex appear to be the re-representation areas from which a more integrated sense of self emerges.[13] In studies with humans, subjective feelings of self, including sadness, anxiety, pain, disgust, sexual arousal, trust, social rejection, and even the pleasantness of music, are associated with increased activity in the anterior insular cortex and, when emotional reactions are activated, the anterior cingulate cortex. When subjects are instructed to think about themselves, these same cortical networks are activated. It appears we also think about the feelings of others by simulating empathic feelings in these same networks. Empathic feelings are thought to help agents understand and predict what others may be thinking and how they may behave by generating similar dispositions in ourselves as we think about what others might do.[14] These feelings also propagate down the interoceptive lattice to modify social and somatic dispositions, as well as activity in the autonomic nervous system.

This brings us back to the case of Phineas Gage. The prefrontal cortex is a region of the brain which is highly developed in humans and is known to be involved in planning and logical reasoning. The ventromedial prefrontal cortex is thought to re-represent interoceptive feelings in the context of planning and reasoning. However, damage to this region disconnects logical decision making from the empathic evaluations in the interoceptive stream. Individuals with such damage are able to learn tasks that require factual reasoning, and they generally score well on standardized intelligence tests, but their ability to plan ahead and make social judgments is impaired. They don't recognize how their decisions will be perceived by others and thus have difficulty making appropriate choices. In short, their empathic feelings and social values are largely disconnected from their reasoning skills.[15]

This was the fate of Phineas Gage. His personality changed radically after his accident. He routinely made crude and offensive comments and then was surprised when others were offended. He promised to deliver work at a particular time, but if he was distracted for some reason, he was not motivated by the social contract to return to the promised task. He made up stories when it suited him. He threw fits when he didn't get his way. In short, he was no longer the responsible and trustworthy person he was before his accident. As Damasio remarked, Gage lost a critical part of his humanity: his sensitivity to the feelings of others.[16] His management strategies became confrontational, and he was soon fired from his job as crew foreman. He had similar difficulty holding other jobs for long. Without a link between the reasoning networks in the prefrontal cortex and the empathic evaluation centers in the interoceptive stream, Phineas Gage simply failed to recognize the social consequences of his decisions.

AN INTEROCEPTIVE COGLEY

Although we are not quite ready to bring Cogley to consciousness, we are ready to provide him with networks to support a rich, interoceptive proto-self and networks where the sensitivities and moods of the proto-self can be re-represented as more complex feelings. Following a bio-inspired design strategy, it seems reasonable to begin by giving Cogley a variety of low-level interoceptive sensors and analysis networks that continuously inform him about his physical status and internal need states. These somatic inputs will serve to embody Cogley's feelings. However, they must be linked to reactivity networks that adjust his output dispositions in quasi-emotional ways to complete the process. For example, if the sensors that monitor energy resources find that his energy is low, they should motivate him to seek recharging opportunities in a timely manner. Sensors that detect a risk of damage should motivate Cogley to withdraw so as to avoid injury. Sensors signaling positive states should motivate Cogley to approach and investigate, and so on.

To make these interoceptive states adaptive, Cogley will need links between his reactivity networks and his planning stream so that his moods can motivate approach and withdrawal dispositions. In addition to these directional dispositions, specific need-states should be connected with the priority setting networks of the planning stream. These connections should promote decisions which resolve the needs. This will cause general approach and withdrawal activities to gravitate toward situations favoring need-relevant investigatory and consummatory end-state activities. Of course, reward signals will be needed to promote learning about the activities which end up satisfying need states. The result will be neither a purely general purpose learning system nor a fully special purpose system, but rather a hybrid system with properties of both strategies and a bias to favor the most urgent interoceptive needs first.

To keep his activity levels linked to daily cycles, like his human partners, Cogley will also need a circadian rhythm module. This module should entrain to daily light-dark cycles and bias his activity so that he becomes more active in the light part of the cycle. There may even be value in having Cogley engage in sleep cycles. Sleep would not only be a good time for Cogley to recharge his batteries, but it would also be a good time for him to replay his recent learning and memory experiences, so as to tune those networks better. Replaying memory events during sleep could result in dream-like experiences for Cogley, if his attention networks are partially active during sleep. This would no doubt give Cogley added insight into what dreams are like for his human partners.

To enable Cogley to anticipate interoceptive outcomes, and to react ahead of time, a *valuation learning* system will be needed to link exteroceptive cues to his interoceptive reactivity systems. This will be Cogley's equivalent of the basolateral amygdala. The system should enable Cogley to learn what exteroceptive cues precede reactivity changes. These changes should motivate exploratory and investigatory behaviors, and some may even come to serve as interim reward cues. They will enable Cogley to learn about cues and actions that lead to rewarding end states. Systems monitoring both upshifts and downshifts in interoceptive states will be needed to inform the memory networks when to store memories and to latch memories in place on recall. These systems will ensure that Cogley's memories and actions will come to be organized around important interoceptive states.

One distinction between the reactivity dispositions that we typically categorize as emotions and those that we categorize as motivations is that those states which we call emotions generally involve signaling.[17] Signaling about emotions makes it possible for emotional feelings to spread to other agents. Therefore, Cogley's emotional networks must also include signaling, and signal tracking networks, so that he will be able to signal about his emotional states, and so that he will be able to recognize the emotional states of others. Both facial expressions and voice quality changes should be used as nonverbal channels for emotional communication. Cogley's recognition systems for these cues should also engage empathic states of interoceptive reactivity. This will better enable him to understand the actions of others by tuning his interoceptive states to more closely match those of others as he observes and thinks about them. If Cogley is to be a highly social agent, then his systems for monitoring social status and social reactivity states should be considered as fundamentally important as his physical condition. This means that Cogley should sometimes give his social needs priority over his physical needs.

These sorts of social signaling systems have not escaped the attention of robot designers. One of the earliest robotic systems to use such social signaling strategies was Kismet, a conversational face robot developed at MIT in the late 1990s. Kismet was designed to recognize and react to movements, faces, and voices in lifelike ways, using eye shapes to signal what his cameras were tracking. Kismet interacted by following movements and by speaking in gibberish, English-like phonemes which sounded like speech but had no meaning. Similarly, Kismet didn't understand what he heard, but he followed the direction of conversational sounds with eye contact, and he would even make and break contact as he took turns speaking and listening. He was also motivated to gibber just to fill in awkward pauses in the conversation. Interacting with Kismet was much like interacting with someone who spoke an unknown language. No higher-order message was communicated, but the interactions seemed surprisingly real because they used nonverbal channels that humans find socially meaningful.

Building on her experience with Kismet, Cynthia Breazeal, a professor of robotics at MIT, began designing a robot called Leonardo in collaboration with the Stan Winston Studio, a model designer of Star Wars fame. Leonardo was named after the Renaissance scientist, inventor, and artist Leonardo da Vinci. Standing at nearly twenty-five inches tall, Leonardo has thirty-two separate movements for changing his facial expressions, and about the same number for making body and arm gestures. This makes him capable of many subtle human-like expressions. In fact, Leonardo is arguably the most emotionally expressive robot ever built. However, he does not resemble a human. Because Leonardo would never be fully human, the developers decided it would be better to give him his own distinct species look. However, Leonardo uses many of the same kinds of facial expressions and gestures that his human partners use.

Similar early processing and preparedness mechanisms will be important for shaping Cogley's social awareness. We have already proposed that, like Kismet and Leonardo, Cogley will be equipped with low-level mechanisms for detecting and producing facial expressions. Six emotional states are commonly signaled in human facial expressions: anger, disgust, fear, joy, sadness, and surprise.[18] These emotional expressions are found across various cultures. Further, like visual perception, it appears that the analysis of emotional states occurs in several stages. Some expressions, such as fear, are recognized in the amygdala prior to conscious awareness. More subtle emotional evaluations require added levels of re-representation to be accurately represented and are not readily recognized from static visual cues alone. For example, humans are not very accurate in assessing emotional states from still pictures of faces. However, allow them to observe the changes in facial expressions over time, and they do better. Add in supporting actions, gestures, and emotional vocalizations, and they are much more accurate. In effect, each of these channels carries only a small amount of information, but the sum of them is highly informative, especially when the agent doing the signaling is well-known.

We have already noted that humans are predisposed to interpret comforting contacts as socially reinforcing, and that even non-touch contacts such as eye contact, attention sharing, and soft melodic voice tones are thought to release oxytocin. Cogley must be designed to react to such signals with his only sociality reinforcement signals to keep him tuned to the emotions of his social partners. Over time, Cogley should even learn to activate social reinforcement signals based on positive social exchanges – in particular, in response to social affirmations and verbal compliments. This will further enhance his social learning. In contrast, frowns, glares, looks of disgust, and loud raspy voice tones should trigger punishing effects which discourage current behaviors. Of course, Cogley will also need to signal his social feelings using these same channels, so that his human partners can recognize and respond to how he feels.

Activating emotional states and reinforcement signals in reaction to emotional cues will ensure that Cogley learns to interpret the emotional states of others and learns to discriminate his own states more accurately. In particular, these signals should promote the formation of associations in higher-level networks which can re-represent interoceptive states in the context of exteroceptive cues and social interactions. To give some idea of how these associations might be recombined to form broader emotional dispositions let's assume that, in addition to the six emotional states signaled by facial expressions, Cogley is sensitive to levels of interoceptive *anticipation* as signaled by anticipatory reactivity states, and that he is sensitive to his level of motivation to seek rewards; let's call this latter dimension *want*. This would mean that Cogley would be capable of representing eight core emotional dispositions: anger, anticipation, disgust, fear, joy, sadness, surprise, and want. As the diagram in Figure 8-2 shows, this combination would be capable of representing a wide range of emotional dispositions when scaled by intensity.

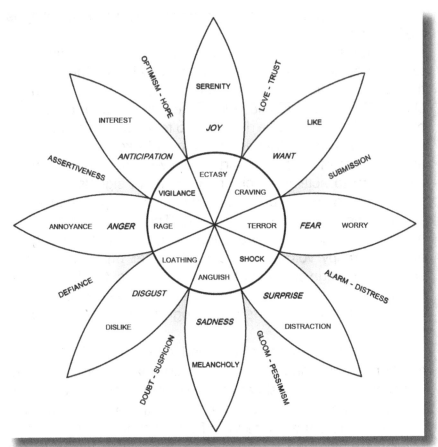

Figure 8-2: A framework for representing emotional feelings.[19]

In addition to these eight core emotions, several higher-order emotions formed by re-representing the core dispositions in dyads are shown in the intervening spaces of the diagram. Re-representations formed by combinations of even more distant core states could produce an even broader array of feelings. For example, *amazement* would seem to involve a combination of surprise and joy, while *longing* would probably include elements of love, want, and sadness. Given that we'll want Cogley to distinguish between different "want" categories and other situational cues, it seems clear that he will need several levels of re-representation networks if he is to learn to recognize the many variations in feelings that his human partners do. Clearly, an ability to form these combinatorial representations will be essential if he is to become sensitive to the range of feelings his human partners take for granted.

Curiously, the human brain appears to represent status and reactivity feelings in largely separate areas. Initially the cortical representations for status project to the insular cortex while the cortical representations for reactivity project to the anterior cingulate cortex. Although there is cross communication between these regions, status and reactivity are subsequently re-represented in largely separate areas of the anterior insular cortex. Given that this separation appears to be important for humans, it seems reasonable to design Cogley with a similar tendency to keep his status and reactivity feelings largely separate. The emotional categories in Figure 8-2 are based mostly on reactivity dispositions, so this combinatorial mapping appears to provide a reasonable beginning for re-representing reactivity states. However, Cogley will need another set of combinatorial mappings to re-represent status feelings. Core elements for status should probably include feelings of wellness versus distress, satiation versus discontent, pleasure versus pain, acceptance versus rejection, and care versus neglect. These are simply initial suggestions. I'm sure they can be improved. However, being able to re-represent these aspects of status in varied intensities and combinations will clearly enable Cogley to form much more human-like feelings of self.

CONCLUSION

The importance of the interoceptive stream cannot be overemphasized. As Damasio notes in the quote that opens this chapter, "were it not for the possibility of sensing body states that are inherently ordained to be painful or pleasurable, there would be no suffering or bliss, no longing or mercy, no tragedy or glory in the human condition." Without these evaluations, the other two streams would not be very adaptive. The planning stream links affordances with potential action plans, but modules in the interoceptive stream determine what motives are activated, what plans are favored, and what actions will be rewarded and reused. The perceptual stream analyzes sensory inputs and interprets perceptions in the context of past memory associations, but the interoceptive stream determines what perceptions and co-occurrence associations are important enough to be stored in memory and which memories are likely to be recalled when those percepts are encountered again.

The interoceptive stream also sets the values that must be weighed in making logical decisions. Without these values, Cogley would lack direction and social insight. With them, he will be more sensitive to his needs and more socially aware. Cogley will also need a variety of nonverbal signaling and detecting systems to enable him to recognize and empathize with the interoceptive states of his partners. In humans, these signals and the reactivity states they engender are essential for interpreting the motives and actions of other agents. Without these signals, there would be no trust, no bonding, no caring, and no motive to help, and without these feelings, human society as we know it would not exist. However, with a functional interoceptive stream in place, Cogley will have such feelings. Empathizing and sharing in the feelings of others will make it possible for Cogley to become a productive member of his social groups. We'll have more to say about this group identification process later.

CHAPTER 9
THINKING IN STREAMS

It should now be clear why separate processing streams are found in the vertebrate brain. Each of the three streams is specialized for a different cognitive task. The planning stream is optimized for learning about affordances, locations, and effectivities and for resolving best-fit action plans. The perceptual stream uses a multilevel analysis hierarchy to resolve complex patterns. It operates more slowly, but it can link perceptions in flexible memory associations that support categorical recognition and that can be reactivated to supplement ongoing perception. The interoceptive stream tracks somatic and emotional status and manages reactivity states. It is these status and reactivity states which guide the processing of the other streams in service of interoceptive needs. It is also these status and reactivity states that come to be re-represented as conscious feelings in states of attention.

Although the initial function of the processes in each stream is to analyze ongoing experience, these processes also provide tools for thinking beyond current experiences. Actions can be simulated in the planning stream. Perceptions can be imaginatively recalled in new combinations in the perceptual stream. Feelings can be reactivated to guide planning or to help interpret the mental state of other agents. Because activity in one stream can trigger reactions in another, creative thinking is a natural product of these interactions. However, there is one additional feature of this circuitry that favors creative interactions. The brake turns out to be a critical factor in the emergence of intelligent planning. The ability to hold a brake on committing to an action plan makes it possible to pause and deliberate for longer time periods before resolving plans.

THINKING IN STREAMS

Animals may constantly be seen to pause, deliberate, and resolve. – Charles Darwin, *The Descent of Man*, 1871

Tom is on the kitchen counter now and is considering a leap to the top of the kitchen cabinet. The distance from the counter to the cabinet top is about four feet straight up, a fairly long leap for Tom from a still position. Were he frightened he could easily scale it, but he is less motivated, and there's a chance that he could misjudge the jump and fail. I can tell that Tom is considering this jump because he's making preparatory movements with his hind legs while fixing his gaze on the target-landing spot. Tabby, a leaner and more agile cat, made this leap a few moments earlier. Tom appears to be following. However, Tom is heavier and has never tried this leap before. He holds his position for about ten seconds, making preparatory movements as he considers the action, and then he stops. He subsequently crosses the counter and effectively completes the jump in two smaller steps, first leaping atop the refrigerator and continuing from there to the top of the cabinet.

In an earlier chapter, we suggested that thinking is what agents do as they consider taking action. Having set out the basic operational properties of the three streams, we can now begin to explain how certain kinds of higher-order thinking emerge from preparatory interactions among processes in the three streams. We have already noted that we recall memories by using backward links from memory nodes to reactivate the original perceptual networks. In effect, we use the same perceptual networks to experience memories on recall as we did when we first experienced them. Similarly, we suggested

157

that thinking about past feelings, or the feelings of others for that matter, involves activating similar feelings in ourselves. It should not be surprising then to find that we think about actions by activating the same networks that we use to plan actions – that is, by activating the planning networks at low levels, and sometimes for extended time periods, as we consider taking action.

In retrospect, we should have known this all along. It takes a special arrangement of networks to process perceptions from elementary features into pattern combinations and object shapes. If perceptual memories were recalled to some other place, then a similar hierarchy of perceptual networks would be necessary to resolve and re-represent the same level of detail. Duplicating such a network when one already exists would be highly inefficient. Natural selection favors simple solutions whenever possible, so it tends to reuse complicated networks rather than building new versions of them. The same strategy applies to feelings and actions. It takes a special hierarchy of networks to re-represent status and reactivity inputs as feelings. Once such a hierarchy is in place, it is highly unlikely that another hierarchy would evolve just to think about feelings. Similarly, once selection processes have crafted networks for planning actions, there would be no value in building yet other networks for thinking about actions. After all, they would need the same kinds of connections between affordances, spatial-motor maps, and motor programs, so they would end up overlapping each other anyway.

Reusing established networks for thinking about perceptions, actions, and feelings is more efficient. However, the implications of this strategy go far beyond the efficiency of simply reusing the same networks for thinking. If we think by simulating actions, imagine events by recalling perceptions and feelings, and think about other agents by empathizing with their feelings, then we should be able to generate even more complex plans and ideas by allowing those separate processes to interact with each other. Simulated actions can promote the recall of new memories. Recalled memories can reactivate new motives and feelings. These in turn can change what actions are considered in simulation, and the process can be repeated. In short, simply by allowing thoughts to interact across the three streams, an agent can create running patterns of simulated actions, imagined perceptions, and empathic feelings.

It may not seem like a big difference at first, but these strategies enable thoughts in the three streams to extend beyond the cues of the present and consider new possibilities. Our goal in this section will be to develop a bio-inspired concept of thinking that will help us design Cogley to become a more creative thinker. The sections are all short and are more hypotheses than proven mechanisms, however I believe they capture some interesting insights into creating thinking.

PLANNING BY SIMULATING ACTIONS

Studies of brain activity suggest that humans represent actions by activating motor-programming circuits at low levels while holding the brake on action execution. When people think about reaching, areas related to reaching programs are activated in the brain. When they think about eating, areas related to mouth movements are activated. When they think about running, areas related to locomotor movements are activated. If we measure the end muscles for those actions, we find that they too show low levels of activation. It seems the brake circuitry allows just enough activation through to put the muscles for planned actions in a preparatory state. This enables an action plan to be broadly represented in simulated activity before committing to it. This extended level of representation has important implications for how decisions are made, because the brain reacts to its own activity.

The process is something like revving up the engine of your automobile while holding your foot on the brake – except that when the motor circuits in the brain are revved up for extended periods, they trigger a broader set of associated memories, reactivated feelings, simulated activities, and anticipated outcomes. This extra processing often enables a more reasoned decision to be reached. Indeed, as Darwin noted, animals are often seen "to pause, deliberate, and resolve" before taking action. Of course, we cannot always see the preliminary movements as easily as in the example of Tom preparing for a leap, but we often see the pauses. In Tom's case it was probably the worry that he might misjump that motivated him to pause. His preliminary movements provided evidence that motor circuits were being partially activated in simulated actions. The fact that he changed his plan suggests that simulating actions sometimes contributes to exploring alternate plans.

A critical feature of this architecture, and one which makes extended planning possible, is the fact that the brake in the basal ganglia can be controlled voluntarily. Some actions, for example, must be stopped with precision. Actions sometimes need to be interrupted for other priorities. Many terrestrial mammals, for example, have emotional circuitry, triggered by fear, which causes them to freeze when they sense danger at a distance. In social animals, dominance is often based on the threats of harm, and the softer side of fear we call worry can also activate the brake at times, causing social agents to inhibit activities in the presence of more dominant agents. In fact, when a group is large and the social hierarchy involves several levels, staying within accepted social bounds may require constant vigilance. Some animals even seem to develop group-sensitive rules that cause them to inhibit unacceptable actions in the presence of higher-ranking conspecifics, lest they be disciplined. In humans, we call such internalized social rules conscience. In our pets, we attribute them to good training. Whatever we call them, learning to inhibit actions is a common characteristic of higher intelligence.

Researchers have long noted that there is a correlation between the complexity of social structures and the intelligence in primates, and they have speculated that this is related to the increased need to monitor status relationships in complex social groups.[1] This appears to be the case for humans. Ongoing concerns about social status constantly motivate them to pause and deliberate before taking actions that might diminish their status or offend another. This suggests that we humans have become more systematic thinkers, in large part because worrying about the consequences our actions causes us to pause, deliberate, and resolve more often. This link between worry and thinking is still largely a hypothesis. However, it provides a logical path by which socially guided inhibitory control can be learned, and it may even help explain why people seem to get a worried look on their face when they are thinking deeply. The hypothesis is simple, but the implications are noteworthy. Social worry appears to increase inhibition and, in doing so, to increase the time that an agent spends deliberating on an action before committing to it.

This suggests that if Cogley is to develop human-like intelligence, then he will need to be sensitive to social criticism. In fact, he will need to be concerned enough about his social status to pause and deliberate before taking actions. In effect, concern over his status should often motivate him to set a brake on acting while he considers the consequences of his actions in more detail, by simulating them for longer time periods. Cogley should be encouraged to do this as he develops. However, rather than trying to train this strategy in full, it would be useful to make simulation due to worry an automatic background activity. Many animals have background vigilance activities that accompany ongoing behavior. One only needs to watch small, foraging birds to notice that they constantly shift between feeding and looking out for predators. In a similar way, Cogley needs to intersperse social worry with his ongoing activities. However, to keep him from growing depressed from all this worry, it will be important to make resolving problems a rewarding experience. Thus, while worry will motivate Cogley to reconsider his action plans, simulations will help him eliminate his worry, and that should be rewarding. Following this strategy Cogley should even come to enjoy the opportunity to resolve problems by strategically pausing and deliberating. It will give him an enhanced sense of control over his decisions.

IMAGINING BY EXPLORING AND BLENDING MEMORIES

The snapshot model of episodic memory helps to explain how capturing a series of overlapping configurations enables connected episodes of experience to be reconstructed. Simply by directing attention to a feature in one configuration, that feature can serve as a retrieval cue that triggers the recall of related configurations. It is then possible to redirect attention to yet another feature and repeat the process. This provides a simple strategy for searching through snapshot memories. For example, suppose that elements of a search image are primed in short-term memory as retrieval cues while episodic memories are being searched. As each new snapshot is recalled, attention would naturally be shifted to the features that are more closely associated with the search goal. As these features are encountered, they become the new retrieval cues. The search is then repeated until a snapshot connected with the target item is found. At that moment, memory modulation circuits would be engaged and momentarily latch onto the memory. This simple process of successive searches with better retrieval cues naturally drives the search toward better fitting associations. It doesn't always work, but it's surprisingly effective.

It is even possible that events originally encountered separately might be recalled together because they happen to share a connection with some common retrieval cue or context. Research suggests that switching attention between two items results in transient periods when the activations of the two items overlap.[2] It follows from this observation that rapidly shifting attention back and forth between two items could extend this overlapping effect. This means that events that never actually occurred together in the real world may nevertheless "occur together" during memory recall. If the joint recall of the events happens to suggest more promising outcomes, then the events may be bound together in new configurations. For example, suppose a chimpanzee has had experience using sticks to manipulate items that were out of reach outside its cage. Suppose the same chimp had experience climbing on boxes to gain access to items that were out of reach in height. Subsequently, this chimp is given a problem in which the boxes alone do not provide enough of a boost for him to reach his goal. He is stumped. Then he happens to notice a stick nearby, and the common association of gaining access to objects that are out of reach brings these two different actions together in an imagined combination. Aha!

The excitement of noticing this chance co-occurrence would be enough to latch the two separate memories together, at least in the moment. If the combination of climbing and poking with the stick helps reach the goal, then stronger memory modulation effects would be activated. The chance overlap of the two ideas would become a more permanent combination, a new insight. In effect, the same mechanisms which enable memories with overlapping events to be connected in searches helps to explain how separate events might overlap and be bound together on recall. Forming insights in this way requires a rich store of memories to serve as building blocks for imaginative combinations. Attention shifting provides a mechanism for exploring those building blocks. Memory modulation circuits provide a means for binding them together. However, insights require experience. The more experience an agent has with similar situations, and the more she explores her memories when stumped, the more likely she will be to find experiences that bind together in productive combinations.

Cogley will need a rich history of varied experiences before he is likely to have the range of memories needed to construct such insights. However, once those memories are in place, then simply by shifting attention among items, new overlapping combinations will occasionally occur. To make sure that Cogley engages in this strategy, we should provide him with motives that cause him to explore memories in times of worry or uncertainty. That is, worry and uncertainty should not only promote pausing and physically looking around, but also looking around" in memory. Finding matches to perceptual inputs is what memory networks do, and looking around will prime a variety of memories. Thus, in times of worry Cogley should look both externally and internally. Over time, this should result in synergistic links between action planning and memory exploration, such that simulating an action will trigger the recall of related memories, and recalling memories will prompt the simulation of new activities.

While imagining events or features in new combinations may occasionally lead Cogley to new insights, it seems likely that much of the time he spends in such memory explorations will not be very productive. We occasionally reach creative insights following this strategy. However, most of the time we spend scanning memories and discovering new co-occurrence associations simply results in associations that seem interesting in the moment but don't end up being very productive. Still, exploring memories is a low-cost activity, so as long as it occasionally results in something useful, the process should be encouraged. If we design Cogley to consider such associations as transiently rewarding, as we humans seem to do, then he should develop the habit of exploring memories at times of uncertainty and latch on to those combinations which evoke interesting interoceptive reactions, especially when they are associated with highly desired outcomes. We have yet to explain the mechanisms that will enable Cogley to manage his attention; it will take us a few more chapters to make that happen. However, as we have noted, attention is the key to managing memories, so once Cogley learns to manage his attention, he should be able to guide his thinking in more imaginative ways.

EMPATHIZING BY REACTIVATING FEELINGS

Empathizing with feeling states adds an important dimension to decision making. As the accident of Phineas Gage revealed, the ability to make sound judgments depends in large part on being able to consider the social consequences of an action. We appear to be able to do this is because we have interoceptive modules, some cortical, some subcortical, which can be reactivated by the social signals of other agents and by memories of our own experiences. By activating similar feelings states, we are better able to evaluate situations and make decisions that take those effects into account. This is partly how our memories help us make better decisions. We can recall outcomes and how we felt, and that influences how we plan. Just as there are mirror-neuron networks in motor regions for recognizing the actions of others, there are also interoceptive networks which are sensitive to social signaling and which can partly mirror the feelings of others; they detect signs of how another agent is feeling and engage similar feelings in us.[3]

Based on such findings, Vittorio Gallese and colleagues have argued that the ability to empathize is essentially a process of partially imitating the feelings of others.[4] If someone else is sad, we share in their sadness as we think of them. If they are happy, we share in their happiness. In a social context this allows an agent to predict when another agent may be approachable or not. Many animals recognize the moods of others and anticipate their actions based on their assessment of those moods. At a planning stage, this enables them to predict when they should inhibit interactions with others and when their interactions might be welcome. For example, Tom has learned to track my moods and routines. He recognizes when I am busy or irritated, both by attending to my actions and my tone of voice. Based on such observations he knows when to keep his distance and when he's welcome to rest on my lap. In fact, many of the vocal cues that signal feelings, especially the tone of our voices signals, appear to be evolutionarily old and to share common characteristics across a broad range of modern species.[5]

On an interpersonal level, sharing in the feelings of others also enables caring agents to form alliances and to work together cooperatively in shared goals. When you notice what another agent is trying to do, and if you also care about that agent, then you are naturally motivated to help support the activities. However, the ability to empathize with the feelings of others introduces one more important property of mind. By empathizing with the feelings of others as we simulate actions, we can often gain added insight into what actions they may be motivated to do next. By recalling memories of their past behaviors, we can sometimes better represent what they know and what they find important in a situation. These are core elements of an ability to simulate the thoughts of other minds, and they are the kinds of thinking which are naturally encouraged by episodes of worry in a social context.

Researchers often refer to this kind of social reasoning as having a *theory of mind.* Obviously it not as much a theory as it is an ability to predict the actions of another agent by observing, empathizing, and simulating potential actions. However, it is a very useful skill. Cogley will need similar observational and empathic skills to help him predict the behavior of other agents. It will take time and practice before he reaches this stage, and we will need to introduce more social mechanisms before Cogley becomes good at predicting the behavior of other agents. However, once he learns to consider the perspective of other agents in his simulations, he will begin to behave as if he has a theory of mind. It won't be a formal theory. Sometimes his predictions will be guided by the emotions they are signaling. Sometimes his predictions will be augmented by observing where they look as they plan. Sometimes his predictions will be enhanced by memories of what they've done in the past. Many times it will be a combination of such indicators that guide his assessments. However, by following this strategy Cogley will be able to use his simulation skills to help him predict the behavior of other agents. For Tom and me, that will qualify as having a theory of mind.

CONCLUSION

The intricacy of mechanisms in each of the three streams strains our ability to comprehend the full functionality of any one stream. Yet when we step back and look at the bigger picture, the basic functionality of the three streams enables us to explain how many higher-order cognitive functions emerge. Simulating actions provides a mechanism for extended planning. Recalling and recombining memories in imaginative combinations provides a way to construct new expectations and to gain insights from past experiences. Empathically simulating feelings, guided by memories and nonverbal signaling, provides a basis for understanding the feelings and motives of other minds, as well as those of our own mind. The ability to hold a brake on committing to a plan helps us to understand how imagined plans, perceptions, and feelings can be run in extended simulations before decisions are made.

These interactions call to mind a process for influencing thought which Antonio Damasio called the *somatic marker hypothesis*.[6] Damasio noted that conscious experience depends not simply on perceptual memories but on the somatic and emotional status feelings and the states of motor and emotional reactivity that those experiences engender. We don't merely observe states in the world, we react to them, and those reactions further influence our interpretation of the situation and our subsequent reactions. Perceptions and action plans are thereby partly grounded in the neural mechanisms that prepare the body to react to the events. In fact, Damasio argues that bodily reactions are some of the earliest "markers" for many feeling states. Among other things, this hypothesis argues that memories are best recalled when motor circuits are partially reactivated, because states of motor activation are part of the markers that get stored with an experience. Damasio suggested that these states can be re-engaged by simulating an action in what he described as an "as if body loop."[7] Our model of thinking in this chapter is consistent with Damasio's ideas. The low-level motor activation states, which occur as actions are being simulated with the brake held on, should provide just the sort of "as if body loop" that Cogley will need to simulate activities, trigger the recall of related memories, and activate associated interoceptive states.

The obvious advantage of having this simulation process is that it will enable Cogley to explore potential plans before he makes decisions. It even introduces the possibility of exploring unlikely scenarios, engaging in fantasies, or speculating about the actions of another agent. This model of thinking is, of course, still incomplete. As of yet we have provided no organizations to bind the three processing streams together in focused self-centered flows, and we haven't provided a steering system to manage the flow of his ideas. These will be needed to make his conscious planning more productive. However, the three streams model provides a rich framework for thinking about how a mind with imaginative cognitive skills might emerge, and that's start. Our next task is to look at the organizations that coordinate ideas from the three streams in coherent self-centered flows. Consciousness and the experience of self-control will emerge from interactions made possible by these coordinating systems.

PART TWO:
CONSCIOUSNESS AND SELF

CHAPTER 10
MAPPING SPACE AND BRIDGING TIME

The three streams architecture provides us with a model of mind that perceives, acts, and feels in largely separate regions of the brain. Extensions of these processes even provide a glimpse into the origins of creative thought. Some occur by simulating possible actions, others by selectively recalling and recombining perceptions, and yet others occur by reactivating past feeling states from memory or by empathizing with other minds so as to make feeling informed decisions. These strategies are obviously important aspects of mind. However, despite the insights which they provide, this model still lacks mechanisms by which these separate faculties can be readily coordinated in adaptive combinations. The first step in understanding this coordination process will be to describe the proprioceptive networks in the brain which coordinate behavior in space and time. In our quest to understand consciousness, this focus on spatial and temporal coordination may seem like a distraction. However, proprioception is a critical component of self-awareness. Although the proprioceptive coordinators do not directly result in conscious experience, they nevertheless play a role in selectively enhancing certain aspects of experience over others. This means that proprioceptive networks are important for guiding the flow of information in conscious minds.

MAPPING SPACE AND BRIDGING TIME

> When we learn to move ourselves, we learn to distinguish just such kinetic bodily feelings as smoothness and clumsiness, swiftness and slowness, brusqueness and gentleness, not in so many words, but in so many bodily-felt distinctions. Short of learning to move ourselves and being attentive in this way to the qualia of our movement, we could hardly be effective agents – any more than a creature who 'does something and then looks to see what moves' could be an effective agent.... An agent who holds sway is a bona fide agent precisely insofar as she/he is aware of her/his own movement, aware not only of initiating it, but aware of its spatio-temporal and energy dynamics. – Maxine Sheets-Johnstone, "Consciousness: A Natural History," 1998

Just as a pre-conscious representation of self-status (the proto-self) emerges from the integration of early status and reactivity modules, a pre-conscious representation of agency (perhaps we should call it the *proto-agent*) emerges from the integration of early proprioceptive modules in the brain. As Maxine Sheets-Johnstone notes, sentient experience depends in large part on systems of proprioception and kinesthesia. These systems enable a mobile organism to sense itself acting in relation to the world. A proprioceptive mind is more than just a collection of sensorimotor reflexes. It's an interconnected organization of spatial and temporal knowers which are capable of coordinating behavior in more adaptive ways. It is aware of external stimuli in the context of internal feedback, of the spatial orientation of actions relative to objects and spaces, and of the contiguity of events and its reactions. Proprioceptive senses are founding elements in an agent's ability to represent itself as a coherent agency that is capable of reacting to the world in space and time.

There are two proprioceptive organizations, sensorimotor meta-networks to be more accurate, which coordinate activity on a broad scale in the vertebrate brain. Both of these meta-networks involve an extensive array of interconnected circuits that sense events and coordinate actions. The first of these is the spatial orientating and targeting system in the tectum, a structure in which visual inputs are especially well represented in most vertebrates. The second of these is the temporal coordination system in the cerebellum, a proprioceptive organization with widespread connectivity. Neither of these meta-networks directly results in conscious experience, and yet without them consciousness as we experience it would be largely deficient. In particular, these networks provide preliminary distinctions which help manage the flow of information in the brain and which coordinate decisions in space and time. While lacking a distinct focus of attention, the information they represent nevertheless primes higher-order knowledge systems and shapes the flow of attention which those systems make possible. Such is the layered nature of conscious awareness.

A Place for Space

Embryologically the brain and spinal cord develop from a neural tube. The lumen of the tube becomes the central aqueduct or canal of the spinal cord and is filled with cerebral-spinal fluid. In the brain the central aqueduct expands to form a series of fluid chambers, known as ventricles. The brain region that lies above the midbrain ventricle is called the *tectum* (see Figure 10-1). The term is Latin for covering or roof. In mammals, the major structural landmarks in this part of the brain are two sets of bulges on the back of the brainstem which are known as colliculi, meaning hills. The larger hills, the superior colliculi, are the primary sites for visual input to the midbrain. The smaller hills, the inferior colliculi, are the primary sites for auditory inputs. Given that visual processing tends to dominate spatial representations for many mammals, the label superior colliculi (SC) is often used as a general label for the mammalian tectum. We will use this convention at times to make contact with the work of other authors. However, because "tectum" is used to refer to this region in non-mammals as well as mammals, we will more often use it to refer this region. The tectum is an early network system for representing the location of the body in sensory space, and that turns out to be critical for making many action decisions.

An important aspect of the networks in the tectum is that they receive both proprioceptive motor information and a wide array of sensory inputs. The motor input is from systems that detect the orientation of the head and body. The sensory inputs come from virtually all fast-acting directional senses, including electrosense in electric fish and infrared heat detection in reptiles. Even magnetic field detection appears to be represented in the tectum, in those vertebrates with that sense. Olfaction, a slower and less directional sense, is notably absent here. A key feature of the organization of the tectum is that each sensory region is mapped .spatially in layered arrays. The strategy of representing sensory and motor information in topographic maps is widespread in the nervous system, but the organization of the tectum is particularly interesting because the maps for many different sensory systems are arranged in registration with each other. This means that spatial representations for different senses are arranged in physically adjacent layers.

This layered spatial alignment facilitates the cross-referencing of directional information from different senses with each other and with motor orientation systems. For example, the registration of visual and auditory maps allows each system to inform the tectum about sensory activity in a given spatial direction and, in turn, to guide motor reactions that align with that direction. In fact, some intervening cell layers respond optimally only to a combination of auditory and visual cues. This enables a target to be tracked using combined auditory and visual information, when neither input alone would be sufficient. The alignment also makes orientating the head and sensory receptors toward sources of information easier. Not surprisingly, predators who depend on their orientation skills for tracking prey, like our feline companion Tom, have a particularly well developed tectum.

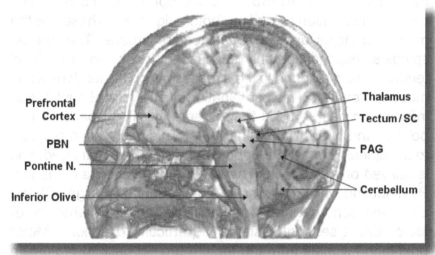

**Figure 10-1: Brain structures described
in this and the following chapter.** [1]

A fascinating feature of the spatial orienting networks is that they have evolved mechanisms which allow them to discover and maintain the alignment between sensory and motor systems. This was elegantly demonstrated by Eric Knudsen and Michael Brainard, when they raised barn owls with prism goggles that displaced their visual field laterally.[2] Barn owls are nocturnal hunters and are well able to localize prey using auditory cues alone. However, their eyes also work well in low light, and they often use a combination of visual and auditory cues to locate prey. The initial effect of wearing the prisms was to create a misalignment between the auditory and visual maps in the tectum so that when a bird heard a sound, it would look in the wrong direction. However, the owls soon developed a new auditory-visual mapping in the tectum which re-aligned the displaced visual inputs with their auditory inputs. Apparently, the tectal networks align themselves by connecting sensory and motor-feedback regions that show strong simultaneous activity, so when the maps are displaced a new alignment gradually develops.

To complete our description of how the tectum is organized, we need to add a little more detail about the motor inputs to the tectum. There are two kinds of proprioceptive senses that provide information regarding motor alignment. These are the muscle spindles and the Golgi tendon organs. The muscle spindles respond primarily to muscle stretch, and the Golgi tendon organs respond to the stresses put on the tendons when muscles contract. This combination of stretch and contraction information can be used to represent the motor patterns that position the head, limbs, and torso. This information projects to the spinal cord, where opposing muscles are managed, and is relayed on to the tectum, where the orientation is managed. By detecting which patterns of motor orientation align with the strongest sensory inputs, the tectum can learn what motor states direct sensory systems in particular directions. Motor orientation signals from the tectum project back to brainstem and spinal motor nuclei which manage motor activities.

The tectum thereby guides motor orientation on a broad scale. In particular, there are projections from the tectum which influence motor networks for locomotion, reaching, head turning, and gaze direction. Other projections map to visual motion tracking areas in the cortex and to spatial-motor-planning areas in the posterior parietal lobe. It is the ability to align motor systems with objects in space that makes actions effective, and the tectum is a core network in this process. When a frog flicks it tongue to catch a passing fly, it's the tectum that provides the orientation and targeting information that coordinates the motor systems controlling the head and tongue muscles with the fly's location in the visual field. When you turn your head to gaze in the direction of a particular cue, the tectum helps align your head and eyes with the spatial mapping of the sensory information. This alignment also changes the gain on sensory inputs, because features at the center of the alignment are processed more intensely.

One orienting activity which the visual tectum manages,

and one which strongly influences attention, is eye movement. Because the visual inputs to the tectum of most land vertebrates come from the rod receptors in the eye, which are distributed broadly across the retina, the tectum has just the sort of wide-field view needed to make targeting decisions for directing gaze. These decisions are important because they determine where the more detailed visual processing region of the eye, the fovea, is directed. Targeting the direction of the head and eyes using this broad-field view is a simple yet highly effective strategy for managing sensory acuity, and it helps to explain how low-level orienting systems can literally bring higher-level systems to a focus. It is also what makes the tectum a major coordinating system in the brain, and a key member of what I call the *attention steering committee.* The orientation decisions of the tectum guide attention.

In addition to its role in guiding attention, the orientation of actions toward objects implies an ability to distinguish between the agency of actions and the target of actions. As Maxine Sheets-Johnstone notes,[3] a distinction between the body and the world, and the ability to coordinate actions between them, is an essential component of self-awareness. Given that the tectum is the first low-level system that makes movement-related body-world distinctions possible, it is a critical system for providing the early representations needed to form a sense of body-centered location. This body-centered information is subsequently re-represented in several areas of the cortex, where it provides the basis for a conscious distinction between the self and the world, and a sense of where the self is located in the world. In fact, this spatial sense of self is so influential that damage or disturbances in the posterior parietal cortex can even result in out-of-body experiences, experiences in which the observer self seems to be located in a different place than where the body is located.[4]

GETTING COGLEY ORIENTED

Though the structure of the tectum is complex, building such a system is, in principle, a tractable engineering task. In fact, a robotic barn owl has been designed which can use visual and auditory inputs to guide the orientation of its cameras. It can even learn to adapt to visual changes imposed by putting displacement goggles over its camera.[5] So emulating the function of the tectum should not present too large a problem. All robotic systems that act in space must have some means of orienting directional sensors, such as cameras, and a means of locating and directing actions toward objects. Many of these systems also use this information to build longer-term spatial maps, which can then be used to plan routes. Cogley will need all these skills if he is to have human-like orientation skills.

We haven't given much attention to the spatial maps in the cortex, but earlier during our discussion of the FARS model for reaching, we noted that there were three well-defined spatial maps in the parietal cortex which represent objects in relation to particular motor spaces. In the macaque monkey, where they have been best studied, these are known as LIP, MIP, and VIP. They are located in the lateral, medial, and ventral regions of the intraparietal sulcus. LIP is the short-term spatial-memory map for objects in the visual field. Research indicates that the items in this map are continuously realigned so as to be mapped relative to the central focus of the eye. MIP is the short-term spatial memory map for objects within the reach motor field. It is presumably realigned with changes in body orientation. VIP is the memory map for features close to the face and is presumably realigned with head movements. These maps are not simply egocentric; they are motor-domain centered. That is, the objects tracked within these maps are items relevant to specific domains of motor action. Cogley will need a similar set of motor maps to manage visually centered, reach-centered, and facially centered actions. These maps will provide him with three core workspaces for thinking about his personal space.

There are, however, some complexities involved in maintaining different spatial maps. For example, if Cogley wants to reach for something that is some distance away, he will have to use his visual-space map to bring the object into reach space. To do this, he will need some representation of where his reach map is located in his visual space. Similarly, bringing an object close to the face for inspection requires a connection between the reach-space map and the face-space map. To make this possible Cogley will need networks that can transform locations in one map space to those in another. These *coordinate-transform* networks are widespread in the motor system. In fact, researchers Yale Cohen and Richard Andersen note that visual transforms tend to dominate the motor regions in the tectum and parietal cortex, such that other sensory modalities, like touch and audition, are routinely remapped into visual coordinates.[6] This suggests that visual coordinates have become something of a standard in the vertebrate brain. The advantage of having a standard coordinate system is obvious when different modalities – say, hearing and vision – must be used jointly to guide an action like reaching. It would be difficult to guide actions based on different sensory modalities if they couldn't be transformed in a common reference system. Visual coordinates appear to be that common system.

The need to keep maps aligned also suggests how out-of-body experiences might occur. If for some reason an agent's visual space map was not properly aligned with his reach space map, then his coordinate transforms would be out of whack. In fact, he could have the experience of calculating that his face and body were in one location while his visually-guided sensory systems were situated in another location. This is apparently what happens in out of body experiences. When such anomalies occur, it seems that the visually centered perceptual map is treated as the primary spatial center of self. Why the visual map? Well, it seems it is the dominant map for the spatial sense of self because most of the modules in the attention steering committee use visual maps. Obviously, Cogley will need a number of coordinate-transform networks to interface between his different maps and keep them aligned so that this sort of confusion won't occur often. However, whenever his maps slip out of alignment, he will feel disoriented, until he can get his bearings". That is, until he can get his spatial sense of self realigned with his other motor maps.

Given that Cogley's vision will depend on cameras, camera orientation will obviously be a critical determinant of his field of vision and, by extension, his attention. However, to mimic human attention processes, Cogley will need two sets of cameras. As noted earlier, in most terrestrial vertebrates the visual inputs to the tectum are from the rod receptors in the eye. The rods cover a wide field of view and are highly sensitive to movements. However, they lack color visions and are less dense, so they produce a lower-resolution image than the receptors in the central fovea. To establish a parallel visual system for Cogley, he will need stereoscopic orientation cameras to provide low-resolution, wide-field, black-and-white information for orientation. Spatial information from other modalities should be mapped so as to be in registration with the spatial information provided by these cameras.

A second set of narrow-field, high-resolution, color cameras should be used to provide more detailed visual information, as the fovea does in the human eye. This high-resolution system will be part of Cogley's attention-focusing system. Given that the other sensory modalities are aligned with the low-resolution visual field for orientation, it seems reasonable to duplicate this alignment strategy for the high-resolution cameras by enhancing the inputs in any sensory modalities which are spatially aligned with the high-resolution field of view. In that way, the circuits for the orientation of the high-resolution camera can also be used to enhance the acuity of the other senses. I doubt that Tom would be very impressed with this design. He can orient his auditory sound collectors, his external pinnae, independently of his head and eye orientation, and even independently of each other. However, humans cannot do so, and for now this arrangement seems like a workable compromise. In addition, it provides us with a simple strategy for managing other aspects of attention. Perceptual processing for all sensory inputs aligned with the focus of the high-resolution visual inputs will be enhanced.

Having a wide field of view of the world will enable Cogley

to use his low-resolution inputs to build rough spatial maps quickly and to use that view to select regions to inspect in more detail with his high-resolution cameras. Following this strategy, whenever Cogley scans a scene, his low-resolution cameras should be used to detect regions of interest, and his high-resolution cameras should be oriented to those regions to get more detailed information. This strategy will enable the detailed visual inputs to be linked to locations in the wide-field view, along with detailed information from other modalities in that field of view. In effect, the wide-field will provide a big-picture view of the world that serves as an organizing context for the detailed information view. Because most motor affordances won't require high-resolution analysis, motor decisions will usually be planned and implemented based on the wide-field view, enabling them to be performed more quickly.

The mix of inputs provided by the low-resolution and high-resolution views will provide Cogley with different strategies for thinking about the world. The low-resolution view will serve as the primary visual input for Cogley's planning stream, and the high-resolution view will serve as the primary input into Cogley's perceptual stream. However, because the high-resolution view will be anchored to the wide-field view, this view should also be able to influence motor decisions by enhancing processing for particular features. In effect, simply by focusing on particular features Cogley should be able to influence his motor-planning networks. Though low-resolution inputs may be adequate for most motor decisions, this will not be the case for Cogley's perceptual stream. High-resolution inputs will provide these networks with more detailed information for perceptual analysis and memory. However, the big-picture framework provided by the low-resolution view also needs to be represented in the context side of the memory networks so that when high-resolution perceptual features are recalled, they can be linked back to relevant big-picture contexts.

TRANSITIONS IN TIME

Some action categories have their own dedicated motor-program generators, but actions rarely occur in isolation from each other. Gait patterns, for example, frequently need to be adjusted to support balance or to change locomotor direction. Reaching for an object requires coordinating shoulder movements with arm extensions and hand closing; if the timing is off, then the grasp will fail. Action sequences are thought to be learned by networks in the basal ganglia. However, the task of fine tuning the timing and intensity of the separate actions in real time depends largely on support from the cerebellum. The complexity and scope of the cerebellar networks are truly amazing. The name cerebellum literally means little brain. However, appearances can be deceiving. Although the little brain accounts for only about 11 per cent of the total volume of the human brain, it nevertheless contains approximately half of all the cells in the brain and, by some estimates, up to 80 per cent of all the neurons in the brain.[7] The presence of such an immense number of neurons in such a small package suggests that the cerebellum is a processing system of great importance. Research suggests that this system is dedicated to the temporal coordination of activity in the brain on a grand scale.

By analogy, compare the brain to the central processing system of a modern microcomputer. In computer systems, the main processor is designed to handle a variety of common input and output operations: memory addressing, if-then branching, register shifting, and integer math calculations, all of which it does in a fairly efficient manner. However, the organization effect comes into play here because the basic operational design needed for managing these functions is not very efficient at performing floating point math calculations. As a result, most modern microprocessors have an auxiliary processor, sometimes called a math coprocessor, which is designed specifically to handle floating point calculations. When the main processor detects a floating point instruction, it passes the calculation task to the math coprocessor. The coprocessor performs the calculation quickly and returns the result to the main processor. This makes the system run much more efficiently. It is my contention that the cerebellum exists for a similar reason. It is the brain's temporal flow coprocessor.

Temporal flow calculations in the cerebellum are used in three major kinds of tasks.[8] These are balance, action integration, and the temporal integration of cognitive information, a process which I call *cognitive flow*. In the case of balance, the vestibulocerebellum uses vestibular and postural information to calculate ongoing changes in body position, anticipate changes in the body's center of balance, and activate the postural reflexes needed to maintain balance as the body moves. In many terrestrial vertebrates, maintaining balance is a calculation-intensive operation. So it is not surprising that it involves a dedicated region of the cerebellum. Real-time action integration is another calculation-intensive task. The spinocerebellum coordinates the actions of the limbs and digits by adjusting the timing and intensity of elemental movements in each action. These adjustments allow combinatorial movements to be assembled into finely coordinated acts. The graceful movements of skilled athletes depend in large part on the ability of the cerebellum to coordinate muscle actions in a smooth and timely manner.

The timing adjustments needed to manage cognitive flow involve yet another region of the cerebellum, the cerebrocerebellum. This region, which involves the large lateral zones of the cerebellar hemispheres, calculates the temporal relationships between a variety of sensory, motor, and interoceptive events and exchanges information broadly with areas of the cortex and many subcortical regions. This effectively primes various brain regions to react in anticipation of common sequential transitions. These anticipatory reactions result in a smooth, coordinated flow of cognitive processing. The coordination of outputs for balance and action integration is largely directed to motor nuclei in the brain stem and bypasses the central thalamocortical pathways. As a result, conscious agents are rarely aware of these motor adjustments, although they may be aware of the fluid coordination of their actions. However, many of the cognitive-flow calculations are directed back through the thalamus to the cortex, and these signals are often in a position to influence the flow of attention.

Temporal Coordination Learning

Learning in the cerebellum is commonly studied using a Pavlovian stimulus-stimulus pairing procedure. One common experimental paradigm is eye-blink conditioning. Suppose a tone is used to predict an air-puff delivered to the eye. The tone is called the conditioned stimulus because reactions to it must be trained (i.e., conditioned), and the air-puff is called the unconditioned stimulus because it triggers a response without training. The unconditioned response to the air-puff is a protective eye-blink. The initial response to the tone is simply to notice and possibly to orient toward its source. Training involves presenting the tone immediately before the delivery of the air puff. In a typical procedure, the tone begins sounding two seconds before the air puff and continues while the air puff is presented. As learning proceeds, the subjects learns that the tone predicts the air-puff, and a protective eye closure, the conditioned (learned) response, begins to occur to the tone in anticipation of the air-puff.

The neural circuitry supporting temporal coordination learning is complicated, but as usual, understanding the basic organization will help us understand the process better. Given that half the cells in the brain are involved in this circuitry, a little extra detail seems in order. Inputs into the cerebellum are provided by neurons characterized by two kinds of axonal connections, *mossy fibers* and *climbing fibers*.[9] The mossy fibers arise from nuclei in the spinal cord and brainstem, in particular from the massive pontine nuclei in front of the cerebellum. These input regions receive signals from nearly every part of the brain.[10] This arrangement guarantees that nearly any sensed experience will be passed along to the cerebellum. In the eye-blink conditioning example, the tone activates mossy fibers. The mossy fibers branch to make synapses with granule cells and Golgi cells, which in turn connect with a complex web of cells. It is estimated that the profuse branching here allows the signals from the mossy fibers to reach approximately one hundred thousand different Purkinje cells. Because the Purkinje cells manage output activity, this arrangement ensures that each mossy-fiber input can act as predictive cue for many different outputs.

The climbing fibers provide information about the output activity that the cerebellum learns to predict. All the climbing fibers arise from the inferior olivary complex. The olive is a large group of nuclei in the brain stem that sit on each side of the medulla (see Figure 10-1). These nuclei receive signals from output command centers such as the vestibular nucleus, the red nucleus, the superior colliculi, and other output controllers in the brainstem. There is a one-to-one relationship between the climbing fiber inputs from the inferior olive and their Purkinje cell targets in the cerebellum. The Purkinje cells, in turn, connect with the output drivers in the deep cerebellar nuclei. These nuclei can activate the activity of each output being monitored by the climbing fibers. Normally, the Purkinje cells inhibit these deep output drivers. However, learned associations with mossy-fiber activity release the inhibition and activate the outputs. It turns out that the output drivers also inhibit further climbing fiber signals.[11] This means that strong climbing fibers signals represent *unexpected activity* – that is, types and amounts of activity that the cerebellum has not yet learned to anticipate well.[12]

In this manner, the climbing fibers provide a kind of weighted "error" signal for activity that has not been predicted by the cerebellum. Research suggests that this error signal enables the Purkinje cell to learn what combinations of mossy fibers are active *immediately prior to* each climbing fiber signal. Whenever the mossy-fiber activity is detected, the Purkinje cell is disinhibited, and that activates the deep motor outputs for the expected action. In contrast, the mossy-fiber activity which occurs earlier in time causes the Purkinje cells to form stronger inhibitory connections. As a result, the Purkinje cells come to suppress activity early and to disinhibit output responses around the time that the unconditioned stimulus is expected to arrive.[13] The strategy of suppressing activity early, and ramping up output activity near the time that a response in anticipated, is what enables motor outputs to be smoothly coordinated with each other.

Think of it this way. The cerebellar complex sits right behind the sensory and motor pathways running along the brainstem and listens in on all the input and output chatter passing up and down this pathway. The pontine nuclei and its associates sample potential cues and relay them to the cerebellum via mossy fibers. The inferior olive provides continuous information about ongoing output activity to the Purkinje cells. Each Purkinje cell learns what mossy-fiber activity best predicts the climbing fiber activity to it and ramps up output activity in timed anticipation whenever those mossy fibers are active. This anticipatory strategy is highly flexible because the mossy-fiber inputs branch in a web of processes which can be connected with a vast number of Purkinje cell outputs. Each climbing fiber, in contrast, provides feedback about one elemental output to a Purkinje cell. Following this strategy, a wide variety of cues can be used to control the timing of very specific outputs.

Although the motor consequences of temporal coordination learning have been emphasized in the eye-blink learning example, the temporal flow predictions provided by the cerebellum are also critical for anticipating other source of activity, including changes in interoceptive reactivity states. This occurs because the cerebrocerebellum receives inputs from a number of interoceptive structures ranging from the cingulate cortex and the extended amygdala to the reticular formation and the autonomic nervous system. Further, the cerebellum directs outputs back to these same systems.[14] In short, the cerebellum is not only involved in anticipating and coordinating motor activity, but it is also in anticipating and coordinating the interoceptive states which motivate and support actions. The autonomic and motivational effects of cerebellar learning are often overlooked. However, interoceptive changes are a critical part of many anticipatory transitions, and the cerebellum learns to anticipate them, just as it anticipates motor activity.

Cognitive Flow

Cortical projections to the pontine nucleus, including those from the prefrontal cortex, enable cortical processes to serve as inputs into the cerebellum. This means that "thoughts" managed in the cortex can function as predictive cues for the flow of cognitive processing. Outputs from the cortex to brainstem command centers, like the red nucleus, also inform the cerebellum about cortically organized action plans. The red nucleus has historically been considered a motor system, however there is growing evidence that it is also active in sensory processing tasks, presumably because sensing typically involves orienting and motor-related tracking activities.[15] These tracking activities are cognitive responses in the sense that they influence attention, and the cerebellum learns to predict them too. As a result, when plans begin to take form in the cortex, the cerebellum can send anticipatory outputs to the red nucleus, and via the thalamus, to broad areas of the cortex. These are effectively conditioned cognitive-flow responses which dynamically ramp up tracking and anticipatory processing activity to support the plans.

The cognitive-flow activity managed by the cerebellum has broad effects on thought and decision making. Research shows that the cerebellum is active during imagined movements, mental searches, and sensory discrimination tasks.[16] Thus, it is not surprising to find that patients with cerebellar damage show deficits in planning successive actions, in making decisions, and in shifting attention.[17] Deficits in cortical anticipation also affect language. Some cerebellar lesions result in grammatical errors, errors in the normal flow of words,[18] and in extreme cases even in a failure to initiate verbal comments following events that would typically trigger comments in normal subjects. These deficits may be so severe that some patients appear to be unable to speak, although their vocal apparatus is fully intact.[19] In these cases, it appears that perceptual information simply fails to trigger the kind of cognitive flow needed to go from seeing to saying.

Taken together, these observations suggest that though the

cerebellum is not required for the immediate focus of conscious attention, it plays a critical role in guiding the temporal flow of attention. Because the cerebellum seems to be optimized for predicting changes in the 100–200 millisecond time frame, it is easily fast enough to influence sequential steps in conscious processing, which are estimated to take 300–500 milliseconds. As a result, the cerebellum must also be considered another core member of the attention steering committee. Without its anticipatory prompting, minds would be more passive, and many basic cognitive tasks – such as forming sequential plans, making logical decisions, and speaking about what we experience – would be impaired. With this cognitive-flow module in place, minds become more dynamic and self-stimulating. Given that half the cells in the human brain are located in the cerebellum, it seems safe to conclude that the cognitive flow it provides is an important part of human thought, and for that manner of the thought of all vertebrates. The cerebellum literally keeps thoughts flowing in anticipatory sequences.

GETTING COGLEY IN THE FLOW

For a bipedal robot, balance is obviously a processing intensive task, but it has been accomplished. Sony's twenty-three-inch-tall bipedal robot Qrio, short for "quest for curiosity," not only keeps his balance while walking, but he can also jog, dance, kick a soccer ball, and even jump slightly off the ground with both feet. Like humans, if Qrio senses that he is falling, he moves quickly to adjust his balance, and if unsuccessful, he anticipates the danger and tries to break the fall with his hands or rump, so as to protect his head. Honda's fifty-one-inch-tall robot Asimo, which is featured in some of Honda's television advertisements, boasts similar skills, including stair climbing. Asimo reportedly uses thirty-four separate actuators, each with its own controller, to give him life-like movements. Neither of these robots uses a self-organizing general purpose temporal coordination system as complex as the cerebellum, but a substantial degree of coordination is required to jog and dance, and that's a start.

Providing a temporal flow coordinator which operates on this scale for Cogley will be slightly more complicated because we'll want to use a general-purpose algorithm for all aspects of temporal coordination. Balance and action integration should probably be implemented together in Cogley. Cognitive flow might be implemented separately. However, the same kind of timing algorithms should be used. Only the input and output paths will need to vary. Cogley's balance networks will need vestibular and postural information as inputs, and he should use this information to anticipate changes in his center of gravity in the next processing cycle and generate well-timed limb and postural changes to compensate and maintain balance. Action integration will require networks listening in on motor commands and anticipating sequential actions. Cognitive flow will require that high-level networks also provide inputs to Cogley's temporal flow coordinator so that it can "listen in" on his cognitive processing and anticipate his shifts in thinking.

The criterion for successful cognitive flow is less obvious. However, a system that learns about often-occurring sequences of its own processing will come to anticipate its own activity and begin to ramp up routine processing strategies in well-timed sequences. Using the flow of motor activity as a model, cognitive flow should have the effect of smoothing the assembly of routine cognitive activities into well-practiced and gracefully coordinated sequences. Obviously the cognitive flow of information should not be so influential that it takes full control of ongoing processing and runs on in isolation from other inputs. Instead, it should simply bias ongoing processing sufficiently to help Cogley anticipate common transitions, so that he is better prepared to handle expected events when and if they occur. It should also be clear that many aspects of cognitive flow may occur in parallel, so they may not all gain attention. However, those aspects of cognitive flow which do gain attention will be the most influential in shaping the flow of his attention. That should help give Cogley's behavior an added quality of connectedness, because what he attends to will influence his reactions in predictable ways.

CONCLUSION

Brains evolved to coordinate actions with objects and events in the world. Two critical domains for such coordination are space and time. Spatial coordination in the brain is made possible by the layered architecture of the tectum, a midbrain structure known as the superior colliculi in mammals. Inputs from all directionally oriented sensory systems pass into the tectum where they are arranged in spatial maps. Different sensory systems are mapped in registration with each other and with motor orientation information that can direct eye, head, and locomotor activities. The registration of the maps makes it easier to target actions using combinations of sensory inputs. A spatial coordination system built on this design will provide Cogley with the ability to orient and track external events using combined sensory inputs. However, it will do much more. It will provide the initial mappings he needs to represent his body in relation to the events in the world, and it will provide a framework for anchoring focused high-resolution sensory inputs into a big-picture view of the world.

Temporal flow coordination emerges from the architecture of the cerebellum and its location in the flow of sensory, motor, and interoceptive activity as that information travels up and down the brainstem. Temporal coordination ensures a smooth flow of motor activity, cognitive processing, and interoceptive states. Given the scope of the functions that are supported by this architecture, it will require a significant amount of Cogley's network resources. However, it will add a dynamic character to his thinking. Action plans will be guided by anticipatory sensorimotor associations. Memory recall will tend to flow in common sequences of occurrence. Feelings will flow in anticipation of changing actions and events. The spatial and temporal coordinators won't result in conscious experience. However, they will provide Cogley with bottom-up mechanisms which will help coordinate the flow of spatial and temporal information, and some of that information may reach consciousness. In this way, they will provide Cogley with a dynamic sense of his own agency. Our next task is to bring part of this agency to consciousness and provide it with some executive guidance functions so that it can learn to manage its cognitive flow better. Only then can Cogley truly be considered to be a conscious, rational agent.

CHAPTER 11
IT BINDS, THEREFORE I AM

The proprioceptive coordinators for space and time don't result in conscious feelings. However, they do provide low-level processes that help coordinate the flow of information in the brain. The tectum provides the connectivity needed to represent the location of the body in relation to cues in the world, to use those representations to track selected cues, and in doing so, to manage the flow of information by managing the orientation of receptor systems. The cerebellar complex provides the systems needed to connect predictive cues in well-timed coordination with common outputs sequences, and in doing so, to manage the moment to moment flow of cognitive activity throughout the brain.

Attention and core states of consciousness emerge from another coordinating system, a global coordinator which is capable of binding features from all three streams together.[1] The primary brain structure responsible for this binding is the thalamus and its connections with the cortex. Attention results from filtering, competitive selection, and interoceptive biasing processes in the thalamus. Consciousness emerges as a feeling of what happens because interoceptive activity is bound with perceptions and actions in states of attention. When interoceptive states gain attention, we call them feelings. Thus, feelings are always a part of attention. Feelings add subjective and self-referential properties to attention. That is the basis for the experience of phenomenal consciousness.

Adding to the core experience of consciousness is a second global coordinating system which guides attention in task-related activities. That coordinator is localized in the prefrontal cortex. Our model of this system is that of a task-related memory, one that is capable of forming working hypotheses about what feature categories to attend to during a task. The idea is that working hypotheses come to guide attention and consciousness in ways that enable conscious agents to attend to task-relevant cues and thus to act more intelligently. It's a simple model of intelligent decision making with profound implications.

It Binds, Therefore I Am

Flexibility carries a cost: Although our elaborate sensory and motor systems provide detailed information about the external world and make available a large repertoire of actions, this introduces greater potential for interference and confusion. The richer information we have about the world and the greater number of options for behavior require appropriate attentional, decision-making, and coordinative functions, lest uncertainty prevail. To deal with this multitude of possibilities and to curtail confusion, we have evolved mechanisms that coordinate lower-level sensory and motor processes along a common theme, an internal goal. – Earl Miller & Jonathan Cohen, "An Integrative Theory of Prefrontal Cortex Function," 2001

The three streams model of brain function includes mechanisms which serve to organize information within three primary domains of cognitive activity: motor decisions, perceptual processing, and interoceptive monitoring. However, with the exception of a few broadly distributed reinforcement signals originating from the interoceptive stream, this model provides little insight into how activities are coordinated across the three streams. Yet many tasks can only be accomplished when a number of functions are brought together in globally coordinated interactions. Competition for better systems of coordination has been an underlying force in the evolution of the brain. A coordinated predator is more likely to be successful than her less coordinated competitors. A coordinated prey animal will be more likely to escape than his less coordinated conspecifics. Attention, consciousness, and intelligence are all products of this competition for better coordination. Attention emerges in a mind which can selectively focus its activity in service of specific perceptions, actions, and interoceptive needs. Consciousness emerges as feelings and values when interoceptive states gain attention. Intelligence emerges as minds acquire the ability to manage the flow of attention in more productive ways.

To begin to understand how these aspects of mind take form, we need to look at the organization of two systems for coordinating activity on a global scale in the vertebrate brain. The first of these is a system for selectively focusing subsets of cues in states of attention. The second is a top-down coordination system which is capable of providing some measure of executive control over attention. As global coordinators, these systems are involved in many aspects of cognition and behavior. I don't claim to provide a full description of their roles or properties. Instead, I will try to essentialize their functions in order to make the descriptions of each of these coordinators more tractable. Consciousness emerges within a process of interoceptively weighted attention made possible by the organization of the thalamus and its connections with the cortex and the interoceptive stream. How this coordinating system gives rise to core states of conscious experience will become surprisingly clear once we understand the organization of the thalamocortical architecture. The executive control networks of the prefrontal cortex provide top-down biases that make conscious minds more sensitive to task-related cues and more capable of managing their attention. By paying attention to task-relevant cues, and by shifting attention as a task progresses, conscious agents come to act more intelligently.

COORDINATING ATTENTION

The vertebrate cortex is a massively parallel, highly interconnected meta-network. However, simulation studies find that massively parallel architectures are inherently unstable. They operate at the edge of chaos in a constant interplay between integration and segregation. Without stimulation they may lie dormant, but one spark of excitation can spread like wildfire across the entire system. Yet thousands of modules of perceptual and motor activities may be activated in the brain each second. To take adaptive action, the brain needs to engage mutually related states of excitation for balance, orientation, perception, planning, and action decisions. Obviously, it needs some way of coordinating these activities while filtering out other sources of activity, if it is to overcome its chaotic tendencies.[2] The strategy that the vertebrate brains have evolved is to form a signaling conduit for information passed to the cortex; the conduit selectively favors processing connected activities while filtering out others. This strategy of selective filtering on a global scale is what we call attention.

The neural organization that makes attention possible in the vertebrate brain begins with the thalamus. Most of the sensory and motor information that reaches the cortex passes through the thalamus. Twenty years ago this structure was considered little more than a relay station that perhaps performed a little preprocessing as a signal was passed on to the cortex. Surprisingly, however, researchers Murray Sherman and Ray Guillery have found that less than 10 per cent of the synapses in the core thalamic nuclei are from the sensorimotor sources that are being relayed to the cortex.[3] The remaining 90-plus per cent are from sources that modulate the transmission of the source signal (see Figure 11-1). Approximately one-third of the modulatory synapses are local connections within the thalamus. These have binding and filtering effects. Another third of the inputs are reentrant feedback circuits from the cortex. These provide top-down guidance over the flow of information through the thalamus. The final third of these modulatory inputs are from low-level interoceptive networks. These guide processing based on internal status and reactivity states. This is not exactly what would be expected from a standard relay. Much more is going on within the thalamus.

Filtering in the thalamus depends on the thalamic reticular nucleus, which covers the top of the thalamus like a thin cap. However, rather than being a homogeneous cap, the reticular nucleus is actually divided into a number of functional sectors.

For example, there are separate regions for vision, hearing, touch, movement, and reactivity functions. Each sector has topographically mapped connections which interact with functionally related thalamic nuclei. Fibers passing in both directions between the thalamus and the cortex send off collaterals to the reticular nucleus. It appears that signals compete for passage through the filter using a reciprocal inhibition strategy, in which stronger signals inhibit the weaker signals in each domain and thereby decrease their ability to inhibit them. This happens because the outputs from the filter are inhibitory and project back to the thalamic neurons which connect to the cortex. The thalamic neurons projecting to the cortex act like a switch. When they fire, they activate stellate neurons in the cortex, which in turn disinhibit the cortical pyramidal cells for those circuits. This effectively *turns on* the corticothalamic loops for the particular features being relayed to the cortex. However, when the connecting neurons are inhibited by the filter, as most usually are, the loops are turned off. Thus, only the strongest set of signals is turned on and can activate the cortex at any time.

One way that a signal can be strengthened is by binding it in synchronous activity with other signals. Early evidence for synchronous processing in the brain was provided by Wolf Singer, Charles Gray, and colleagues, who reported widespread cases of synchronous firing activity in the sensory cortex.[4] Initially working in the visual cortex of cats, these authors noted that cells for different visual features became active in synchrony when a distinct cue (such as a bar of light with a specific size, orientation, and velocity) moves through the visual field. The synchrony occurred even though the processing regions for different features (e.g., size, orientation, and movement) were spatially separated in the visual cortex. This suggested that synchrony is a key mechanism for binding features into coherently processed combinations. The mechanism is thought to work because features that have broader synchronous support tend to outcompete those which are less well connected.

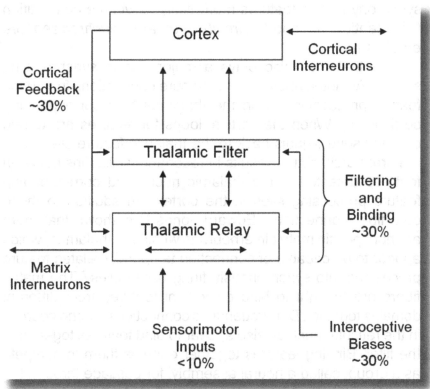

Figure 11-1: Modulatory effects on thalamic information flow.

Subsequent work has found that neural synchrony is also part of the process that underlies attention. For example, when different visual targets were presented to each eye of a cat, and the cat was required to react to one target to gain reward, synchrony among the neurons representing the chosen stimulus pattern increased in intensity as attention shifted, while the synchrony among neurons representing the non-attended features decreased.[5] In effect, synchrony appears to provide a general mechanism for binding coincident features together. If the features are often coincident, like the redness and roundness of an apple, then Hebbian learning mechanisms produce more permanent bonds that result in added synchrony due to associative connections. However, mere coincidence appears to be sufficient to induce short-term synchrony, and synchrony makes features more likely to win the competition for attention. In short, *circuits that are synchronized are emphasized.*

The thalamus also plays a major role in synchronizing signals. As noted above, there are core connections from each feature-processing site in the thalamus to a corresponding cortical site. When the cortical loops for features are turned on, they send signals back to the thalamic feature sites. This reentrant architecture results in resonant oscillations between features detected in the thalamic nuclei and corresponding feature processing sites in the cortex.[6] In addition to these core fiber connections, Edward Jones has shown that there are nonspecific matrix interneurons within the thalamus, which appear to be capable of connecting temporally related feature processing into synchronously firing assemblies.[7] The matrix fibers are thought to bind coincident features from different domains together. Dendrodendritic connections among neurons in the reticular nucleus also appear to bind features together in the filter. Binding features together enables them to compete as a group, called a neural assembly, for passage through the filter. The details of this process are still sketchy, but it appears that all the corticothalamic loops for the group are turned on in synchrony. This results in resonant firing patterns among all the features.

Extended patterns of thalamocortical resonance were initially assumed to be just another form of neural synchrony. However, research suggests that as the thalamic bindings extend across functional regions, they result in rhythms with more complex phase alignments, apparently because different functional regions have their own intrinsic rhythms and timing patterns.[8] Simulation studies in large-scale networks have shown that the alignments tend to propagate in patterns with characteristic sequences and delays, resulting in polysynchronous rhythms rather than simple synchrony.[9] The general idea is that, like synchrony, these polysynchronous alignments enable connected feature assemblies to outcompete unaligned competitors. However, the polysynchronous connections enable larger patterns to bind together while the components still vary somewhat locally. This presumably allows the organization to vary more flexibly as the composition of the assembly gradually changes.

More work is needed to explain how these polysynchronous assemblies bind and propagate as they do. However, electrical couplings among adjacent neurons, matrix cross-connections within the thalamus, the harmonic coupling of rhythmic patterns across systems, and intrinsic processing rhythms within different subsystems are all thought to contribute to the process.[10] Multiple binding frequencies have been found.[11] Short-range bindings generally involve higher frequencies, such as those in the gamma (30–60 cps) and high-gamma (80–200 cps) range, while bindings among more distant neural structures tend to produce couplings at lower frequencies, beta (14–30 cps), alpha (8–13 cps), and theta (4–7 cps) rhythms. The form of the assemblies gradually shifts and flows as new features join with the resonant assemblies, and other features are lost. Most of the transitions occur gradually, apparently because they involve extensive interconnections that are slow to be assembled and disassembled. However, the tendency of these assemblies to form gradually morphing chains of activity seems to provide a good match to the shift and flow of conscious awareness over time.

As we have noted, a sense of status and readiness to act emerges from activity within the interoceptive stream. However, what is not often recognized is that this activity also plays a major role in the flow of information through the thalamus. It turns out that thalamic processing is bound to core regions of interoceptive stream via projections from the parabrachial nuclei and adjacent regions in the brainstem.[12] The circuitry is immensely complex and the role of these interoceptive inputs remains to be verified. However, functionally it seems clear that to be adaptive these connections must promote attention to the sensory and motor signals which are bound to important interoceptive activity. Natural selection simply could not bring value information into the thalamus and not use it to promote more adaptive decisions. Further, the confluence of interoceptive inputs with sensorimotor information is exactly what is needed to add *a feeling of what happens* to attention. Given that less than 10 per cent of the synapses in thalamic relay nuclei involve neurons from the signal modality being relayed to the cortex, while about 30 per cent involve interoceptive inputs, it appears that interoceptive associations have over three times more influence on what gains attention than the activity of the primary sensory or motor signals.

This explains why sensorimotor events are always accompanied by feelings. Associated interoceptive activity is always a major part of what gains attention. An interoceptively guided mind is a coordinated agency that perceives not only the external world but also its own reactions to what it perceives and how it acts. In effect, attention is immersed in a constant flow of information about the status of the body and its states of reactivity. As Gerald Edelman notes, this naturally introduces a self-referential aspect to experience.

> A ubiquitous set of inputs to the dynamic core is continually received from bodily and brain systems concerned with motor behavior and homeostatic control. These inputs to the core are not only among the earliest but are also among the most persistent, and they provide a fundamental basis for subjectivity or the self-referential aspects of consciousness.[13]

Because primary consciousness involves interoceptive feelings, an agent with this level of consciousness experiences an ongoing sense of how its status changes as it perceives and acts. This awareness of status enhances the sense of self-agency and naturally motivates actions and perceptions which improve self-status. Given that this initial stage of self-awareness results from the resonant binding of feelings with perceptual and motor activity, I call it the *resonant self*. As Rodolfo Llinás notes, the self is a unique and highly adaptive consequence of binding features in the interoceptive domain with those processed in the exteroceptive domain.

> *This temporally coherent event that binds, in the time*
> *domain, the fractured components of external and*
> *internal reality into a single construct is what we call*
> *the "self."* It is a convenient and exceedingly useful
> invention on the part of the brain. It binds, therefore
> I am![14]

Although Gerald Edelman refers to the feeling-bound awareness of experience as *primary consciousness*, it is more often referred to as *perceptual or phenomenal consciousness*. Antonio Damasio refers to these initial stages of consciousness as *core consciousness*.[15] However, in recent work he seems to have extended this term to include several components of consciousness.[16] I will therefore use the term phenomenal consciousness to refer to the primary stage of consciousness. As Damasio characterizes it, the lower levels of the interoceptive stream effectively integrate interoceptive inputs into what he calls the *proto-self,* a highly interconnected summary representation of somatic status and reactivity. The dispositions of the proto-self are broad and pre-conscious. The phenomenal experience of consciousness emerges when selected aspects of proto-self activity are then re-represented in the context of attention to sensorimotor activity. This re-representation effectively results in what Damasio calls the *feeling of what happens.* A conscious agent thus feels the effects of attending to how it perceives and acts, and it can change how it perceives and acts in response to its feelings. As Damasio notes, this gives a conscious agent an added adaptive advantage.

> And what was that advantage? It was the possibility
> of connecting the very core of life regulation with the
> processing of images.... Core consciousness is the
> door to a revelation of regulatory values, the passage
> into the possibility of constructing in the mind some
> counterpart of the regulatory value hidden in the
> brain core, some new and more open way of sensing
> the life urge and the means to hold on to life.[17]

When we combine the ideas of Gerald Edelman, Rodolfo Llinás, and Antonio Damasio together with the architecture of the thalamus proposed by Edward Jones, Murray Sherman, and Ray Guillery, we find a simple yet elegant account of the mechanisms underlying consciousness. Attention is managed by a filtering system that is weighted by interoceptive activity. When interoceptive states gain attention, they are interpreted as feelings. A sense of self emerges as attention networks re-represent interoceptive states in the context of perceiving and acting. With such a neural organization in place, an agent is always in contact with his subjective reactions to his own experience. "I" am the emergent agency that experiences feelings in the context of reacting to the world. I understand the world in a more intense and more personal way because of those feelings. I am therefore more sensitive to how my decisions affect me. That's the adaptive advantage that consciousness provides.

There is one more aspect of the resonant self that deserves special mention. As we have already noted, social status and reactivity are also a core part of the interoceptive lattice. In fact, the circuits which represent social status and reactivity appear to reuse many of the same neural pathways used for representing somatic status and reactivity. As a result, social care and acceptance produce feelings of well-being that energize an agent much like somatic feelings of wellness do. In a similar way, social rejection and loss tend to produce feelings that mimic the distress and hurt of physical pain.[18] And because social states also contribute to interoceptive status and reactivity, social values can also influence what gains attention. In short, social feelings have evolved to be core elements of conscious awareness. They have a major impact not only on what sensorimotor features gain attention, but also on an agent's sense of self. In fact, for highly social agents, social feelings may sometimes be more important than somatic feelings of self.

A FEELING OF WHAT'S COGLEY

The remarkable thing about the neural architecture that leads to attention and consciousness is that although its organization is complicated, the basic logic is surprisingly clear. If we are to give Cogley a feeling of what happens, then he will need a central attention complex in which attention filtering is influenced by associated feelings of status and reactivity. To give proper priority to these interoceptive associations, we should arrange for the feelings to have over three times more potential to impact the transmission of sensory or motor relay signals than the signal itself. In effect, Cogley should constantly be interpreting his perceptions and actions based on the value of associated feelings. If Cogley is to have feelings of self as rich as those of his human partners, then his interoceptive networks must integrate low-level status activities, including homeostatic needs, emotions, and empathic feelings, and he must be able to re-represent them in more complex combinations. In short, he must have a broad array of status and reactivity inputs to provide him with the range of feelings typical of his human partners.

Following a biologically inspired design strategy, the features required for Cogley's attention architecture seem rather straightforward. Sensorimotor activity will have to be relayed to high-level perceptual and motor-planning networks through a multimodal central attention channel. Processing in this channel should be segregated into modality and functionally specific submodules and then relayed to corresponding modality-specific, high-level networks for further processing. However, the relay signals should be subject to filtering, so that only those with the highest activations tend to win the competition. High-level perceptual and motor networks should also have reentrant connections with the central channel submodules such that processing can be further refined by resonating activity between the higher-levels modules, Cogley's version of a cortex, and the central channel submodules.

In order to make Cogley's attention algorithms selective, a set of modality-specific filtering submodules should be situated between the central attention channel and its links to higher-level processing networks. These filtering modules should provide inhibitory feedback for signals in the same modality. The inhibition should be directed to competing inputs in proportion to the activity of each input. Stronger signals should inhibit weaker ones and diminish their ability to inhibit them, thus driving attention to a focus. At the same time, cross-connections in the central channel should enable related features to bind together in mutually supporting assemblies, which can compete for passage through the filter as a group. As a result, those assemblies which can recruit the most robust activity will tend to win the competition for passage through the filter, and less active assemblies will be suppressed. To ensure that Cogley's selective attention gives priority to feeling-centered information, we'll also need to add connections between low- and mid-level interoceptive modules and each sensory and motor relay module. These will cause assemblies with stronger interoceptive bindings to be favored in the competition for attention.

To establish cross-modal connectivity within the central attention channel, Cogley will need a matrix of cross-connections that are capable of binding activity in different central channel submodules in resonant cross-modal assemblies. These bindings should be based on the coincident timing of inputs and the tendency of learned associations, including interoceptive associations, to activate coincident activity. Because competing assemblies will inhibit each other, it seems likely that competing resonance states, to the extent that they form, will tend to occur out of phase with each other. A simplified thalamic model of this sort has shown that connections between the core relays and the high-level modules (the cortex equivalent) can result in resonant activity, whereas a diffuse set of matrix connections can serve to cross-connect submodule features in larger assemblies.[19] Cogley's temporal flow module, his version of the cerebellum, would naturally be expected to guide temporally bound features in meaningfully changing sequences as this process continues.

Because artificial neural networks are digital, they will lack some of the natural rhythmic properties that biological neurons have. However, firing rhythms can be simulated. Processing rhythms are prominent characteristics of many brain regions, and it seems likely that many of these rhythms result, at least in part, from interactions with local inhibitory circuits. Such timing circuits can readily be added to Cogley's networks. Resonant oscillations between Cogley's central channel modules and his high-level processing modules will also be essential for guiding attention via top-down processing. It seems likely that several kinds of resonant bindings can be managed within Cogley's architecture. A computational model of activity in cortical columns with such binding potentials has shown that oscillations tend to emerge at precise thresholds, after critical durations of activity.[20] When such simulations are conducted in networks with long-range connectivity, and the flow of signaling exceeds a critical threshold, the networks have been shown to ignite in large-scale synchronous binding states. Similar shifts in binding patterns have been found to accompany shifts in consciousness in humans.[21]

Building broad network systems with resonant binding and ignition properties will no doubt be challenging. However, the fact that these properties have already been simulated in neural network models demonstrates that it is possible. Cogley's networks will not bind exactly like biological networks. However, that doesn't appear to be critical. Once we can arrange for Cogley's networks to bind on multiple levels, in polysynchronous assemblies linked by a central conduit architecture, and once we can arrange for that architecture to use a competitive filtering process, prioritized by interoceptive associations, then Cogley's attention should shift and flow along topics and in time courses similar to those of his human partners. Because the assemblies with the strongest interoceptive reactions will tend to win the competition for attention, they will constantly connect his perceptions and actions with important physical and emotional feelings. These will provide Cogley with a sense of his own status and reactivity as he perceives and acts. He will, in short, experience a feeling-centered resonant-sense of self.

Research suggests that it takes 300–500 milliseconds for a new assembly to stabilize and gain attention. Assuming a 30–40 Hz thalamocortical cycle, this time implies that it usually takes 8–12 cycles for the focus of attention to stabilize. These numbers should provide us with guidelines for adjusting binding properties in Cogley's networks so that his shifts in attention will tend to occur in similar time frames. In order to have alternative assemblies that can quickly gain attention, it would also be useful to allow several of the next strongest assemblies to be maintained in low-level resonant states. These alternative assemblies will effectively be on the brink of attention and be able to gain attention quickly. This will enable Cogley to shift his attention among several situationally relevant assemblies in a flexible manner. We'll return to this idea and describe an architecture to support these alternative assemblies in the next chapter.

The states of consciousness provided by the current design will not be equivalent to the rich states of consciousness that we humans usually experience. They will be limited to feeling-centered perceptions, such as the awe we experience when we notice a beautiful sunset, the pleasantness that accompanies good food, the joy that accompanies an attractive smile, or the anxiety that occurs when startled by a sudden noise. Still, we have reached a critical milestone. With this architecture in place, Cogley will experience an ongoing feeling of what happens as he perceives and acts. He will be phenomenally conscious. Subsequently, as Cogley experiences perceptual changes in the context of planning actions and recombines memories in imaginative forms, his feelings will take on more importance because they will also come to shape his plans and guide his subsequent imagination and learning. Because his high-level interoceptive networks will support several levels of re-representation for feelings and will have extensive connections with his semantic memory networks, as Cogley learns to re-represent his feelings, those refined discriminations will become part of his memories. The quality of his conscious experiences will grow more meaningful as past feelings are recalled in the context of ongoing perception. However, our discussion of these extended forms of consciousness will have to wait. First, we need to describe a second global coordinator, one that will help Cogley guide his attention in more intelligent ways.

AN EXECUTIVE COORDINATOR

The prefrontal cortex, commonly abbreviated as the PFC, is a second major global coordinating system in the brain. It was part of this coordinating system which Phineas Gage damaged in his accident. As Gage's subsequent course of life illustrated, the PFC is a critical brain structure for reasoning. In fact, theorists have characterized the role of the PFC variously as one of extended planning, logical decision making, attention biasing, rule learning, goal recognition, and consequence evaluation. This variety of functions has caused the PFC to be viewed more generally as an executive control module.[22] However, despite agreement that it supports many executive faculties, exactly how the PFC performs these varied functions remains controversial. Our goal in this section is to summarize the connectivity and functions of the PFC and to propose a systems-level model of how the PFC exerts executive control, so we can provide similar functions for Cogley.

Some authors interpret the differing anatomical connections in the PFC as evidence that the different regions must have qualitatively different modes of operation. Others have suggested that the differing connections may simply imply that different regions have different contents, while the basic processing algorithms may be the same. Of course, both of these perspectives may be partly true. Content differences on a broad scale may give rise to qualitatively different functions. For example, the dorsolateral PFC connects to the occipital, temporal, and parietal cortices, which suggests that this region plays a role in managing spatial information and early action planning. Adjacent areas in the dorsal PFC have connections with premotor structures such as the supplementary motor area, the premotor cortex, the rostral cingulate, and the frontal eye fields. This suggests that this region participates in the supervision of action planning. The ventrolateral PFC connects with the visual, auditory, and somatosensory cortices, suggesting that it plays more of a role in guiding perceptual processing. A rostro-caudal hierarchy of connections in the lateral PFC appears to bias cognitive control via progressively abstract higher-order representations, suggesting that this region uses different levels of abstraction.[23]

The PFC also plays a role in emotional and social decisions. The medial PFC has major efferent connections with dopaminergic systems in the ventral striatum and appears to be important for managing motivation and mood states. The ventromedial region has broad connections with the hypothalamus, the amygdala, and the interoceptive cortices, which suggests it plays a role in representing goals and values. Recent evidence suggests that the hierarchical organization of the lateral PFC is also involved in managing social decisions and that inputs from other parts of the PFC also contribute to social processing.[24] For example, inputs to the anterolateral PFC from the ventromedial region appear to be critical for linking interpretations about goals and values with social decisions, while the ventrolateral region seems to be critical for representing social norms (both obligatory and prohibited norms) during decisions.

Based on the patterns of connectivity described above, it seems clear that the PFC has the potential to influence a wide variety of functions in the brain. In fact, it has links with prominent regions in all three processing streams. However, despite the obvious regional differences in connectivity, there are also many common features found across regions which suggest that the mode of operation in each region may be similar. For example, each region of the PFC integrates information from multiple high-level areas. The sensory inputs come from perceptual association areas rather than from primary sensory areas, and many of the association areas are sites of multimodal convergence. Similarly, the motor connections are with premotor "planning" areas rather than with primary motor areas. In addition, the different regions of the PFC are systematically interconnected, allowing the outputs from some regions to influence others. This suggests that the PFC may act as a kind of supernetwork composed of a number of specialist subnetworks. It seems likely that it is the ability of this supernetwork to coordinate information across multiple domains which gives it an executive character.

Obviously, it would help if we could simplify our thinking about how this supernetwork exerts its executive control. Fortunately, a number of theorists have proposed models for how the PFC does it job.[25] Most of these models suggest that the PFC is best characterized as a working memory system which is capable activating and sustaining task-relevant information. The idea is that the multiple regions enable the PFC to learn about representational states for a number of functions at the same time, while internal links within the PFC serve to bind those representations together in task-specific configurations. The sustained activation of feature representations during a task is proposed to account for the working memory function of the PFC, whereas the activation of motivational regions in the PFC is proposed to hold a particular feature set on task until it has been completed and the motivation is released.

The idea that motivational links hold the PFC on task follows from studies of the effects of damage to the PFC. One of the classic signs of PFC damage is the inability of subjects to stay on task. Subjects with PFC damage are easily distracted by irrelevant or previously relevant cues. This level of distractibility would be consistent with a diminished ability to sustain working memories and their links to motive states while inhibiting others. A similar inability to sustain connections between situational recognition, logical decisions, and social feelings could explain the many instances of impulsive behavior and poor social judgment which are typical of patients with PFC damage, as they were for Phineas Gage. It seems we're on the right track. A working memory model in which cues bound to interoceptive reactions guide attention and hold an agent on task appears to account for many of skills that the PFC supports and for the deficits that occur following damage to the PFC.

There are a few other properties of the PFC that deserve mention. The PFC has strong connections with subcortical regions of the planning stream. Further, it seems clear that the storage of working memories in the PFC is guided by dopamine reward signals. The subcortical connections and the reward-driven memory processing are thought to account for the task-related nature of working memories. Given that dopamine is known to be released by predictive cues as learning progresses, it follows that working memories could come to form memory configurations for sequences that lead to reward. Given that the PFC has robust reciprocal connections with the temporal flow networks in the cerebellum, working memory configurations would even be expected to change in sequential steps as a task progressed. Research also indicates that the PFC has direct projections to the filtering networks in the cap of the thalamus, which would seem to provide it with a more direct means for biasing attention.[26] The mix of task-specific working memories, hierarchical filtering biases, trial-and-error reward learning, and temporal flow learning appears to provide an ideal metasystem for learning about complex actions sequences and biasing attention to those sequences. In fact, connections between the PFC and the cerebellum seem to be ideally suited to generate the sequential shifts in attention that would be required for multistep tasks.

KEEPING COGLEY ON TASK

So how might a similar working memory complex be implemented in Cogley? A simple start would be to add a supplementary set of rapid-learning networks that spontaneously sort through high-level input patterns, looking for the dominant features in each domain they sample and locking on to those that precede increasing changes in reward density. The regional domains in this executive supernetwork should each receive high-level inputs from functional association areas. Spatial map areas and spatial-motor association areas should serve as high-level inputs into a spatial executive network to form Cogley's equivalent of the dorsolateral PFC. High-level motor decision areas should provide inputs into Cogley's executive motor network. High-level interoceptive areas should provide inputs into Cogley's executive goal state network, and so on. Each executive area, in turn, should have reciprocal connections to their source areas, as well as projections to functionally related areas in the central attention channel filtering networks. These projections should enable the executive networks to bias the features at the center of attention.

Randal O'Reilly and colleagues have proposed that learning in such an executive complex should be triggered by an *adaptive-critic* gating mechanism which is sensitive to both rising and falling reward densities.[27] The idea is that performance learning would initially depend on the standard trial-and-error learning algorithms in the planning stream. However, as rewards are encountered, they should cause the different domains of the task-memory complex to lock on to the high-level features that best predict the rise in reward density. This adaptive side of the algorithm can then serve as a sort of inferential learning tool for forming *working hypothesis* (my term) about what features are most important for a task. By enhancing the activation of this configuration of features on subsequent trials, the working hypothesis could then bias attention to feature categories which can guide the performance of the slower trial-and-error learning algorithms.

As long as a working hypothesis is successful, the configuration of cues that is enhanced by this algorithm should not change. However, if something happens to ramp up reward density, then the hypothesis should be updated to emphasize cues that accompany the increase in reward density. In contrast, if the reward density ramps down too far, then the critic side of the adaptive-critic algorithm should reset the executive complex so that the configuration will be resampled on subsequent tries. In effect, a run of successes should cause a hypothesis to form and be refined, and a run of failures should result in a new hypothesis being adopted. Amazingly, this process is not much different from the strategy of runs and tumbles that *E. coli* use to track a goal gradient. Runs continue with the direction that works. Tumbles reset the direction. Following this strategy, the executive complex should develop inferences and hypotheses which can be refined with experience. This model has been extended by some impressive computational models which show how these executive hypotheses might come to include hierarchical biases.[28] With these additions, Cogley should be able to form hierarchically structured hypotheses.

To hold Cogley on task, the executive complex should be designed so that interoceptive needs tend to keep a set of task-related working memories active until a goal is reached or the task is otherwise aborted. Once a task is completed and the motive is reduced, the configuration should become inactive. However, when the same task is resumed later, the working hypothesis from the last episode should be the starting hypothesis for the next one. In effect, the working hypothesis should be learned, refined, and carried forward. Given that these hypotheses are used to guide learning, it follows that as routine actions become automated as habits, this system would no longer be needed to guide their learning. This would free up the executive modules to support the learning of more complex tasks. Of course, some tasks like playing chess may be too complex to automate fully. Executive hypotheses may always be a required for success in these more cognitive tasks.

There are some limitations to this hypothesis-forming strategy for guiding attention. Hypotheses established in this way are not formally tested, so in some cases, they may end up being incomplete heuristics, or merely vague intuitions. It is also clear that some intuitions, especially those involving statistical predictions, sometimes result in systematic biases. To overcome these biases, it is helpful to learn to apply rule-based calculations in some situations, rather than always relying on heuristics.[29] Thus, in the long run it will be important for Cogley to learn to test his hypotheses more formally. However, as an initial strategy for guiding attention, learning executive biases based on what cue sets are associated with increasing success in a task seems like a good starting place. Testing potential plans by simulating their outcomes while hypotheses are active will also provide Cogley with a simple, first-level strategy for evaluating and refining his hypotheses. Cogley's hypotheses may not always be correct, but the ability to form hypotheses, even if they are wrong, will make him a more human-like agent.

CONCLUSION

The thalamus has traditionally been characterized as a relay for passing information to the cortex, but it has proven to be a surprisingly complex relay. It does contain a number of bottom-up sensorimotor inputs which can be relayed to the cortex. However, it also functions as a central conduit which is capable of binding multimodal activity in resonant assemblies, as a competitive filter for those assemblies, as an interoceptive gain controller, and as a target for returning cortical influences. It is this combination of processes which makes attention and consciousness possible. Attention is the global focus of processing which emerges from filtering and bindings within the thalamus. Consciousness is the experiential effect of having sensorimotor attention bound to and guided by interoceptive activity. Because associated interoceptive states are part of what gains attention, feelings are always a part of conscious experience. That is why Damasio describes the initial stage of consciousness as the feeling of what happens. It is the ongoing awareness of perceiving, acting, and feeling which results in a connected sense of self. As Rodolfo Llinás characterizes the effect, "It binds, therefore I am." And we know this architecture occurs in animals like Tom, because many of the studies on visual attention by Gray, Singer, and colleagues, and on the structure of the thalamus by Sherman and Guillery, used cats as subjects. We share a similar thalamic architecture with Tom.

Consistent with the organization effect, we have also described a way of thinking about the role of the executive modules in the PFC. The idea is that the working memories in the PFC create what are effectively working hypothesis, selective memory biases for task-related cues which facilitate attention, learning, and task performance. Following O'Reilly and colleagues,[30] we have proposed that this task memory is guided by an adaptive-critic learning algorithm. This strategy allows hypotheses to develop and shift depending on their success. Because the PFC forms hierarchical hypotheses and interacts with sequential learning algorithms in the basal ganglia and with the temporal flow algorithms in the cerebellum, this model can guide task performance in sequential steps. Admittedly, the essentialized model proposed here oversimplifies the complexity of the PFC. However, there is something very satisfying about this essentialized hypothesis-capturing process. It will enable Cogley to develop intuitions, to infer relationships, and to form task-related hypotheses much like his human partners do. These hypotheses will guide his attention based on the features his executive networks currently assume are important for succeeding in a task. In short, they will enable Cogley to become a more intelligent agent.

Chapter 12
The Steering Committee

The thalamus serves as the gateway to the cortex, but it's a gateway with some very special properties. As we have noted, cross-connections between nuclei within the thalamus appear to bind associated neural functions together in rhythmic assemblies. However, the cap of the thalamus serves as a filter which ensures that only the most active assemblies can pass through to the cortex. Reentrant connections from the cortex add top-down influences, which also bias the selectivity of these circuits. It is in this context that competition among assemblies taking form in the thalamus results in states of selective attention. Because interoceptive states are always part of the assemblies that gain attention, ongoing experiences are always accompanied by associated feelings. This feeling of what happens is the essence of phenomenal consciousness.

However, as remarkable as this phenomenal stage of conscious experience is, simply being able to react to the world with associated feelings is not always very adaptive. To make this process more adaptive yet, other support systems are needed to guide attention to the most relevant information and feelings in the moment. The organization that makes this possible is what I call the *attention steering committee*. We're all familiar with committees, so the idea that a committee of brain modules helps to manage attention provides us with a metaphor for thinking about how a modular agent-system might steer its own attention. However, to expand this metaphor, we'll first have to introduce a few new committee members, including one new member that will act as the chair of the steering committee.

THE STEERING COMMITTEE

> The cortex is a very highly and specifically interconnected neural network. It has many types of excitatory and inhibitory interneurons and acts by forming transient coalitions of neurons, the members of which support one another. 'Coalitions' implies 'assemblies' – an idea which goes back at least to Hebb – plus competition among them.... On the basis of experimental results in the macaque, some researchers suggest that selective attention biases the competition among rivalrous assemblies.
> – Francis Crick & Christof Koch, "A Framework for Consciousness," 2003

The strategy should be obvious by now. We're building a model of conscious mind in stages. The brain organization required to explain the full range of conscious experiences is simply too complex to take on in a single step. That said, we've reached another critical transition point. The thalamocortical architecture provides an organization within which sensory, motor, and interceptive processes can be bound together in self-referential states of attention. The prefrontal cortex adds task-related hypotheses that can partly guide attention. However, that architecture is still not sufficient to explain the range of experiences that influence attention. We need a more complex steering committee. Fortunately, most of the knower modules needed to populate this steering committee have already been described. Learning and memory modules are obviously important, but several new modules will also be needed. We're often not aware of many of these steering committee functions, at least as separate modules. However, once we come to recognize how each committee member contributes to the decisions process, we will have a much better understanding of how to design a self-steering agent.

A simple strategy for managing attention was proposed by neuroscientists Robert Desimone and John Duncan in a review article published in 1995. They noted that neural subsystems analyzing features on many different levels in the brain have an intrinsic bias to process mutually supporting features, while suppressing unconnected features. We characterized this process earlier with the aphorism, "Circuits that are synchronized are emphasized." Desimone and Duncan noted that these supporting interactions enable feature assemblies to form and to compete for attention within the thalamocortical architecture, and this in turn suggested to them how attention might be managed. They argued that by coupling systems which are capable of enhancing processing with other sources of activity, the systems should be able to bias the likelihood that assemblies containing those features would win the competition for attention. This strategy for managing attention has come to be called the biased-competition model.

Desimone and Duncan argued that both top-down and bottom-up mechanisms are often involved in biasing attention. Learned top-down processes are thought to provide some measure of conscious control for biasing attention, whereas low-level emotional states often provide unconscious bottom-up biases. One obvious process with learnable top-down controls is the working memory in the prefrontal cortex. As we suggested in the previous chapter, simply by enhancing activation for particular features, working memory processes appear to be able to bias attention for some feature-sets over others. However, it turns out there are many other processes for biasing attention. One simple behavioral strategy is to change head orientation, because the direction of the head influences the acuity of auditory and visual processing. Yet more focused control of attention can be achieved by changing the direction of gaze to sample particular visual features, because items at the center of gaze are processed in more detail.

In addition to the strategies for biasing attention via working memory and gaze direction, there are a number of other clever mechanisms which enhance features based on other sources of saliency. Our goal in this chapter is to explore these mechanisms in enough detail that we can provide Cogley with a broad set of tools for managing his attention. Once we can define a variety of mechanisms for biasing attention, we'll be able make Cogley a versatile self-steering agent. Of course, he will only be a partially self-steering agent. You and I and Tom have the capacity to steer our own attention at times. However at other times our attention is guided more reflexively. Cogley will also be subject to reflexive changes in attention. However, he should be able to guide his own attention in many situations. We'll begin our search by looking at the attention mechanisms involved in managing gaze direction.

GAZE-GUIDED GAIN MODULATION

There are several reasons that gaze is a particularly effective strategy for influencing attention. The first and most obvious is that gaze determines where the dense receptors in the fovea are focused. Unlike the broad field of inputs which the tectum receives, the receptors in the fovea are tightly packed and are all routed through the thalamus to primary visual areas in the cortex. It is these more detailed inputs which provide the primary visual information used in the perceptual stream. A second reason why gaze direction has a major influence on attention is that visual coordinate-transform networks appear to have become something of a standard in the brain.[1] This means that gaze has the potential to influence processing activity in many sensory and motor regions. In fact, research shows that simply looking at a particular finger causes activity in the motor regions for that finger to increase.[2] Similar effects occur when gaze is directed at particular perceptual features. This processing enhancement effect is referred to as *gain modulation*.

In addition to the acuity and gain modulation effects which gaze produces, there is another feature which ensures that gaze exerts a strong bias on attention. It turns out the there is a special nucleus complex in the thalamus which enhances the binding of features via resonant coupling with visual activity in the cortex. In mammals, this is the pulvinar nucleus complex, a region that is particularly well developed in primates.[3] The pulvinar is what researchers Murray Sherman and Ray Guillery refer to as a *higher-order* thalamic relay, because it is driven by preprocessed inputs, rather than by basic sensorimotor inputs as is typical of first-order relay nuclei.[4] In the case of the pulvinar, the driving inputs come from visual processing areas in the cortex, including gaze planning regions. In our steering committee metaphor, the pulvinar can be considered to be the chair of the committee. It brings the members of the committee together, summarizes the discussion, and guides the policies that influence attention.

To explain this role in more detail, we'll have to go beyond the committee metaphor and look at the brain systems that manage both gaze direction and feature saliency. Many of them interact directly with the pulvinar. The control structure for managing gaze direction is truly fascinating. Low-level control begins with the spatial orientation networks in the visual tectum, known in mammals as the superior colliculi. Because the retinal inputs to the tectum are from the magnocellular ganglion cells, which interface with rod receptors all across the retina, the tectum receives a wide-field view of the visual scene. As we have already suggested, this wide-field view is an ideal platform for selecting feature areas for more detailed processing. Because the tectum has direct connections with the motor nuclei controlling head and eye movements, it is the ideal low-level system for orienting head direction and targeting the visual saccades which redirect gaze. This is why the tectum must be considered a primary member of the attention steering committee.

Complementing the bottom-up direction of gaze by the tectum, there is a motor area in the frontal cortex of mammals dedicated to selecting gaze targets based on top-down information.[5] This area, which has come to be called the frontal eye field (FEF), has connections with the pulvinar and with visual-perceptual feature processing areas in the ventral stream.[6] These connections enable feature saliency to influence gaze planning. Not surprisingly, the frontal eye field is also closely interconnected with the circuitry in the tectum. In addition to this eye field, there is another gaze control region in the dorsomedial frontal lobe which is connected with the frontal eye field. This region is known as the supplementary eye field (SEF). The supplemental eye field lies adjacent to the supplementary motor area, a region that has been implicated in early stages of stimulus-guided motor planning. For this reason, it appears to be ideally situated to promote gaze shifts to update feature and location information as part of ongoing motor planning,[7] a motor planning process that I refer to as *look ahead.*

There is also an eye field in the parietal cortex, known in humans as the parietal eye field (PEF). This area is functionally homologous to the region in the macaque monkey which was labelled the lateral intraparietal sulcus (LIP) in the FARS model for reaching. In humans, the equivalent functional area lies slightly beyond the lateral intraparietal sulcus landmark, so the region has been given a different label although its function is homologous to that of LIP. The parietal eye field provides a short-term memory map for objects in visual space. This memory acts as a constant reminder of where to make visual contact with recently encountered objects and affordances. Research indicates that this memory is mapped relative to gaze direction such that the center of the map is always at the center of gaze. In fact, the locations of objects within the map are automatically updated whenever gaze is shifted.[8] This arrangement appears to ensure that visual coordinate transforms are always linked to current gaze direction.

In addition to the biases on attention that result from shifts in gaze, attention is also influenced by *gaze planning.* This aspect of attention is sometimes referred to as covert attention. As it turns out, gain modulation effects begin when a saccade is planned.[9] This happens because potential target areas in the cortical eye fields and in the tectum become active *prior to* gaze shifts. This rise in activity results in spreading gain modulating effects even before gaze is shifted. With respect to look-ahead motor planning, this implies that merely considering an action can influence the flow of attention, even if the action never occurs, and gaze is never overtly shifted. However, simulated actions often do result in real gaze shifts, because gaze shifts are a natural component of action planning. This is part of the reason that the eyes often shift as someone thinks. Eye movements are signs of shifting attention, and many people develop habits of eye-shifting that provide hints as to how their attention is shifting.

Other aspects of the visual scene also affect gaze planning. For example, visual scanning has been shown to depend on regions of high-color intensity or on busy areas with high contrasts and sharp edges.[10] Reaching and pointing are also known to influence gaze, apparently because the orientation networks in the tectum which direct arm movements can also be coupled with gaze. As a result, pointing can bias gaze and gaze can bias pointing.[11] Semantic memory is known to influence attention, such that subjects tend to scan regions where informative features were previously found.[12] Gaze is also partly controlled by learned strategies. For example, periodically fixating on the rear-view mirror while driving or checking the location of teammates on a basketball court during play are learned activities that essentially become habits of attention with experience.[13] In short, the systems managing gaze are broadly interconnected and highly flexible, and whatever influences gaze direction and gaze planning influences the flow of attention.

A GAZE-CENTERED COMMITTEE

In addition to the eye fields, there are several other cortical regions which influence visual attention, and there are several models about how these various systems work together. On reviewing these, Stewart Shipp proposed a composite model in which the pulvinar complex is the central player.[14] The basic idea is that the pulvinar serves as a higher-order thalamic relay module, which biases attention by summarizing the saliency of inputs from other modules. Inputs into the pulvinar come from a variety of sources and are linked together based on their location. Recall how the registration of spatial maps in the tectum enables signals in one modality to supplement those in another? An impressive part of Shipp's description is how this registration is extended in the pulvinar. The fact that the maps are in registration means that activity in any one region can enhance processing activity in other aligned regions. In fact, in the quote below Shipp essentially equates the gain modulation effects resulting from this aligned activity with the searchlight analogy of attention proposed by Francis Crick.[15]

> The key point is that the connection zones made by the areas along the ventral visual pathway all have registered visual topographies, which overlap and fuse to form the global primary and secondary visual maps in the pulvinar. Frontoparietal signals for covert attention to a particular location (a corollary of saccade-planning signals) can then be relayed into this map via the tectum. This is figuratively a 'beam of attention', extending along the line corresponding to a specific visual locus within the ventral pulvinar. Activity in the beam then sets off conjugate cortical activity in the corresponding loci of all the visual maps along the ventral visual pathway – like a set of subsidiary spotlights emanating from the master one within the thalamus.[16]

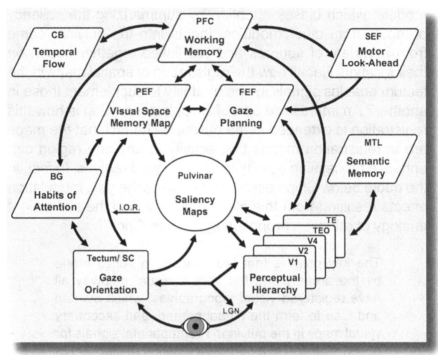

Figure 12-1: An overview of the attention steering committee.

Figure 12-1 provides a functional overview of this model. The details of the circuitry have not all be been worked out, but I believe we know enough about the basic organization to understand its essential functions. Let's begin with a brief walkthrough. There are two paths for inputs from the eye (shown as left and right paths at the bottom). Wide-field, low-resolution visual inputs from across the retina project to the Tectum/SC. These provide bottom-up influences on feature saliency for guiding gaze direction. High-resolution focal inputs project via the lateral geniculate nucleus (LGN) in the thalamus to the perceptual analysis regions in the visual cortex (labelled as V1, V2, V4, TEO, and TE). These perceptual regions also have visually mapped connections with the pulvinar and with the frontal, supplementary, and parietal eye fields. The frontal eye field (FEF) is proposed to bias attention based largely on perceptual saliency. The supplementary eye field (SEF) is proposed to bias attention based on motor planning. The parietal eye field (PEF) is proposed to bias attention based on active spatial locations. The tectum is strongly connected with both the frontal and parietal eye fields.

Extensions to the model include working memory from the prefrontal cortex (PFC), which provides task-related sources of bias. Temporal flow learning in the cerebellum (CB) provides sequential biases that guide shifts in working memory and influence the sequential flow of attention more generally. Feature saliency biases in perceptual regions of the visual cortex can promote semantic associations in the medial temporal lobe (MTL), which further refine feature saliency. Habits of attention, learned by the networks in the basal ganglia (BG), are proposed to influence both top-down processing as guided by the prefrontal cortex and to influence bottom-up processing guided by the tectum. An inhibition of return function (IOR) is proposed to ensure that gaze routinely shifts to new sites. In humans, this function appears to depend on connections between the parietal and frontal eye fields and the tectum.

Shipp's composite model is fairly complex, but the insights it provides are critical for understanding how a versatile system of self-guided attention emerges. The saliency weights in the pulvinar complex are not tied to any single source. Rather they summarize functional biases from a large variety of mechanisms, including learning and memory systems. These summary measures are anchored to spatial maps in the pulvinar, which are maintained in registration with maps in the tectum and in the cortical eye fields. Because activity in the pulvinar can bias activity in other thalamic relays via matrix cross-connections, any region that influences the pulvinar can bias attention. Thus, all brain regions which interact directly with the pulvinar must be considered core members of the attention steering committee.

This is truly an impressive system for coordinating attentional biases all across the brain. Each subsystem has its own particular emphasis, but no single system is in control. It's the collective influence of the committee, as summarized in the pulvinar, which determines what features are most likely to be bound together in assemblies in the thalamus and win the competition for attention. Working memory adds yet another level of control, which helps keep the process focused on task-related features. Not shown, but equally important, contributors to saliency are the interoceptive connections, which bind thalamic features with activity in the interoceptive lattice. These connections ensure that saliency is also influenced by interoceptive associations.

LATTICING VISUAL ATTENTION

Having described how a committee of brain modules can bias attention, it's time to consider how this model might account for certain experiential effects of attention. As noted above, a key feature of the pulvinar is its tendency to map saliency in alignment with the maps in the tectum and with maps in the cortical eye fields. It appears that the spatial-memory pointers in the parietal eye field play a critical role in this process. In fact, activity in the parietal eye field appears to be essential for holding separate perceptual features together, for tracking bound objects as they move, and for coordinating activity in the perceptual stream with activity in the planning stream. For example, patients with Balint's syndrome, a condition resulting from bilateral damage to the spatial map areas in the posterior parietal cortex, have difficulty representing the location of objects with respect to themselves. They report seeing only a small part of their visual field, the focal view of the fovea, at any one time, and when they shift gaze the features outside the current focused view fade from awareness, apparently because they are not linked to broad-field spatial memory locations which keep them active when they are out of central focus. Quite literally, out of sight means out of mind for these patients.

Research suggests that spatial representations in the parietal eye field serve as reference points for anchoring perceptual features to locations in wide-field visual space. For example, when a Balint's syndrome patient known as RM was shown letters of different colors or sizes, he could focus on them and readily name each letter and say what colors were present.[17] However, he had great difficulty deciding which letter was a particular color, even after studying the letters for ten seconds or more. Similar problems occurred when RM was asked which letter was larger. He could recognize and report whether the letter sizes were the same or different, but he couldn't connect the different sizes with specific letters. Other studies have found that RM also has difficulty attending to one object while trying to ignore another.[18] Without the support of the spatial representations in the posterior parietal cortex, he is simply unable to anchor features together or to hold his attention on one group of features at a time. In short, spatial locations in the parietal eye field and their links with the pulvinar appear to be essential for binding feature elements into more complex perceptual units.

To get a more personal appreciation for the importance of spatial location in anchoring features together, consider the objects shown in Figure 12-2. Without counting, estimate how many squares there are on the right side of this figure. Then, estimate the number of squares on the left side. If you respond like most people, you will find that the number of squares on the right, which can be individuated based on differences in location, can be easily assessed visually, a process known as subitizing. However, because the squares on the left cannot be as easily individuated by location, they don't form separate representations, and their number cannot be readily assessed by subitization. Instead, they must be counted. Counting, in this case, requires focusing on a non-central feature, like a corner, and shifting attention from the corner of one square to that of another as you count.

Figure 12-2: How spatial separation contributes to object individuation.[19]

Visual researcher Zenon Pylyshyn initially called the mechanism for tracking spatially separated objects "Fingers of Instantiation" because they worked much like pointing fingers. However, he later adopted a less metaphorical term, *visual indexes*.[20] Alan Leslie and colleagues proposed the term *object index* for this property, noting that the indexes were "sticky" – that is, capable of binding features together.[21] These index models are closely related to Anne Treisman's *object file* concept.[22] The object file metaphor suggests that attending to a visual object is something like opening a file in a computer program. The idea is that the file name serves as an index, or handle, for locating the file information. However, once the file is located and opened, its contents – that is, information about associated object features – also becomes accessible.

Indeed, studies suggest that visual perception follows an incremental process in which units of possible interest (Pylyshyn terms them proto-objects) must be detected *before* they can be identified. That is, the early stages of visual processing must first detect and then track units of perception in order to bind them together with features. It is this process which ultimately enables then to recognize objects. Pylyshyn notes that we even refer to objects in this incremental manner when we talk about them, often beginning with vague demonstratives such as "this" and "that," which are devoid of conceptual information. For example, we may initially pose the question, "What is that?" and direct attention to an unknown proto-object before we later recognize it. This is reasonable once we recognize that memory processes are slow and that we need to be able to track an object, sample it features, and maintain processing for a half second or more before it can be linked with semantic memory associations.

If we think about this indexing process in the context of the attention steering committee, it becomes clear that there are several stages of attention, depending on which members of the steering committee are dominating processing decisions at any time. The spatial orientation system in the tectum/SC serves as the low-level targeting system for gaze direction.

Targeting contributes to the early stages of visual processing by identifying interesting features for more detailed analysis. Pre-attentive targeting requires a spatial mapping sufficient to direct gaze, but the mapping at this level only supports head and eye orientation. This leaves the focus of processing vague. A more complex mapping effect, a sticky index, develops when proto-object locations detected by the tectum are re-mapped in the pulvinar , in registration with locations in the parietal eye field. The sticky index adds short-term memory tracking and feature binding properties to the proto-object. This is the beginning of top-down attention. It enables the object to be tracked and described on a feature level. Object recognition can then be determined by retrieving feature associations from semantic memory. Each of these latter steps contributes new properties to what gains attention as the scope of the processes bound with activity in the pulvinar expands (see Table 12-1).

The hierarchy of attention described here also provides an instructive framework for thinking about some of the differences between attention and consciousness. Attention is an incremental processing effect which results from orienting, filtering, binding, tracking, saliency biases, and executive control biases on perception. Consciousness, in contrast, is an incremental feeling-centered awareness that emerges as interoceptive associations, with different levels of focus and gain attention. Under some conditions, conscious feelings can be largely dissociated from the top-down aspects of perceptual attention.[23] In fact, given that consciousness and attention both emerge in incremental stages, it is often possible to find aspects of awareness that do not correspond to higher-order states of attention. However, interoceptive bindings are always part of the weightings that guide the hierarchy of attention. As such, it seems important to recognize that the acuity of conscious feelings can differ from the acuity of perceptual attention without forcing us to disassociate the two.

Brain Regions & *Functions*	Processing Characteristics	Recognition Level
Tectum/SC Orientation	Low-level features detected. Gaze scanning is activated.	Vague awareness of something.
PEF, Pulvinar Saliency Mapping	Proto-object is tracked. Sticky index is assigned.	What's that moving over there?
Ventral Stream Feature Analysis	Features bound to index. Object can be described.	It's gray and white, and looks furry.
MTL Semantic Memory	Features linked to memory. Object is recognized.	Oh! It's Tom on the hunt again.

Table 12-1: A hierarchical model of visual attention.

Perceptual attention is most intense when saliency biases are strong, feature processing is focused, and both task and semantic networks are engaged in coordinated activity. Conscious feelings, in contrast, are most intense when strong feelings gain attention. As a result, conscious feelings are not dependent on the saliency of top-down perceptual biases, or how well coordinated perceptual and semantic associations have become. For example, if you are isolated in a dimly lit and unfamiliar location, and you are subsequently startled by a piercing scream, the perceptual features at the center of your attention may be only vaguely focused. However, the bottom-up processes that activate your feelings in the moment are likely to be intense. The point to note is that even in the absence of well-focused percepts, feelings may dominate attention and conscious experience.

OTHER BIASING MECHANISMS

Research suggests that the higher-order thalamic drivers, such as the pulvinar, evolved from bottom-up relays and occasionally still work that way. Interestingly, the higher-order drivers often have reciprocal connections with the zona incerta, a reticular-like region in the subthalamic area. The exact function of this region remains uncertain, but it is known to have subsections for somatosensory, visual, auditory, motor, and interoceptive reactivity, and it appears to play a role in coordinating motor activities, such as orientation and locomotion, with autonomic reactivity and sensory processing.[24] There is evidence that the zona incerta is involved in switching higher-order thalamic drivers between bottom-up and top-down modes.[25] When the pulvinar is operating as a bottom-up relay, it relays information from the tectum. In this mode, orienting inputs from the tectum provide directional alerting cues which promote visual scanning and early stages of object tracking. However, as soon as object tracking begins, it seems that pulvinar switches to a top-down control mode. In the top-down mode, cortical inputs have the dominant influence on attention.

In addition to the pulvinar, there are several other thalamic nuclei which function as higher-order relays. For example, in mammals with whiskers, such as our feline friend Tom, the posterior medial thalamic nucleus appears to adjust the saliency of whisker inputs based on top-down cortical activity. Again, switching between bottom-up and top-down control is apparently managed by the zona incerta.[26] The zona incerta receives a duplicate set of incoming whisker signals and passes them on to the thalamic whisker relay. These inputs to the thalamic relay are inhibitory and function something like a noise cancellation circuit to reduce the gain for transient whisker inputs to the relay. However, when an animal actively whisks, top-down motor signals to the zona incerta inhibit the noise cancellation effect. As a result, whenever an animal is not whisking, the gain of the whisker inputs through the thalamic relay is held low. However, as soon as active whisking begins, the gain goes high and whisker inputs become more salient.

The strategy of gating sensory inputs based on sensing-related motor activity is yet another method for adjusting the saliency of particular features to bias attention. Another set of higher-order thalamic relays which appear to serve a gating and biasing function is a diffuse complex in the thalamus known as the midline and intralaminar nuclei. For some time, these nuclei were thought have diffuse attention-gating functions. However, they appear to be more specific.[27] The driving inputs for this complex come largely from the pallidum, the output region of the planning stream in the basal ganglia. This suggests that they play a role in coordinating attention processes in the thalamus with planning activity in the basal ganglia. Van der Werf and colleagues divide the midline and intralaminar complex into four major functional groups.[28] The posterior group appears to be poised to bias attention to cues involved in managing motor activity. The dorsal group appears to be involved in biasing attention based on reactivity states. The ventral group appears to be less involved with motor activities and more involved in biasing attention for sensory and memory processing. The lateral group appears to help bias attention in coordination with executive attention processes.

Kimura and colleagues suggest that the posterior group shifts between top-down attention biasing and bottom-up guidance.[29] When this group functions as a bottom-up relay, it appears to be driven by midbrain motor modules. These include the orienting outputs from the tectum, locomotor biases driven by brainstem motor modules, and the emotional behavior patterns managed by the periaqueductal gray. This pattern of input suggests that the motor dispositions of the proto-self are also early drivers of attention in support of more refined motor decisions. This is exactly how a hierarchal attention process would be expected to operate. In fact, this organization provides some insight into how the top-down control of attention must have evolved. In animals with more limited cortical control, the bottom-up inputs are still largely the primary drivers of attention. Apparently, the top-down nuclei were gradually taken over by high-order cortical processes as they became more important in decision making.

Our model of the attention steering committee has been dominated by visual processing networks. However, we have also noted that gaze-centered coordinate-transform networks are interconnected with many sensory and motor-control areas. The interaction of the pulvinar and the gaze control circuitry has wide-ranging effects on attention, even beyond visual processing. To put this in a broader perspective, it should be noted that the tectum is not simply a visual center; it coordinates multiple sensory inputs for orientation decisions. In effect, other modalities can influence gaze direction. This also appears to be true for the pulvinar. For example, though the pulvinar complex is largely visual, the medial pulvinar nucleus receives auditory information by way of the medial geniculate nucleus in the thalamus. Similarly, somatosensory inputs reach the anterior pulvinar nucleus. As a result, the pulvinar complex appears well situated to bias attention beyond visual channels.

If this model is roughly correct, it implies that there are probably similar saliency mappings for auditory and somatosensory features which can also be bound to spatial locations. Consistent with this hypothesis, research suggests that separate sound features are only recognized as a single combined sound if they can be grouped together by location.[30] Non-visual features are also processed in a hierarchical manner beginning with a vague pointer, "Did you hear that?" and moving on to more localized pointers. In effect, it seems logical to assume that like the tectum, the pulvinar complex is a multisensory processing center. In fact, many of the components of the attention steering committee have multisensory inputs. Although the cortical eye fields are not explicitly multisensory, the very fact that there are coordinate-transform networks, which link other sensorimotor maps to eye coordinates, means that other sensory modalities can also influence decisions in the eye fields, and vice versa. It's hard to imagine a steering committee that could be more sensitive to multiple sources of saliency. The nexus of attention is managed by a committee that is dominated by visually-guided gain modulation processes, but visual coordinate-transform networks and motor-centered gating mechanisms link the visually centered focus of attention with other domains of sensorimotor activity.

A SELF-STEERING COGLEY

There are still aspects of the attention steering process which we don't understand well. However, our goal is not to build an exact replica of the human brain, but rather to use what we know about its architecture to provide Cogley with a bio-inspired model of its essential functions. In thinking about how this organization might be implemented, several points come to mind. The first and most obvious is that attention is dominated by visual processes, gaze direction, visual gain modulation, and visually dominated top-down biasing mechanisms. A second point is that many of the systems which have the potential to influence the flow of attention are part of the planning stream. These include the frontal, supplementary, and parietal eye fields, which plan gaze direction, and the motor decision processes in the basal ganglia, which guide habits of attention. A third point to note is that several of the systems which manage attention are subject to modification by interoceptive feedback. These include working memory in the prefrontal cortex, habit learning in the basal ganglia, temporal flow learning in the cerebellum, and semantic memory associations in the medial temporal lobe.

Fortunately, many of these have already been described. The new additions are largely those needed for gaze control and saliency mapping. To implement similar functions in Cogley, we'll need a saliency mapping module in the central attention channel and three spatially mapped eye-control modules to manage gaze direction. The master eye-control module should be sensitive to perceptual inputs and should provide top-down biases to the saliency module, as well as to the orientation module. A supplementary eye-control module will be required to encourage the master module to implement look-ahead gaze checks as part of action planning. The third eye-control module is the visual-space memory eye-control module. This module has already been described. However, we have now introduced a few new requirements. It should be capable of sustaining activity for the location of at least six separate objects at the same time, and a background algorithm should ensure that the map is updated after each shift in gaze so that it is always configured in eye-centered coordinates. In fact, it seems appropriate to keep the maps for all of Cogley's eye fields synchronized around the center of gaze.

In order to provide Cogley with the ability to subitize, we should add a dedicated subitizing network, with a counter function, to his visual-memory space map. This counter should be able to differentiate from zero to up to six spatially separate objects, with seven and above coded as many. The counting function should be able to use a set of feature cues as a guide and be able to identify the number of feature matching sets in the field of view. For example, Alex, the African Grey parrot trained by Irene Pepperberg, was able to selectively subitize following queries such as, "How many blue keys?" and report the correct number, even when there were distracting green keys and blue rocks on the same tray.[31] This implies that the subitizing network can use a distinct search image. In addition to the numbers one through six, Alex also recognized when none was a correct answer. I don't believe he was ever taught a word for "many," however it seems likely he could have learned this concept. We'll want Cogley to have similar small-number perceptual skills.

Cogley's perceptual analysis networks should also have eye-centered mappings so that high-activity features can be relayed into the saliency module map. These should also be mapped into Cogley's master eye-control module so that it can provide feature-sensitive top-down guidance for selecting gaze direction. The saliency module should also be bound to the sensory and motor relay modules in Cogley's central attention channel by matrix cross-connection links. These bindings should have the effect of extending visual gain modulation effects in the saliency module broadly across the central attention channel. Although the saliency mapping complex should be dominated by visual inputs, it should also include auditory and tactile saliency maps, connected in registration with the visual map, so that attention can be managed in other domains.

Given that the saliency mapping module is to be located within the central attention channel, there is one more potential complication that must be resolved. Thus far we have proposed that the filter in the central attention channel should suppress activity in competing assemblies, so that only one assembly can gain attention at a time. However, we have also noted that it seems useful to provide workspace for up to six separate candidate assemblies at a time. The idea is to enable new assemblies to form in the background and subsequently compete for attention. This should enable attention to be shifted quickly between established assemblies. Obviously, this will not work if the inhibitory filters in the central channel suppress the indexing and maintenance of all alternative assemblies. Exactly how this is managed in the vertebrate brain is unclear. However, given evidence that the pulvinar is capable of maintaining more than one map, I propose that Cogley should use a two-level saliency mapping strategy. One map, the pre-attention map, should be used to map saliency for up to six separate assemblies, each with their own attention indexes. These indexed assemblies should be maintained with low-level resonant connections. The assemblies in this index map should then compete for attention in a winner-take-all strategy, and the winning assembly should then project to a second map, the active attention map, in which stronger resonant connections allow the assembly to pass the filter in force and dominate attention.

We should extend this design by adding the requirement that links for the new winning assembly should always be transferred into the active attention map *before* the pointers to the old assembly are disabled. This will guarantee that the two maps will temporarily overlap and influence each other. By allowing reciprocal connections between the active map and the pre-attentive map during this transition, overlapping associations may also influence the composition of the pre-attentive assemblies. We have already noted that switching attention between two items results in transient periods when the activations of the two assemblies overlap.[32] Initially, we suggested that this could support the blending of memories into new insights. However, it actually introduces a general-purpose way of blending any ideas that reach attention. Allowing features to overlap as attention is switched should make it easier for Cogley to blend ideas together in creative ways.

The use of coordinate-transform maps to generate gain modulation effects appears to have become a general strategy in the vertebrate brain. Obviously Cogley will need a number of coordinate-transform networks to coordinate activity between his different sensory input maps. In particular, auditory and tactile inputs should be transformed into visual coordinates so that action decisions can be managed by multimodal inputs. These networks should be arranged so that activity in any one location of the sensory map will result in enhanced activity in the corresponding location of the visual space map, and vice versa. This will enable other senses to influence visual processing, and visual processing to influence them. Let's assume that the coordinate-transform networks in Cogley's low-level orientation module have wide-field, gain-modulation effects, whereas those connected with Cogley's visual-space eye-control module have more focal effects. This should enable Cogley to gain modulate his attention on both global and focal levels across sensory modalities.

Using whisking as a model, it seems appropriate to provide Cogley with an action-gated saliency module, a homologue for some of the gating effects we attributed earlier to the zona incerta. By default, this module would hold sensory gains at low levels but should ramp up the gain when particular sensing activities, such as active touching, are engaged. Following this action-gated saliency strategy, it also seems reasonable to gate the inputs from various motor maps. This could be accomplished by ramping them up when actions involving those maps are being considered, and ramping them down when other actions are being considered. In effect, whenever Cogley is planning a reach, inputs from his reach-space map should be enhanced. However, when he's planning behaviors unrelated to reach space, then the saliency of inputs from that map should be reduced. Connecting action programs to particular spatial maps would be an automatic way of raising the saliency for action-related cues and thereby biasing attention to appropriate locations and programs for specific actions.

Given this action-gating strategy, it also seems practical to provide Cogley with a general-purpose mechanism for coordinating his attention with ongoing decisions in the planning stream. To do this, Cogley will need a general-purpose, cross-stream binding complex in the central attention channel. Given that there are several channels through the basal ganglia, let's assume that this is a multi-module complex with at least one module supporting each channel.[33] The driving signals for this complex should be output channels from Cogley's mid-level action networks, his version of the pallidum. Matrix connections within the central attention channel should enable output decisions to bias attention to task-related cues, an effect that should promote task-focused attention. A second goal of this complex should be to enable Cogley to develop more efficient habits of attention. The idea is that as actions become automated as habits, they come to bias attention to relevant cues through these gating modules. In many cases, these habits of attention may gradually take over the role of the higher-level working memory networks in biasing tasked-related attention, freeing up the higher-level networks for tasks that have not yet been automated.

The proposed architecture is still rather sketchy, but it will provide Cogley with a flexible system that is sensitive to many aspects of perceptual and motor experience and one which can prioritize its processing based on multiple sources of saliency. Gaze directing and planning circuitry will provide this architecture with general-purpose tools for biasing the flow of attention and will enable the overall system to acquire a degree of self control. Higher-order gating relays will provide added mechanisms for selectively biasing attention to specific activities. To make this control both self-centered and adaptively meaningful, these tools must be guided by learning and memory networks, which are sensitive to interoceptive needs. In this way, the feelings of the resonant self will not only follow attention, but they will feed back to influence the skills and interests of many steering committee members and thereby change the way Cogley attends to events as he gains experience. Following this strategy, Cogley will gradually learn to steer his attention more efficiently.

CONCLUSION

According to the biased competition model, attention is managed by selectively enhancing processing for features that favor some assemblies over others in their competition for attention. In addition to working memories, a primary top-down process for biasing attention in the modern vertebrate brain is gaze planning. The frontal eye field appears to bias attention predominately for perceptual feature saliency. The supplementary eye field appears to bias attention in support of motor planning. The parietal eye field biases attention for recently active locations. All three eye fields are involved in managing gaze decisions. Gaze direction is effective at biasing attention in large part due to the broad influence of gaze-centered, coordinate-transform networks. These transform networks are found in the tectal and parietal motor areas, where they convert between other sensory modalities and visual coordinates. A mechanism that appears to leverage these gain modulation effects is the visually dominated saliency mapping complex in the thalamus, known in mammals as the pulvinar. A broad collective of other networks can influence the saliency evaluations in the pulvinar. Prominent among these are the working memory complex in the prefrontal cortex, the decision systems in the basal ganglia, the spatial orientation system in the tectum, the temporal coordination networks in the cerebellum, and the semantic memory processing regions in the temporal lobe. Collectively, these modules function as what

I have metaphorically termed the attention steering committee, and they guide the flow of attention.

Having outlined the organization of this steering committee, it seems worthwhile to take a step back and consider the emergent properties that this committee makes possible. We have already noted that a central attention process which binds perceptual and motor activity with interoceptive feelings results in a mind that experiences a feeling of what happens as it perceives and acts. We labelled this experiential state the resonant self. The architecture described in this chapter provides the resonant self with a committee of processes which can bias its ongoing activity. As this dynamic self-steering system gains experience, its ability to manage the flow of its own experience grows because the feelings which accompany attention feed back to influence what the members of the committee learn and remember. This is the kind of maturing, self-guided attention process that we expect from conscious agents. Once we provide Cogley with a steering committee which is guided, in part, by past learning and memories, then his past experiences will increasingly come to influence what he notices, learns, and remembers. This means that, like you and me and Tom, Cogley's history of attention will become an integral part of his personal identity. It will influence who he is and what he does.

CHAPTER 13
THE UPWARD SPIRAL

Attention is not an all or none process. It emerges in stages by building on the results of earlier processes. Pre-attentive targeting processes are always operating in the background, evaluating sensory inputs for potential value and promoting the investigation of some features over others. Without this preliminary screening, later systems would have no basis for deciding where to look or what potential objects to track. Adding to these initial biasing strategies are saliency mapping circuits from yet other sources, and gating processes that bind feature processing with action planning. It is this layered strategy of selecting, refining, and re-representing which enables pre-attentive features to be represented as proto-objects, proto-objects to be re-represented as feature-bound objects, feature-bound objects to be linked with contexts, and contextual associations to prime recognition and situational memories.

Given its close relationship with attention, it should not be surprising to find that consciousness is also not a single-stage phenomenon. As we note in this chapter, consciousness emerges in layers of awareness. Once the feeling of what happens emerges, other processes supplement and refine those feelings. When orientation to sensory cues is re-represented in the context of spatial maps, a *feeling of where I am* emerges. As Pavlovian learning mechanisms begin to anticipate sequential dependencies, a *feeling of what might happen next* takes form. When action planning comes to predict changes, a *feeling of being able to change what happens* takes form. Semantic memory adds feelings of familiarity, continuity, and recognition to ongoing perception. Episodic memory adds the ability to re-experience selected episodes of past experience. In effect, the feeling of what happens is expanded in an upward spiral of representations that add new dimensions to what can gain awareness.

THE UPWARD SPIRAL

My view is that both consciousness and attention occur in levels and grades, they are not monoliths, and they influence each other in a sort of upward spiral. Low-level attention precedes core consciousness; it is needed to generate the processes that generate core consciousness. But the process of core consciousness results in driving higher-level attention toward a focus. – Antonio Damasio, *The Feeling of What Happens*, 1999

Consciousness seems less mysterious when we can identify the organizations that make different aspects of experience possible. However, consciousness experience takes many forms. Although the architecture of the thalamus enables the phenomenal states of conscious experience to emerge, the *feeling of what happens* is merely the first stage of self-awareness. As our discussion of the attention steering committee has shown, there are a number of other neural modules which contribute to the subsequent organization and flow of neural assemblies as they gain attention. Sensory orientation, cognitive flow, action planning, sequence learning, semantic memory, and working memory all add to the qualities of experience which gain consciousness. Many of these properties have already been described. In that sense, this chapter can be considered something of a summary. However, our goal here is to place the varied properties of consciousness in a broader framework, first to gain insight into how more complex qualities of consciousness emerge, and second to provide a more systematic set of categories and labels for talking about the various stages of conscious experience.

Attention to the feelings that accompany perceptions and actions gives rise to an initial sense of self, a state of connected awareness, which we have termed the resonant self. The organization which makes the representation of self possible begins with the architecture of the thalamus, although a fuller sense of self involves many other regions of the brain. Damasio suggests that the subcortical components of the interoceptive stream – in particular, the PBN, the PAG, the hypothalamus, and the amygdala – provide pre-conscious inputs regarding an organism's status and reactivity states. He terms this level of representation the proto-self. He argues that it is the re-representation of changes in the proto-self, in the context of sensorimotor activity, which leads to an emergent sense of self. Further, he suggests that this re-representation minimally requires connectivity through the thalamus and probably extensions into the interoceptive cortices.[1]

The strategy of providing low-level sensors and channeling their input through a competitive central attention conduit is what we'll need to arrange for every internal state to which we want Cogley to be capable of consciously reacting. This strategy also provides us with a model for thinking about how consciousness evolved in vertebrate brains. Many simple action patterns (e.g., tracking and feeding in primitive vertebrates) are largely released through feature detectors in the tectum and/ or pretectum and are guided by orientation responses directed by the tectum. However, some decisions are supplemented by channeling information through the thalamus, where it can be combined in larger assemblies, via attention processes, and routed through decision algorithms in the basal ganglia. This strategy enables more complex feature sets to serve as sign stimuli. It also provides a means for more complex interoceptive priorities to influence action decisions while the networks in the basal ganglia assemble more complex action patterns.

Over evolutionary time, more information has come to be chaneled through the thalamus, and more decision making through the basal ganglia, whereas the tectum has come to serve as a first-level orientation system for managing gaze direction and guiding attention. The thalamus also has become the standard path for managing the flow of information to the cortex. Cortical processing appears to have begun as a means of influencing action decisions with olfactory information, because the early cortical networks were dominated by olfactory inputs. Olfactory information is not very directional and does not project to the orientation networks in the tectum. However, olfaction provides a very important sense for identifying items. In the course of evolution, it appears that information from the other senses was gradually relayed to the cortex to add to the olfactory information, and to further refine item identification. As this happened, the thalamus began to serve as a central relay channel for passing information to the cortex, and for binding it with other sources of information in states of attention. The extent of the thalamocortical control no doubt evolved gradually, and there is much that we have yet to learn about its evolution.[2] However, given that the basic vertebrate brain architecture has been in place since jawed fish evolved, some 440 million years ago, it seems likely that most vertebrates, even primitive cartilaginous fishes, share aspects of phenomenal consciousness.[3]

And what kind of events would evolution have selected for enhancement by conscious attention? It seems inevitable that it would have begun by enhancing interoceptively motivated adaptations which could benefit from guidance by more complex sign stimuli and more varied action patterns. As William McDougall noted long ago, instinctive reactions work best when they gain attention and keep an agent focused on relevant cues and motives.

> Absorption of the organism in any particular task or mode of activity is what we call in ourselves "attention;" and that general excitement whose indications we observe in the animals, when their instincts are strongly excited, is what we call in ourselves "emotion." We may therefore define "an instinct" as any innate disposition which determines the organism to perceive (to pay attention to) any

object of a certain class, and to experience in its presence a certain emotional excitement and an impulse to action which finds expression in a specific mode of behavior in relation to that object.[4]

McDougall's characterization of instincts as interoceptively guided dispositions of attention and action places the emergence of consciousness in a much broader perspective. If, as we have suggested, consciousness emerges from interoceptively guided states of attention, and if instinctive action patterns are enhanced by the same process, then it seems likely that all animals whose instincts are guided by interoceptively weighted attention experience some aspects of a feeling of what happens as they perceive and act. Those who wish to limit conscious experience to humans or to human-like animals may find this conclusion troubling. However, I'm simply following the logic of the organization effect. The borderline between unconscious reactions and reactions enhanced by phenomenal consciousness appears to be phylogenetically much older than generally thought, because the neural structures that bring such properties into existence are themselves quite old. It is difficult to characterize these early stages of conscious awareness in terms of human experience. However, it seems likely that they roughly parallel the early stages of proto-object attention, vague and only partially connected feelings, but with the potential to bootstrap more focused feelings. As Antonio Damasio notes:

When core consciousness began, millions of years and many species ago, we [our vertebrate ancestors] were very far from the current sophistication of modern consciousness, very far from the ease with which we can describe, using language, the reasons behind our actions, past or intended. However, when core consciousness began, we were on the right track and we transcended the critical threshold. We were telling ourselves, without using any words, the answer to the question we never asked, that yes, there was an individual perspective to our percepts, and yes, there was an individual ownership of images, and yes, it was all tied to life.[5]

Graded stages of self-awareness seem inevitable given the layered nature of cognitive representations and the gradual evolution of the thalamocortical architecture. Indeed, the gradual assembly of layered organizations is apparently recapitulated in the stages of our own gradual neurocognitive development. The human foetus passes through stages of physical development, including one in which it has gill-like structures, and it seems quite likely that it passes through graded stages of awareness as the networks supporting consciousness take form. The human brain is still largely immature at birth, and the conscious experience of newborn infants is quite primitive. In fact, the conscious experience of the adult shark must necessarily exceed that of the human infant, at least for the first few months of a human infant's life. An adult shark is capable of much more sophisticated perceptions and decisions than a newborn human. However, although the consciousness of infants may be primitive at its start, it is capable of being extended across years of development. Sharks do not share in very much of this incremental development.

Although phenomenal consciousness may have begun as a means of extending the complexity of the cue combinations which guide instinctive actions patterns, once interoceptively guided attention was in place, other processes began to rely on it. We have already noted that features that gain attention show increased processing activity. Even simple kinds of learning mechanisms tend to work better when cues and their consequences are enhanced by attention. Further, evidence suggests that learning about more complex configurations and contingencies, even those seemingly as simple as trace eye-blink conditioning, may not be possible without attention and memory processes to bind features together and bridge gaps in time.[6] Similarly, it seems clear that many categories of perceptual recognition, and the concepts they support, could not be assembled without selective attention mechanisms to guide their organization and bind them with verbal labels.

Gary A. Lucas

The feeling of phenomenal consciousness is simply a starting point in this process. It provides a framework in which other aspects of awareness can add to the qualities of conscious experience.

FEELINGS OF LOCATION, ANTICIPATION, AND AGENCY

Location

In addition to the feeling of what happens, which emerges with the resonant self, there are several others kinds of feelings that are generally considered core elements of self awareness. The attention steering committee has several visually guided committee members. These members are capable of biasing attention via gaze direction, and they map gain-modulation effects in relation to visual fields. This visual bias begins in the tectum; is remapped and expanded by the frontal, supplementary, and parietal eye fields; is re-mapped again in the saliency networks of the pulvinar complex; and is enhanced by broadly distributed gazed-centered visual coordinate-transform networks. As a result, what is promoted to attention by the steering committee is routinely experienced in the context of where gaze is mapped in relation to the world. This is to say, we often experience a feeling of where we are, a visually dominated locus of observation, as we experience the feeling of what happens. I call this spatially located sense of self the *observer self.*

The observer self probably accounts for the popular idea that there is an internal agent, a homunculus, watching and managing our experience. The classic argument against an internal homunculus is that such an arrangement would seem to require an infinite regress of homunculi – a pre-homunculus to manage the homunculus, a pre-pre-homunculus to manage the pre-homunculus, and so on. However, thinking in terms of an incremental spiral of processes neatly sidesteps this problem. The pre-processing networks need not have complete control over the next processing level. They simply need to provide processes which contribute to the later organization. The early

250

proto-homunculus effects, such as the functions of the tectum, provide little more than reflexive biases in orientation. However, because the modules at each higher level re-represent the information from previous levels and then add new features to it, the knowers at each higher level become increasingly cognizant. This process recurs to the point that some systems represent various subsets of self-awareness. The central nexus of attention, the level of organization on which you and I begin to emerge, binds these subsets together and may sometimes even focus on their individual influences. In fact, the observer self is one of the most persistent and recognized subsets of self-awareness. When other aspects of self lose alignment with the observer self, an agent typically feels disoriented until he or she regains their bearings. That is, until the other aspects of self can be realigned with the observer self.

The good news is that it doesn't appear that we will need to make any major changes to Cogley's design to enable him to become aware of his observer self. We have already provided him with a visual-space memory module, his parallel to the parietal eye field, and we have linked it with his central channel saliency maps, his parallel to the pulvinar complex. These mechanisms all involve visually centered spatial maps. We have also proposed providing Cogley with visual coordinate-transform networks so that information in his other senses can be linked to, and biased by, visual orientation. This should provide him with a coordinated awareness of himself as an observer. Given that his visual space maps will be directly involved in managing his attention, this means that the feelings of the resonant self will typically be linked with inputs mapped in relation to the observer self. In effect, these two primary aspects of self will typically be aligned so that Cogley will normally experience himself as a feeling agent with a distinct perceptual locus.

Anticipation

We have previously described two Pavlovian learning systems in the brain which specialize in anticipatory learning: a valuation learning process managed largely by connections starting in the amygdala, and a temporal coordination process managed by the cerebellum. Although these systems have somewhat different learning rules, the learning processes in both these systems are considered types of Pavlovian conditioning because both mechanisms react to the sequential pairing of events, and because they often operate in tandem. One obvious effect of Pavlovian learning is a shift in motive states and feelings. To be sure, motives and feelings constantly shift and flow as an agent encounters new events, either because the events in the external world change or because changes in orientation and movement change what is perceived. However, anticipatory learning enables an agent to feel changes even before events occur. Experientially, this can be characterized as a *feeling of what might happen next*, a forward-looking feeling of what happens in which feelings change in anticipation of subsequent events.

Complementing the anticipatory learning for interoceptive changes, which occurs in the amygdala, the anticipatory mechanisms in the cerebellum are optimized for learning about the timing of sensorimotor and cognitive reactions. The cerebellum monitors the flow of information up and down the brainstem and learns to use cues to time reactions by ramping up output activity just prior to the time they are needed. A large part of the cerebellum is dedicated to coordinating the timing of motor activities. Well-timed postural adjustments are essential for maintaining balance. Adjusting the timing among various skeletal-motor outputs helps make actions more fluid and efficient. Most of the anticipatory motor signals project down the spinal cord and may never gain consciousness. However, in humans, about one-third of the anticipatory activity in the cerebellum projects back through the thalamus to the cortex. These projections tend to ramp up gaze orientation, perceptual processes, motor planning, and other cognitive reactions in preparation for subsequent events. Because these anticipatory signals pass through the thalamus, they have the potential to reach consciousness.

Anyone who has listened to a musical recording several

times knows that they soon come to anticipate succeeding passages or tracts, even to the point of beginning to hum some passages in anticipation. When the sequences are repeated perfectly, as when playing a recording, this happens quickly. When the sequences generally follow the same pattern, the learning is a little slower, but common events soon come to be anticipated to the point that we look for them and think about them in anticipation. Such shifts in attention are most obvious on occasions when the event has not yet occurred, but we nevertheless find ourselves thinking that it might and planning accordingly. These anticipatory shifts in sensorimotor attention are a major aspect of cognitive experience, and it is no accident that half the cells in the human brain are located in the cerebellum.

Agency

Another prominent feature of conscious experience is an awareness of the self as a planning and decision-making agent. This sense of agency emerges as the experience of planning and decision making comes to predict changes in feelings. As we have characterized the process, action planning in the cortex begins with the detection of affordances and their locations in motor spaces. Candidate actions are then disambiguated based on spatial relationships between affordance locations, starting body positions, and potential action effectivities. The leading action candidates are subsequently evaluated based on various connection weightings added in the striatum. The transition from planning to execution occurs when one action candidate becomes dominant, the brake in the basal ganglia is released, and the activity of the dominant action module in the pallidum is passed on to motor-program generators in the brainstem.

Because motor planning can proceed without conscious guidance, unless the brake is held on to sustain the planning phase, many routine decisions may be resolved before a developing plan ever gains consciousness. We know this because a spike of activity reliably occurs in the supplementary motor area (SMA) about 550 milliseconds prior to the execution of voluntary acts, and yet awareness of the action does not occur until some 350 milliseconds *after* the spike.[7] Given that the SMA is a primary cortical target for returning motor signals from the pallidum, its rise in activity appears to be correlated with the process of resolving an action plan in the basal ganglia. The decision activity in the pallidum projects back to the cortex through the thalamus, where it can be bound in sensorimotor assemblies and gain attention. We have already noted that it takes 300–500 milliseconds for new associative assemblies to be organized within the thalamocortical architecture. The 350 millisecond time span from the onset of the readiness potential to conscious awareness, which is routinely found in these studies, is consistent with this time course.

Given that motor movements don't begin until about 550 milliseconds after the spike of activity in the SMA, these times imply that subjects usually become consciously aware that they are initiating an action some 200 milliseconds before the effective muscle movements begin. Studies have shown that this is sufficient time for the processing that accompanies conscious awareness to trigger inhibitory states, which can activate the brake in basal ganglia and inhibit the action, or at least delay it while it is being further evaluated. This appears to happen because the effectivities of actions are linked to feelings states, which in turn can activate conflict resolution circuits in the anterior cingulate and inhibitory reactions in the medial prefrontal cortex.[8] Thus, conscious awareness is not simply informed about the plan; conscious processing adds another layer of control to the decision process. Further, if the agent extends the planning phase by holding the brake on longer, he can consciously explore the later stages of the decision process more thoroughly by simulating actions and their possible consequences before committing to them. Sustaining the planning process enables a broader range of conscious processes to contribute to decision making. Recall the example of our feline companion, Tom, simulating jumping movements prior to leaping and changing his plan after the evaluation phase.

In addition to assembling action plans based on available affordance and potential effectivities, it is also the case that we often guide action plans by consciously shifting attention to particular affordances, locations, or goals, prior to taking action. In effect, there is often a spiral of automated and consciously considered processes which serve to initiate and guide the planning phase. The aroma of the coffee draws our attention and activates anticipatory motive states. Learned associations soon prompt a glance at our cup. That biases motor planning toward drinking-related grasps while the effectivities of the action are further evaluated. Activity in the SMA occurs as the plan is brought to resolution in the basal ganglia. We are not conscious of every step in the process, but rather of progressive shifts in attention. We smell the coffee. We notice the cup. We recognize an action plan taking form. And we taste the coffee. Each shift in attention takes 300–500 milliseconds to resolve, and unless there is some reason to rush, each may be sustained for a half second or more before the next shift in attention occurs.

Voluntary planning often emerges from a mix of preliminary decisions and attentional biases, many of which result from prior states of attention. The SMA operates something like a project leader in this process. It helps bring the right set of components parts together, but it doesn't decide what to do; it simply reacts to the dominant plan. Part of the planning it reacts to may even include sequential planning processes. Task-related memories supplemented by temporal flow circuitry in the cerebellum even prompts sequential planning steps. This mix of predictors enables an organism who can anticipate changes to learn how to encourage them. Because many of these processes gain attention, the decisions are usually attributed to the planning aspects of self, the *agent self.* The agent self not only experiences thoughts and feelings about what might happen next, but in the context of resolving action plans that predict changes in feelings, it experiences a *feeling of being able to make things happen.* That is, it comes to experience anticipatory feelings associated with its plans, and those feelings further motivate and guide its planning.

AGENT COGLEY

So what additions will Cogley need to experience conscious agency? We have already proposed a planning stream complete with high-level planning networks, Cogley's version of the cortical planning networks, and a mid-level action resolution system with a candidate weighting process and competitive selection algorithms to determine best-fit action candidates. This mid-level system is Cogley's version of the basal ganglia. We have also proposed that Cogley have a braking mechanism in this stream which tends to be activated at times of worry so that he can simulate action plans for longer times when he is uncertain about what to do. Further, we have suggested that Cogley have look-ahead eye fields, which can bias his attention for action planning. This will enable Cogley's planning circuits to recruit added attention in support of ongoing actions.

To complete this motor coordination linkage, Cogley will also need high-level, preparatory motor-planning networks that can disambiguate action plans based on stimulus cues and sequential dependencies. He will need a pre-motor coordination network, his version of the supplementary motor area, to help coordinate his pre-motor planning with activity in the mid-level action resolution networks. The selection algorithms in the mid-level networks should provide feedback to the pre-motor coordinator as the winning candidate is resolved so that the high-level and mid-level networks remain synchronized. This feedback should return through motor modules in Cogley's central attention channel, where it can be bound with related activity in other channels via the cross-connecting matrix circuits.

One goal of these added connections is to enable motor plans to be bound with other activity in the central attention channel, so that Cogley can become aware of how his planning influences his subsequent feelings and learn to change his feelings. The time course of this feedback should be sufficient for an action plan to gain attention before the brake is released. This will ensure that the spread of activation in the central attention channel is capable of activating inhibitory associations in time to prevent a plan from going to completion if, after consideration, it appears to lead to unwanted outcomes. This will also enable Cogley to review plans prior to taking action, and it will give him a dimension of veto power over plans that appear ill-advised. Of course, if Cogley holds the execution brake on for longer time periods, then he will be able to consider his plan in more detail and take even more conscious control over his decisions.

The observation that agents can exert a degree of veto power over an action in progress is sometimes sarcastically described as *free won't*. This phrase is meant to suggest that veto power is not the same as free will – that is, consciously choosing to initiate an action. However, the fact that on some occasions Cogley's actions will only be controlled consciously by late-stage inhibitory reactions after a plan is already in progress does not mean that earlier planning processes cannot influence his decisions on other occasions. We have already described several early processes that influence action planning. For example, we noted that the gaze planning circuitry in the supplementary eye field can bias attention in support of upcoming motor planning, a process that we labelled *look ahead*.

When we introduced this concept, we suggested that looking ahead was an essential part of action planning, because the planning circuits periodically need to update the spatial affordance maps that support various actions as they resolve possible choices. However, given this circuitry, it is also possible to influence motor planning by consciously looking ahead first. That is, the influences can flow in either direction. Attending to spatial locations and supporting affordances tends to bias motor plans toward those locations and affordances, even if no action was planned with them before looking ahead. For example, if the smell of coffee gains attention and triggers your motivation to drink coffee, and if that motivation causes you to glance at your cup, and if those initial processes encourage motor plans that reach for the cup, then you are exercising free will, not free won't. You are consciously encouraging certain action plans to take form. You are still dependent on many of the less conscious parts of the planning stream to assemble the plans. However, those systems are predictable, and as a conscious agent, you learn to manage them.

While we are considering the link between gaze planning and general motor planning, there is another aspect of the gaze control circuitry that needs to be mentioned. It turns out that gaze control is such an important process for managing attention that it has its own separate processing channel in the basal ganglia. Rather than returning decision outputs to the SMA, the circuits in this channel return information about gaze decisions to the frontal and supplementary eye fields. Having a separate decision pathway for gaze means that gaze planning can occur in parallel with planning for other motor actions. To provide similar skills for Cogley, we'll need to add a separate channel for gaze control in his mid-level motor control networks. This parallel control channel should return outputs through the central attention channel to Cogley's eye fields. Providing a separate gaze-planning path will ensure that Cogley's gaze planning will not have to compete with other motor-planning tasks. That should make it easier for his steering committee to use look-ahead and gaze-planning strategies to manage his attention and commit intentionally to actions.

Perspective Taking

There is one additional aspect of agency which emerges from an awareness of movement in personal space. Parts of the superior temporal sulcus of the human brain, a fold in the temporal lobe below the auditory region and above the visual region, are sensitive to biological movements. The recognition of particular movement patterns is also linked to corresponding motor-program regions in the parietal cortex, so as to enhance action recognition by activating the motor programs needed for simulations and the mirror neuron networks needed to predict effectivities. Given that movements are salient events, the locations associated with them are also often tracked. Animate movements in particular are key features for recognizing and localizing other agents, and it appears that we detect and track our own actions via the same networks. We appear to be able to differentiate self-generated actions from the actions of others because self-generated movements are correlated with privileged knowledge about the timing and execution of motor programs, with proprioceptive feedback, and with concurrent changes in feelings of self in the insular cortex. The actions of other agents do not result in such synchronized feedback. Thus movements associated with synchronized feedback are attributed to our own agency, and non-synchronized effects are attributed to external sources.[9]

The linkage between movement detection, motor programs, and location also appears to be part of what enables agents to reason about the actions of others. As we discussed earlier, we tend to think about the mental states of other agents by simulating their feelings and possible actions. One skill that appears to facilitate perspective taking is to simulate the actions of another agent while focusing on the objects and cues that they might encounter. This effectively results in taking their perspective. It is not fully clear how we do this. However, one possibility is that we learn to use coordinate-transform maps to link our own agent-self perspective with the activity and spatial orientation of other agents as we simulate their actions. This would enable us to not only to emulate their feelings but also to attend more closely to the cues and affordances that are likely to influence their decisions. I believe we should be able to implement this effect in Cogley by enhancing the saliency of features located in the direction of gaze of an agent as Cogley thinks about them. This gaze-of-another bias should enhance Cogley's ability to take the perspective of another agent and reason about the actions. In effect, it will give him a more predictive theory of mind.

Maintaining coordinated representations for all these separate aspects of self does have occasional side effects. For example, it can result in certain alignment irregularities.

Ordinarily the feelings of the resonant self are closely aligned with the perceived location of both the observer self and the agent self. However, the locations of the observer and agent selves can sometimes become disconnected. When that happens, the visually dominated observer self tends to remain linked with the feelings of the resonant self, whereas the motor aspects of self-agency become disconnected. This results in various disorientation anomalies, such as mistaking one's position or size, or sometimes even in out-of-body experiences in which the agency of the body seems to be located in a different place from the observer self. The fact that we can adopt the perspective of another agent and plan actions around cues in other locations may even contribute to the tendency for the observer and agent selves to slip out of alignment. That is, we may sometimes view our agent self as if it were another agent, even when we should know logically that it is not.

Recent research suggests that a part of the brain at the temporal-parietal junction, commonly known as the TPJ, is critical for integrating vestibular activity, biological motion detection, and multisensory feedback processed in the superior temporal sulcus. The TPJ appears to align these input areas with sensorimotor activity in the parietal cortex.[10] This alignment plays a critical role in discriminating between the actions made by self or others, precisely because the actions of others are not closely aligned in time with motor decisions, sensorimotor feedback, and interoceptive changes detected in the insula. Obviously, Cogley will also need a TPJ-like agency attribution network to help him discriminate between his own actions and those of others. This network will also be important for ensuring that the perceived location of Cogley's actions will tend to be aligned with the location of his observer self. Out-of-body experiences and other anomalies seem to occur most often when motor feedback is weak or when it is misrepresented. To keep the observer and agent aspects of self from drifting out of alignment, it would be helpful to ensure that local movement tracking, motor feedback, and interoceptive status changes are well represented in Cogley's agency attribution network.

This should ensure that Cogley's self-generated actions will be readily recognized as his own, and that they will be properly aligned with the location of his observer self.

FEELINGS OF HISTORICAL CONTINUITY

Declarative memories add yet other feelings to the upward spiral of conscious experience. As we have characterized them, declarative memories result in two distinct types of co-occurrence associations, semantic memories and episodic memories. Semantic associations result from co-occurrence associations, which link perceptual items with spatial, task-related, emotional, and social contexts. Linking perceptions with functional contexts has a profound effect on experience. Items are recognized in meaningful ways because they are connected with recurring scenarios, tasks, and locations. This creates a sense of familiarity and continuity for items and their contexts. In fact, the process is often so fluid that features recalled from past experiences may subsequently be bound with current perceptions so as to change what is perceived and expected in the current situation. In effect, a reflective mind always experiences more than is present in the immediate situation. This dual level of experience is what Gerald Edelman characterized as *the remembered present*.[11] It literally places the interpretation of ongoing perceptions and actions in the context of past experience.

Episodic memories formed as networks in the hippocampus bind perceptual items and their contexts into situational configurations. As we have characterized the process, these associations capture momentary events as snapshot-like combinations of co-occurring features. Because persistent features tend to be bound in many configural snapshots, this strategy often enables an agent to reconstruct connections between events by attending to key retrieval cues, and then noting which other configurations are activated in memory. Often, the first snapshot simply provides the features needed to retrieve more relevant snapshots. Assuming each shift in attention takes about a half second to stabilize, it is easy to understand why it may take several seconds or more for past episodes to be vividly reconstructed. However, once episodes of past events are reconstructed, it becomes clear why they provide such a powerful addition to conscious experience. They enable an agent to re-experience selected episodes of past conscious experience.

We've already proposed that Cogley be equipped with memory networks which are capable of forming both semantic co-occurrence associations and snapshot-like situational co-occurrence associations. We have also suggested that his memory recall should depend on two conditions: a retrieval cue to trigger the recall of situationally relevant co-occurrence associations, and a competitive selection algorithm to ensure that the associations with the strongest interoceptive links tend to win the competition for recall. In addition to these memory recall effects, we proposed that Cogley be equipped with a pre-attentive saliency mapping network capable of managing up to six working assemblies. This pre-attentive map was proposed to provide a workspace where new assemblies could be organized and compete for transfer into the active attention map. Thus, several alternate memory events may reach this pre-attentive stage as memories are searched.

We also proposed that attention to new assemblies should result in their saliency map being transferred into the active attention map, *before* disabling the pointers from the last assembly. This was to ensure that the two maps would temporarily overlap, making it easier for Cogley to detect related features and bind them in combined assemblies. To encourage these memory-searching skills in Cogley, it seems appropriate to arrange his memory system so that finding a match to similar perceptual cues is mildly rewarding, even if the match is never associated with any subsequent problem resolution. This will ensure that Cogley routinely explores his memories. One result of this similarity matching bias would be a tendency for Cogley to find similarities, symmetries, repeating patterns, and partial matches to be points of interest. The first-order goal will be to encourage Cogley to explore similarities that might lead to useful insights. However, for auditory cues this should also cause Cogley to find repeating rhythms, phrases, melodies, or rhymes to be interesting. Thus, an interest in similarities will not only make his thinking more human-like; it should also promote some musical and artistic interests in Cogley. [12]

To facilitate memory recall, it also seems reasonable to provide Cogley with a means of reducing conflicts with ongoing perceptions when he is searching for memories. When humans explore their memories, they tend to deemphasize perceptual inputs while focusing attention on retrieval cues in memory. For visual processing this often results in looking up or away and failing to hold focus on objects in the visual field. If we add similar looking-away and defocusing strategies to Cogley, then he should be better able to explore his memories. In fact, at times Cogley may want to completely shut down his external visual inputs by closing his eyes (his camera irises) to help him focus better on internal processing tasks. Humans use this strategy on occasion, and Cogley should have this option as well. Looking away, defocusing, and closing off his visual inputs will provide Cogley with behavioral strategies that reduce the saliency of external cues when he focuses on his memories.

Using the past to interpret the present will add an element of familiarity and continuity to Cogley's experiences. When Tom encounters a familiar setting, he knows intuitively what he should expect. He is calm and confident, readily slips into routine activities, and tends to investigate small changes. In contrast, when Tom encounters a novel situation he is uncertain. He moves slowly and explores cautiously. Being able to reflect on his memories will provide Cogley with a similar dimension of familiarity and novelty, which should give rise to similar feelings of confidence or uncertainty in him. Using the past to reflect on the future will also have a profound effect on how Cogley's sense of self matures. Because episodic memories represent experiences formed in earlier episodes of attention, their recall effectively enables past episodes of attention to influence attention. This adds a historical character to the experiences of the resonant self. This memory enhanced experience of self is the *historical self.*

The historical self is not simply a curiosity. It's a defining feature of a developing self. The historical sense of self emerges from a spiral of interactions as experience builds on past experience. Allowing memories to influence how he interprets his world will provide Cogley with a similar historical sense of self. What he perceives will be supplemented by meaningful associations from his past. His expectations will be guided by episodes from past scenarios. His feelings will be enhanced by feelings associated with past perceptions and activities. What he is able to learn will depend in part on the skills and concepts he has previously acquired. As his experiences grow his historical sense of self, his concepts, his skills, and his personality will also grow. We'll have much more to say on the historical nature of self in later chapters.

CONCLUSION

Consciousness emerges in an upward spiral of processing as a series of increasingly cognizant knower networks represent and then re-represent experience in the context of perceiving and acting. The transition from biological awareness (the tuning of sensory systems to ecologically important cues and reactivity states) to acquired awareness (simple associative connections between cues and reactivity states) to attention-guided awareness was a gradual process. Indeed, as our earlier description of the cognitive lattice hierarchy illustrated, these early processing states still precede conscious awareness. The earliest stages of interoceptive representation result in proto-conscious status and reactivity dispositions. Feelings emerge as changes in those representations are integrated and re-represented with changes in perceptual and motor experiences in resonant states of attention. An observer sense of self emerges as resonant feelings come to be associated with visually centered perceptual maps. Anticipatory feelings emerge as perceptual experience comes to predict changes in the resonant self. A sense of self-agency emerges as an agent comes to anticipate changes in feelings as he plans actions. A sense of historical continuity takes hold as past episodes of experience are used to guide the interpretation of current experience. These experiences may seem to be little more than a patchwork of associations, but they all add to an agent's personal sense of self. If our essentialized model is roughly correct, then these same kinds of representations will also contribute to Cogley's sense of self. He will be a conscious agent who experiences his world in a rich context complete with feelings of spatial location, anticipation, agency, and his own memory-based sense of historical continuity.

CHAPTER 14
ATTENTION MANAGING

Consciousness emerges in an upward spiral of representations and re-representations that each add new properties to the feeling of what happens. In addition to this spiral of properties, there are a number of attention-managing processes, which in their simpler forms don't seem to result in new kinds of awareness per se, but nevertheless influence the overall flow of attention and the experience of consciousness. One of these involves a sustained focus of attention. Another involves time sharing attention. Yet another involves strategically shifting attention in the performance of a task. Still others involve shifts in attention which are motivated by social contingencies. Labelling all these effects distinct states of consciousness is something of a judgment call, and some may quibble with the categories. However, these attention-managing strategies contribute to both the flow of consciousness and to reasoning. Reasoning is arguably a significant addition to conscious experience. Therefore, it seems important to describe the added experiential qualities that these attention-managing strategies introduce, provide labels for talking about them, and consider what kinds of organizations are required to make them work, because we'll want Cogley to have similar attention-managing skills and a similar potential for conscious reasoning.

ATTENTION MANAGING

The modulation of conscious states by attention is likely to occur via input to the cortex from the basal ganglion loops as well as from the gating of core responses through the activity of the reticular nucleus of the thalamus. During behavior dedicated to learning, core activity influences the development of automatic motor sequences by sending signals to and receiving signals from the basal ganglia. In this way, depending on context, neural areas underlying conscious and nonconscious activities can interact to enhance attention or develop automaticity. – Gerald M. Edelman, "Naturalizing Consciousness," 2003

Although there may be times when Cogley may want to defocus his immediate perceptions and take time to explore his memories, there will be other times when it will be to his advantage to stay focused on a task. Many agents become more vigilant when in danger or when the risk of harm is high. Similarly, they become more focused when a reward is highly valued or when decisions must be made quickly. This influences the flow of conscious experience. I will refer to this tendency to stay focused on immediate cues as *task-focused consciousness*. Task focus appears to result from two attention-managing strategies: an increase in attention to immediate cues, and a decrease in attention to memory recall. This kind of focus seems to occur more often during tasks that are well practiced, presumably because they can be performed with minimal need for the kind of extended planning which might lead to memory searches. Given that the prefrontal cortex is thought to play a role in suppressing attention to irrelevant cues and in holding attention on tasks, it seems to be a likely candidate for the brain region that helps suppress attention to memories during tasks that require vigilance, although exactly how we learn to do this is less clear.

Consider the process of learning to drive an automobile. Initially this task requires a learner's full attention. In fact, if the vehicle is equipped with a manually shifted transmission, the task may exceed a new operator's attentional capacities. However, over time most of the actions involved in driving come to be so automated that an experienced driver can negotiate her way home from work for thirty minutes in rush-hour traffic and yet arrive home with little memory of the trip. All the routine tasks – shifting gears, stopping at signals, changing lanes to avoid congestion, adjusting for the proximity of other vehicles – and all the shifts in attention required to support those activities, are automated to the point that routine decision algorithms can select the best fit actions within each microsituation and complete the immediate task without encountering anything that would require reflective memory searches. The decision algorithms simply react to the immediate cues with well-practiced habits.

Task-focused consciousness is particularly confusing because the activity that goes on during this state is often not subject to memory recall. Given the inability to recall such experiences, some people have speculated that they must be going through them unconsciously. However, activities such as rush hour driving are not performed unconsciously. If we interrupt the driver at any point in the trip and ask her about the traffic, or we comment on the actions of a particular vehicle, it is clear that she is vigilantly attending to the relevant details of the situation and is highly engaged in her task. The observation that such activities are not readily recalled does not imply that they were performed unconsciously. Rather, it suggests that when someone attends vigilantly to a task, she doesn't reflect on her activities often and therefore doesn't capture the many configural cross-connections needed to make the experiences easy to recall. Indeed, if we query the driver with more specific retrieval cues like, "Do you remember the green car weaving along the center line on County Road?," we often find that she is able to recall a number of things about the trip, although the different events are not well connected in memory.

Before we dismiss task-focused consciousness as a lesser state, we should note that it is often a highly productive state. In fact, the ability to engage in sustained periods of vigilant attention is often prized for what it allows an agent to accomplish. For example, there are times when a craftsperson seemingly becomes locked into a task for extended time periods, sometimes even hours. During these periods the person may make hundreds of subtle comparisons and perceptual judgments which guide the work. This suggests that the craftsperson is accessing semantic memory associations. However, the memories are task centered and don't involve episodic associations that would link the work activities with other aspects of his life. In fact, his attention is often so task focused that when he finally surfaces from the task, he needs to reorient: "What time is it? Where are my keys? Has anyone left a message?" Again, this suggests that periodic shifts in attention and active memory storage are necessary to create a sense of connectedness among different aspects of our lives. When such reflections don't regularly occur, our experiences are less well connected. I should also note that states of highly focused attention are not unique to humans. No one is more focused on his task than Tom when he is on the hunt. And yes, when Tom becomes engrossed in hunting activities, he often loses track of time and comes home late.

If Cogley can hold his attention focused and avoid reflecting on memories when he needs to be vigilant, then he should also become engaged in periods of task-focused consciousness at times and fail to store memories. From a design point of view, the important thing will be to establish a strategic balance between vigilant attention and reflective attention. If my analysis is roughly correct, and the tendency to shift attention and reflect on ongoing events is part of the process that encourages the formation of robust linkages among configural memories, then we will want Cogley to reflect on his experiences much of the time and only shift into vigilant non-reflective attention when a task demands it. In humans, this seems to be something of a personality dimension. Some of us are simply more reflective than others. Some of us stay more task focused. There is probably no correct balance. However, if Cogley fails to reflect at some minimum level on his experiences, it seems likely that his thinking and his sense of self will become largely compartmentalized. All humans do this to some extent, so we'll have to accept that Cogley will at times. However, it's an outcome that we don't want to go to extremes.

A few chapters ago we noted that the nuclei in the thalamic midline-intralaminar complex appear to act as higher-order relays, which help bind thalamic processing with various kinds of activity. This is exactly the kind of support system that would be needed to bias attention to task-related features and to encourage the formation of habits of attention. We also noted that a few of the nuclei in this complex send outputs to memory areas. Given that there are times when attention is focused on memories rather than actions, it seems likely that the reverse strategy could also be implemented by these nuclei. We do not know exactly how these biases are managed, but the basic strategy seems sound. Therefore, I propose we provide Cogley with a set of central attention-gating modules to encourage task-focused attention and defocus memory processing when vigilance is needed, and to support the defocusing of immediate sensorimotor attention and encourage a focus on memory processing when memory recall needs to be at the center of attention. This will help Cogley shift between vigilant attention and reflective attention, depending on task demands.

TIME-SHARED CONSCIOUSNESS

As we have characterized it, the attention steering committee includes a number of brain networks which serve to bias attention in various ways. Particularly important in this committee are modules related to gaze-planning activities including look-ahead motor planning. Look-ahead shifts in attention are an important multitasking control strategy. If one is walking and talking at the same time, it is essential that the circuits managing walking are able to gain a look at the path ahead from time to time. Otherwise, you would run the risk of crashing into objects as soon as you began a conversation. Of course, breaking attention up in time slices only works well when the tasks don't interfere with each other. If vigilant attention is required, as is the case when driving a vehicle in rush-hour traffic, then time sharing may result in a significant degradation in performance. One only needs to consider the well-known effect that cell phone usage has on driving skills to recognize the problems that multitasking produces when more than one task requires a large share of attention. However, human agents are often motivated to employ time sharing strategies even in such demanding situations.

Recall now that it takes 300–500 milliseconds for the thalamocortical circuitry to focus attention and establish the extended resonant states that give rise to conscious experience. As we have noted, motor plans reach consciousness in about 350 milliseconds, and memory processes seem to take a little longer. Recall also that the declarative memory circuitry has a tendency to sustain recent activity at high levels for several seconds and at somewhat higher levels for about a minute. As a result, once declarative memories are engaged, they tend to be slow to disassemble. In contrast, the planning stream is faster at making decisions and shifting attention for motor activities. Apparently due to the time differences, the motor planning circuits can often steal small time slices of attention without disrupting the stability of declarative memory processes to any great extent.

If motor activities take over gaze control for more than a few seconds, then the organization of declarative memories may begin to shift. However, if the interrupting task does not activate other memories or strong feelings, then prior memory patterns may not be significantly degraded. They can usually be restarted, *as if there was no break*, even after longer interruptions. It even seems likely that many automated motor activities, especially those which occur as background activities, can pass through the filter and operate in parallel with processes involving visual attention. As long as the motor activities do not compete significantly with features at the center of visual attention, there would be no competition. In fact, for many well-learned motor activities, the occasional need for look ahead may be the only shared activity that competes with visually guided attention.

Time-shared attention seems to work best when the interrupting activity is a procedural task which introduces lesser demands on memory processing. However, it is also apparent that humans can and do learn to multitask between activities that both involve declarative memory processing. For example, we may interrupt a conversation to handle an incoming phone call and then return to our conversation. It seems that as long as the interrupting task is short, and the memory demands are moderate, then the original task can usually be reengaged, although sometimes it may take a few seconds to recover. Hmm. Where was I? The process is not always fully successful. However, when there are environmental cues or a partner to help reactivate the previous memory links, it works surprising well. Given that we have proposed a similar architecture for Cogley, he should also be capable of time sharing between memory-directed attention and motor tasks, without losing his train of thought. It may take him some time to learn to rescan his memory after an interruption so that he can reengage prior tasks with ease. However, once he can do this, then Cogley will be capable of experiencing extended episodes of time-shared attention.

EXPLORATORY CONSCIOUSNESS

Having developed some appreciation for the time-sharing capabilities of the central attention networks, we are now in a position to consider another aspect of higher-order consciousness: the experience of being a reasoning agent. Some assume that all reasoning requires the use of language, but as we have pointed out, language is often not necessary for disambiguating plans or recalling memories. As Darwin noted, in problem situations many animals "pause, deliberate, and resolve" before taking action.[1] In fact, we have suggested that they do this in large part by activating motor programs while holding the brake on committing to action so that they can explore their plans in more detail. We have also noted that memories can be explored by attending to an initial feature as a retrieval cue and then using recalled features to retrieve yet more related memories. Further, we have argued that monitoring the emotional signals of other agents and empathizing with their feelings can help us think about how they think. Yet none of these attention-shifting strategies requires language.

In effect, thinking can be facilitated simply by encouraging creative activities in the three streams to interact with each other in more systematic ways. Shifting attention between simulated plans, recalled co-occurrence associations, and feelings can have the effect of building associative assemblies which are sensitive to features derived from all these strategies. Memories can be retrieved by features brought to attention using one strategy, and then linked with those called to attention by another. Action plans can be guided in part by memories of past experiences, but they can also be influenced by their social consequences. Simulated action sequences can set expectations about scenarios or suggest alternatives to current plans. In short, the ability to shift attention among task-relevant features and to use a variety of processing strategies before committing to action is one that often improves decision making. In fact, such shifts in attention are natural ways to resolve uncertainties, and good thinkers learn strategies that encourage them.

Two neural structures in the attention steering committee that are proportionally larger in humans than in their primate ancestors, and which have been implicated in reasoning, are the prefrontal cortex (PFC) and the cerebellum. By our analysis, the PFC functions largely as a task-related working memory, which guides attention during learning and performance. It is known to be important for sequential and conditional planning.[2] In fact, it appears to learn what kinds of features are important in a task and then sets attentional biases – *working hypotheses* as we characterized them – which promote attention to those features. It is also known to be capable of learning about conditional cues and being able to suppress attention to distractors. The PFC appears to be an ideal candidate for managing the strategic shifts in attention required for higher-order reasoning. The cerebellum is the member of the attention steering committee known to be involved in learning contiguous timing relationships between cues and reactions to them, including changes in cognitive flow. In effect, it appears to be an ideal candidate for helping to guide shifts in attention during sequential stages of a task. Given that there are strong reciprocal connections between the PFC and the cerebellum, this combination appears to be well situated to learn about conditional dependencies and shift attention in stages as a task progresses.

Habits of Executive Attention Shifting

We have already proposed that Cogley be designed so that uncertainty and worry will tend to inhibit him from taking immediate actions, at least when drive states are not too high. Instead, we proposed that these conditions should promote action simulations, memory explorations, and empathic considerations as a means of better resolving action plans. In addition, we have suggested that Cogley should be sensitive to social status, such that worry about his social standing should routinely encourage him to pause, deliberate, and resolve before taking action. To ensure that Cogley learns to engage these strategies, we have also proposed that resolving problems should be rewarding. In fact, we suggested that simply finding matches in memory should be mildly rewarding, so as to encourage memory searches, even if they do not end up being productive. Over time, such a system of rewards should have the effect of building habits of attention, perception, and memory searching.

Given that we have proposed that Cogley be equipped with task-related memory modules, which are capable of forming working hypotheses to guide his decisions, it is reasonable to expect him to develop problem-solving hypotheses that promote actively shifting attention as a means of generating new ideas and avoiding embarrassing oversights. However, to improve Cogley's reasoning strategies, I propose one more addition to his decision architecture. When we talked about the importance of gaze shifting as a means of biasing attention, we noted that gaze shifting had become such an important strategy that a separate processing channel in the basal ganglia has been dedicated to controlling gaze, so that gaze decisions don't have to compete with other motor decisions for processing time. This observation led us to add a separate channel in Cogley's mid-level action networks for gaze control. For the same reason, I propose that we add several additional mid-level control channels to support habits of executive attention shifting.

I propose this strategy because research using macaque monkeys has revealed three largely separate learning channels in the basal ganglia which appear to support executive attention shifting. One of these circuits returns outputs to parts of the dorsolateral prefrontal cortex.[3] Another returns outputs to the lateral orbitofrontal cortex. The third returns outputs to the medial orbitofrontal cortex and the anterior cingulate cortex.[4] The channel which returns outputs to the dorsolateral PFC also receives inputs from the posterior parietal cortex. It appears poised to bias attention based on spatial cues.[5] As we have noted, spatial cues are critical for binding features together with sticky indexes in the pulvinar. The dorsolateral PFC appears to be the working memory region that learns what categories of spatial features are important for this. A meta-analysis of recent neuroimaging studies suggests that activity in the medial orbitofrontal cortex is largely involved in the evaluation of reward value and motivational states, whereas activity in lateral orbitofrontal cortex is more involved in the evaluation of punishers and inhibitory relationships.[6] Thus, these channels appear to be poised to manage habits of attention based on excitatory and inhibitory dependencies respectively.

We currently have a limited understanding of the impact of these added attention biasing channels. However, given that the PFC learns to manage attention to task-related cue biases, it seems likely that these channels optimize learning by favoring attention hypotheses that have proven reliable over time. I believe this strategy works for several reasons. First, the networks in the basal ganglia are good at learning to connect actions in sequential chunks, so they make an ideal system for learning about sequential biases that could guide attention in more complex tasks, especially as they interact with the temporal flow networks in the cerebellum. Second, the basal ganglia can often automate routine action patterns, offloading work from higher-order planning networks and speeding up routine decisions. This suggests that working hypotheses can often be automated as habits that encourage shifting attention to relevant cues without the constant need for higher-level reasoning about such decisions. Providing Cogley with separate mid-level attention biasing channels should help him develop such habits of attention.

A second reason that I believe this strategy has merit is that human experts seem to react to problems in exactly this way. They have habits of attention which automatically lead them to attend to task-relevant cues and to shift attention at critical times. Even when they take on new problems, they employ these habits to make their problem solving more efficient. Of course, these strategies don't always work. However, when working on problems in their field of expertise, experts can often tune in to relevant features quickly by using such strategies. If Cogley is to become an expert at anything, he will need a similar ability to develop habits of attention, habits of perception, habits of inhibition, and habits of motivation. Having a number of attention-biasing channels in his mid-level learning networks to automate these attention-biasing strategies should help Cogley become more of an expert in the tasks that he engages most often.

SOCIAL CONSCIOUSNESS

Social feelings often have strong interoceptive weightings, especially for agents who live in complex social communities, and in such settings they often come to bias attention in special ways. In fact, most of the categories of consciousness which we have thus far described include a social side – that is, social aspects which come to be re-represented as higher-order social feelings. There are socially guided feelings of what happens. There are feelings of anticipation for social outcomes. There are feelings of social agency and reflective evaluations based on social memories. Although none of these states is exclusively social, their social side nevertheless results in a spiral of representations which ultimately enables some unique states of social consciousness to emerge. Two of the most important of these higher-order states of social awareness are *social self-consciousness* and *social reasoning*. Self-consciousness is the enhanced sense of self that occurs when an agent becomes concerned (i.e., worried) about how the attention of another agent is being, or may be, directed at them. Social reasoning is the ability to make predictions about the behavior of other agents, a skill that is often described as having a theory of mind.

Social Self-Consciousness

Although we are accustomed to thinking about social interactions as largely within-group status interactions, interactions outside one's social group also have significant effects on social awareness. Nothing makes an agent feel more self-conscious than to recognize that he or she is being watched by a predator. One low-level mechanism that promotes this concern is the tendency of the tectum to detect eye-spots, which, when they are sustained and when they occur in pairs, tend to be interpreted as staring eyes. When signals about staring eyes reach the amygdala, they produce feelings of concern. Group living often provides a degree of protection from predatory encounters, and yet within-group interactions, group living can be similarly intense and include its own forms of predation. Discipline, competition, and thievery are common in social groups and have major effects on an agent's concerns for his or her personal status. As we have noted, these experiences are salient, in large part because changes in social status have direct effects on the reactivity states of the proto-self.

The sense of self that emerges from this complex of social interactions is perhaps best described as a *feeling of how others may react to me.* It results in what we can describe as a *socially aware self.* This aspect of self is particularly sensitive to the social-signaling of other agents. When an agent feels its status is high, the socially aware self is confident and even emboldened by attention. When an agent feels its status is low, the socially aware self is anxiously self-conscious. The interesting thing about these experiences is that although they don't appear to require any new kind of neural architecture, they nevertheless seem special and profoundly important. This apparently occurs because social signals activate domain-specific social feelings and values that markedly influence how agents interpret their status. This is why we consider social self-consciousness to be a special category of consciousness.

Interestingly, a curious test of self-awareness resulted from the tendency of agents to be concerned about how they look. Some highly visual social animals worry about how they look; they even attempt to groom themselves, if their appearance seems to be degraded. A popular test of this aspect of self-awareness is to place a small blemish, a mark, on the face or body of an animal without it knowing, and then allow the subject to observe themselves in a mirror.[7] Attention to the mark when seen in the mirror, especially attempts to remove or groom the mark, is interpreted as evidence of self-consciousness. Among primates, humans, chimpanzees, and orangutans generally react quickly to the mark. These species are therefore thought to experience a socially aware sense of self.

The assumption of the mirror test" is that subjects who are self-conscious, like humans, should react to seeing the mark by trying to remove it. However, most of the gorillas and monkeys tested in this paradigm fail to inspect the mark.[8] Some interpret these failures to mean that those species lack a self-concept and suggest that only a few human-like primates are capable of this level of self-awareness. However, this interpretation assumes that all animals should consider their visual appearance to be important and be motivated to react to the mark when they see it. However, the differences might simply mean that the mirror test is not always an effective way of evaluating self-recognition. Individuals of many species – gorillas and cats, for example – may not be very concerned about their visual appearance. In contrast, researchers have found that some visually aware non-primates, dolphins, elephants, and even magpies react to a mark when seen in a mirror.[9] This suggests that this form of self awareness is not an exclusive property of primates, or even of mammals.

As we consider the mark test as a measure of self-concept, I should point out that there has long been a mirror sitting at floor level near the central hallway in the house Tom and I share. This is a location that Tom walks by many times a day. When he walks from the main room to the hallway, he passes directly in front of the mirror and really cannot help but notice his reflection. However, due to the angle of the mirror, he must glance to the side to get a good look at his reflection when he approaches from the hallway. Because I often see him make these glances as he passes, it is obvious that he notices the image in the mirror. Sometimes, he even pauses to gaze into the mirror. If I make eye contact with Tom via the mirror when he happens to be looking at it, he usually doesn't hold the eye contact for long; instead he typically turns and approaches me. This indicates first that he recognizes I have made eye contact with him through the mirror, and second that he knows I am not beyond the mirror.

Figure 14-1: Peyton's first encounter with the mirror cat.

Figure 14-2: Peyton eleven days later. What problem?

This kind of mirror recognition takes some learning. When a young cat first encounters its reflection in a mirror, they often treat the reflected image as a stranger. For example, when a new male kitten, Peyton, first saw himself in the floor mirror at twelve weeks of age, he adopted a threat display as if to frighten the stranger (see Figure 14-1). However, this threatening display soon attenuated. For the next few days Peyton seemed a little spooked by his reflection, but he never threatened it again, and soon he came to ignore it. Then, after about a week or so, I noticed that Peyton would occasionally come close to the mirror and look at himself without any concern that the cat in the mirror was a stranger (see Figure 14-2). A second male kitten, Casper, never threatened the mirror image. However, early on he was observed looking into the mirror while pawing behind it. Within a few weeks, this behavior also disappeared, and he too was soon admiring his reflection in the mirror. Sometimes I even catch Casper watching me through the mirror.

Tom's most interesting mirror use occurs when one of his toys ends up near the mirror. As he engages in play with the toy, he becomes interested in watching the reflection of the cat and toy in the mirror. Now, I do not claim that Tom recognizes himself in the mirror as you or I would. There is no clear evidence for that. He never uses the mirror for grooming in any obvious way. However, he never reacts to the image as if it is a stranger. Given that the reflection moves whenever Tom moves, the networks in his temporal-parietal junction (TPJ) have surely learned that his behavior causes the image, which just happens to look like a cat, to move, and he seems to find all that interesting. Paying attention to reflections is obviously a prerequisite for mirror recognition of self. I like to think that Tom is three-quarters of the way to self recognition. He recognizes other agents in the mirror. He knows that what he sees is not located behind the mirror. He even seems to understand that he controls the movement of some of the reflections in the mirror. However, it probably never occurs to Tom that the image he sees in the mirror is what he looks like to others, so he's unlikely to be concerned about finding a new spot on his face.

Given these observations, it seems that if we want Cogley to react to the mirror test as humans do, rather than as Tom does, then we'll probably need to use a two-stage approach. First, Cogley will need to be capable of recognizing that the reflection in the mirror is related to his behavior. To do this, he will need to be able to map the changes in the object in the mirror with changes in his movement, much as Tom does. We've already noted that tracking self-movement seems to involve the representation of animate movement by networks along the superior temporal sulcus and their connections to the TPJ. We have also proposed TPJ-like agency attribution networks for Cogley. These networks should enable Cogley to learn that the activity of the image in a mirror is correlated with his behavior. However, to ensure the Cogley continues to attend to his image after he learns it is not another agent, I believe that he will also need to pay attention to certain qualities of the image he sees.

To do this Cogley will need to be sensitive to judgments about the normalcy, smoothness, and overall symmetry of facial and body features, and find such symmetry attractive, as humans do. In fact, he should probably consider these features to be associated with higher status. In contrast, he should find odd and asymmetric features less attractive on himself and on humans, and he should consider spots and blemishes as things that should be groomed and minimized as much as possible. These links between symmetry and higher status, and between blemishes and lower status, should lead Cogley to be concerned about his appearance, and it should motivate him to notice and react to unexpected physical changes, such as marks secretly introduced by curious experimenters. In effect, they should make it likely that he passes the mirror test. Of course, the goal is not to get Cogley to pass the mirror test simply by using some trick. It is to make him overly concerned about his appearance, so that he will react to the appearance of new blemishes like his human partners do.

Social Reasoning

As with other aspects of consciousness, social reasoning develops in an upward spiral of processing and has its roots in a number of low-level emotional signaling modules. Detecting and interpreting the moods and feelings of others is an important first step in predicting their behavior. We have already proposed that

Cogley have facial signaling and detection modules for anger, disgust, fear, joy, sadness, and surprise, as well as mechanisms for assessing emotional tones of voice. Further, we have talked about the need for Cogley to have status and reactivity networks for re-representing feelings in higher-order combinations. We have also proposed that Cogley have several types of modules for detecting other agents. Some of the simplest of these are eye-spot detectors in the tectum. More sophisticated yet are the biological motion detectors which are found in regions along the superior temporal sulcus in humans. These detectors are sensitive to biological movements, including eye movements and locomotor patterns, and provide inputs into the TPJ which are used in attributing actions to self versus others. Cogley will need similar motion-analysis networks to help him identify the locations and actions of other agents and TPJ-like agency attribution networks to discriminate self versus other actions.

When features such as eyes, facial expressions, emotional tones, and spontaneous movement are bound together in a common location, they come to be recognized as an animate agent. An observer doesn't simply perceive a gruff tone and glaring eye contact. She recognizes these as signs of a social agent, and she expects that agent to react by drawing from a repertoire of behavioral and cognitive dispositions. Many animals sense the moods of other agents and routinely use them, together with orientation and preliminary movement patterns, to make predictions regarding what they may do next. However, empathizing with the moods of another agent is primarily a strategy for evaluating their *desires*. If conscious minds are also a product of what they attend to, as we have proposed, then a more complete theory of mind will need to consider what features other agents notice and remember as they take action. Learning to predict the behavior of others based on what they notice and remember results in a theory of mind that is also sensitive to the *beliefs* of others. A theory of mind that can consider both the beliefs and desires of another agent is arguably a more powerful strategy for social reasoning.

So how should we design Cogley so that he can develop a

theory of mind that is sensitive to the beliefs of other agents? We have already proposed that Cogley follow eye movements to track the attention of others. One adaptation that makes tracking eye movements easier in humans is the white sclera surrounding the darker iris and pupil in the eye. Most animals, including other primates, have a darker colored sclera, presumably because the lack of contrast helps them avoid detection by potential predators. For humans, however, the benefit that the white sclera provides for guiding attention apparently outweighs the risks. Humans not only have a white sclera, but they have developed an eye shape in which a wide area of the sclera is exposed, especially in the horizontal dimension.[10] Seeing a dark iris move against this white background makes it easier for humans to notice and follow the gaze of their partners. We'll also want Cogley to have similar high-contrast, eye-like pointers to help his partners track his attention, even though his visual sensors will presumably be cameras that won't need colored irises.

In addition to having eyes that facilitate visual pointing, humans also attend closely to the eye movements of other agents. Many nonhuman animals notice when other animals looks *at them*, and they notice the objects with which other animals interact. However, humans follow the eye movements of caretakers and peers. Tracking eye movements enables them to become more sensitive to what other agents notice. Perspective taking, the integration of gaze direction with predicted actions, appears to depend on regions in the right inferior parietal lobe which represent actions in personal space.[11] Knowing where an agent is looking and what features he can see makes it easier to take his perspective. We have already proposed that Cogley should have networks for tracking the location of other agents, and an ability to use location transform maps, as a means for enhancing features that may be important to the agent being observed. What I am suggesting now is adding coordinate-transform networks to Cogley which will enhance the saliency of cues in the direction of another agent's gaze, whenever he attends to it. This should greatly improve Cogley's ability to take the perspective of another agent.

Language, of course, is the other specialization that helps humans discover what cues gain the attention of others. Human partners routinely comment on what they notice and what they are thinking. This ongoing chatter helps agents learn to track what others know. Cogley will need a similar link between his language, attention, and perspective-taking networks if he is to develop a human-like theory of mind. Fortunately, we have already proposed how language can be mapped into Cogley's sensorimotor schemas. All we need to do is ensure that perspective taking enhances the sensorimotor schemas related to features in another agent's visual field when their actions are being evaluated. Once Cogley takes the perspective of another agent, then their language will be more likely to map into schemas related to that perspective. It will take time to learn how to attend efficiently to what another agent notices, but with these additions, I believe Cogley will have the skills he needs to acquire a theory of mind that can be sensitive to what other agents know.

Some argue that in the absence of language, animals must all be behaviorists – that is, that they must lack internal representations for mental states. Although reasoning about other minds may begin with simple associations, such as the links between facial expressions and emotional states, once an agent begins to represent those states by reacting empathically with similar feelings, he is representing critical features of another mind internally. When he reasons by simulating actions while considering the perspective of other agent, and occasionally supplementing those simulations with memories of their past behavior, then he is no longer using simple associations. He is generating internal models for how another mind might react. Agents without language may have more difficulty learning to consider false beliefs. However, many animals, Tom included, hide in ambush and understand when they are hidden and when they have been detected. These are seeds from which an understanding of false beliefs can develop. If Cogley can build his reasoning about how other agents behave on such skills, and if he can occasionally supplement his reasoning with language-guided schemas, then he will be on the path to developing a theory of mind that is much like that of his human partners, although it may take him several years of practice to get really good at it.

CONCLUSION

Attention and consciousness emerge in a spiral of processes which begin with core feelings. Sometimes we focus on perceptual valuations, sometimes we focus on motor activity, sometimes we reflect on memories, sometimes our attention is guided by the activity of other agents, and sometimes we talk to ourselves. The flow of experience is varied and often changes from moment to moment. Different tasks engage different levels of awareness with varying affinities. However, except in cases of vigilant attention, it rarely seems that we are locked into or out of any single category of conscious experience for long. The very sense of appreciating an event with a depth of consciousness is one of experiencing it from many perspectives and on many levels. Cogley will need a similar spiral of representations and attention-management skills so that he can experience similar categories of consciousness.

Table 14-1 summarizes the categories of consciousness we have proposed in the last two chapters, including labels for the various stages of self-awareness which we proposed for each category. To be complete, the table also includes the sign-guided category, which we will introduce in the next chapter. In humans, sign-guided attention includes language. Note that although this set of categories can be characterized as a roughly interconnected hierarchy of experiences, this characterization does not imply that the highest levels of the hierarchy are the most important, or even that they usually dominate consciousness. In fact, it's just the opposite. The core states of consciousness are arguably the most important and most persistent. There is no consciousness without a feeling of what happens. The other states of consciousness simply add other aspects of experience to the feeling of what happens.

CORE ASPECTS OF CONSCIOUSNESS

Phenomenon:	Phenomenal Consciousness
Sense of Self:	Resonant Self
Characteristics:	The feeling of what happens (Damasio, 1999).
Phenomenon:	Spatial Consciousness
Sense of Self:	Observer Self
Characteristics:	The feeling of where I am.

LEARNING-GUIDED ASPECTS OF CONSCIOUSNESS

Phenomenon:	Anticipatory Consciousness
Sense of Self:	Anticipatory Self
Characteristics:	The feeling of what might happen next.
Phenomenon:	Agency
Sense of Self:	Agent Self
Characteristics:	The feeling of being able to make things happen.

MEMORY-GUIDED ASPECTS OF CONSCIOUSNESS

Phenomenon:	Reflective Consciousness
Sense of Self:	Historical Self
Characteristics:	The remembered present (Edelman, 1989).
	Two subtypes: Semantic Reflection and Episodic Reflection.
Phenomenon:	1. Semantic Reflection
Characteristics:	An awareness of contextual and categorical co-occurrences.
	The feeling of familiarity (recognition) or novelty (uncertainty).
Phenomenon:	2. Episodic Reflection
Characteristics:	An ability to explore situational co-occurrences and reactivate selected scenarios of past experience.
	The feeling of re-experiencing selected episodes of attention.

Phenomenon:	Task-Focused Consciousness
Sense of Self:	Focused Agent Self
Characteristics:	The facilitation of actions by focusing attention and suppressing episodic memories.

Phenomenon:	Time-Shared Consciousness
Sense of Self:	Multitasking Self
Characteristics:	The interactive sense of agency that results from multitasking.

Phenomenon:	Exploratory Consciousness
Sense of Self:	Reasoning Self
Characteristics:	Reasoning by simulating actions, exploring memories, and empathizing with feelings.
	Reasoning begins with the tendency of agents to pause, consider, and resolve. These skills grow as agents learn to manage their own attention more strategically.

AGENT-GUIDED ASPECTS OF CONSCIOUSNESS

Phenomenon:	Social Self-Consciousness
Sense of Self:	Socially Aware Self
Characteristics:	The feeling of how others may react to of me – an enhanced awareness of self that occurs when an agent worries about being the subject of attention from other agents.

Phenomenon:	Social Reasoning
Sense of Self:	Social Reasoning Self
Characteristics:	The feeling of what others might do – reasoning about the behavior of other agents by empathizing with their feelings, simulating their actions, and exploring memories about them.

SIGN-GUIDED CONSCIOUSNESS

Phenomenon:	Narrative Consciousness
Sense of Self:	Narrative Self – Uses signs to guide attention
	Autobiographical Self – Uses signs to organize memories of self
Characteristics:	In producer roles, narrative agents use signs to guide the attention of others. In consumer roles, narrative agents attend to the signs of others and allow the signs to guide their attention.
	With practice, multitasking agents may even learn to play both producer and consumer roles and sign to themselves.

Table 14-1: Categories of consciousness.

It would be nice if we could associate each category of

consciousness with a particular brain region. However, the very fact that the brain represents and re-represents perceptions, actions, and feelings in an upward spiral of processing means that even the simplest feelings are normally expanded and enhanced by cortical processing, in particular by activity in the anterior insular and anterior cingulate cortices. In addition, because the developing mind comes to depend on this hierarchy of processing, when the interoceptive cortices are damaged in adults, then processing in the lower-order systems is often largely compromised. However, as Antonio Damasio notes in his recent book *Self Comes to Mind*,[12] children born without a cortex seem to experience some aspects of phenomenal consciousness. They react with obvious pleasure and pain. They even develop preferences for certain musical passages and react to them with apparent joy. Thus, it seems clear that some rudimentary feeling of what happens can develop without a cortex, especially if the subcortical connections of the thalamus are intact. Still, if we want Cogley to experience more than vague feelings, then he will need higher-order re-representation networks that mimic the functions of the interoceptive cortices. Further, he will need an integrated steering committee to expand his sense of agency and reasoning, including systems that support language recognition and production. Only then will he be able to experience a full range of human-like consciousness.

CHAPTER 15
REACHING UP WITH SIGNS

The upward spiral is a metaphor for the nested processes of representation and analysis that bootstrap iterative cycles of attention into increasingly complex states of awareness. Phenomenal consciousness emerges in this spiral as changes in interoceptive moods are re-represented in the context of perception and action in states of attention. This gives rise to a core sense of self, which we have termed the resonant self. Other aspects of self emerge as other systems add to the perspectives available to the resonant self. Gaze-centered orientations and spatial mappings result in a localizable observer self. Pavlovian learning mechanisms contribute to an anticipatory sense of self. When action planning guides anticipation, a feeling of agency emerges. Memories place the experience of self in a connected historical context. Attention managing processes result in the experience of a reasoning self.

In the social domain, concerns about status and the actions of others lead to feelings of self-consciousness and promote interests in the thoughts of other agents. It is in this context that agents learn to use signs to manage the attention of other agents. Evolutionarily, it appears that sign-guided attention began long ago with simple signs, such as alarm calls and intention movements. These signs were subsequently extended in some species by more standardized gestures and vocalizations. In humans, these extensions led to the vocal signs used in language. Systems of signs, especially language, set the occasion for yet other ways of managing attention and conscious experience. The narrative self is the feeling of agency that emerges when an agent learns to use signs to manage attention. The autobiographical self emerges when narratives are used to guide the recall of memories. If Cogley is to share in these aspects of self, then he must learn to use signs to manage his attention and the attention of others.

REACHING UP WITH SIGNS

> The fact that gestures run the gamut from non-symbolic to symbolic – and emerge along with the first linguistic skills – is strong evidence that children's ability to communicate symbolically is not tied specifically to language but rather emanates from a more fundamental set of social-cognitive skills. – Michael Tomasello, *Constructing A Language*, 2003

I t is my custom to grill chicken for dinner once or twice a week, and it has become a household tradition to share the leftovers with the cats. In fact, the tradition has evolved to the point that I fix extra chicken for them. Both Tom and an older female cat, Calamity, eagerly look forward to these occasions. However, the household has a changed a little. There are two young males, each about three months old, living with us now. When I place the plate of leftover chicken on the kitchen floor this evening, Calamity comes running to the plate, but so do the young males. The boys seem to like this new tradition, but Calamity does not. The boys are newcomers, and she has not yet adjusted to having them around. When they crowd in on the leftovers, Calamity backs away. She looks at the plate of chicken, looks back at me, and then walks to her usual feeding place, where she nods toward her bowl. I immediately understand what she wants. I take her share of chicken from the common plate and put it in her bowl. She looks back at me briefly and then happily consumes her chicken.

A few days later I'm visiting the home of an acquaintance with a young toddler. Anne is fourteen months old and is happily playing with her toys when I arrive. However, I'm a stranger to her. Soon after I enter the house, she grows anxious, moves toward her mother, and raises her arms. Her mother immediately picks her up, and all is well again. Meanwhile, Anne's older brother, John, is playing catch in the backyard. When his ball ventures into the neighbor's yard, he runs after it and quickly discovers the neighbor's poodle is out. The poodle doesn't like it when someone suddenly invades his territory. He backs his ears, raises his lip in a snarl, and growls. John heeds the warning and slowly backs away. Fortunately, the neighbor hears the commotion and arrives to quiet the dog and return the ball. Such events are routine occurrences. However, what may not be so obvious is that in each of these cases, *signs* were used to manage the behavior of other agents. In Calamity's case it was a learned sign, the head nod toward her bowl, which she used to encourage me to put food there. In Anne's case reaching up was a learned sign that encouraged her mother to pick her up. In the dog's case, snarling and growling were biologically prepared signs that signaled a threat of harm and, in doing so, prevented unwanted approach.

The curious thing about the signs in these examples is that although some were learned and others have evolved, they were all acquired in a similar process of selection as predictive signals. The ethologist Nikolaas Tinbergen noted that biologically prepared communication signals are often derived from preparatory or abbreviated movements.[1] Oskar Heinroth, another early ethologist, referred to such actions as *intention movements*, because they often betrayed to the observer what the actor was motivated to do. In the example described above, the dog's snarl, a protective lifting of the lip, signaled that a biting attack might soon follow. To avoid damage to its lip, a dog doesn't actually need to lift the lip until it begins to bite. However, by evolving a pattern of lifting the lip early, the snarl has become a warning sign.

Tinbergen noted that as predictive movements evolve into signs, they often became stereotyped and more demonstrative, a change that has the effect of making them more likely to be noticed. He referred to this process of standardizing and emphasizing signs as one of *ritualization*. The ritualized signs that Tinbergen described were selected by natural selection, but it is also the case that animals develop stereotyped versions of learned signs in a similar way. For example, chimpanzee infants often develop individual ways to signal when they want to nurse, which their mothers soon come to recognize.[2] In most cases these signs begin as instrumental attempts to nurse: pulling, positioning, and rooting activities, which over time develop into more stereotyped signs. Similarly, young chimps often develop a number of gestural signs to encourage social play: tickling and mutual grooming activities among their peers. Again, these are often partial or preliminary components of the activities they signal and are soon recognized by their peers. However, these nursing and play signals are not standardized. Different individuals tend to develop their own favorite form of signaling.

Some learned interactions, however, occur so consistently that they do result in standardized signs. Elizabeth Bates notes that the reaching-up gesture that human toddlers use to signal that they want to be carried begins with the holding response that children use to secure themselves to a caretaker when they are being carried.[3] Young chimpanzees have a similar holding response and react with similar reaching-up gestures as they try to attach. However, once this reaching up pattern is well formed, it often occurs in anticipation of being carried. Sometimes the child may even misread the situation and reach up when the caretaker isn't intending to pick them up. Because caring parents recognize reaching up as an indication that the child is anticipating being carried, rather than disappoint the child, they often pick up the toddler. In this manner, the child begins to learn that reaching up can actually encourage her caretaker to pick her up. Once the child learns how to signal that she wants to be carried, she often begins to emphasize the signal by arm waving or adding vocalizations, which make the reaching up sign more likely to be noticed.

Because this reaching up scenario provides such a nice illustration of how anticipatory interactions bootstrap the alignment of producer and consumer agents so that a sign comes be used to express intention, I have come to use this phrase as a label for the process. As Irene Pepperberg has shown, whenever you treat an agent as if they are using a sign intentionally, you create a situation in which a socially aware agent *can learn to use a sign intentionally.*[4] For example, Calamity's head nod began with the anticipatory inspection of her bowl to check for food as she waited for me to feed her. However, I often reacted to her inspections as a sign that she wanted food, and I tried to get the food to her more quickly. As I did this, Calamity soon began to use the head nod as a way to encourage me to put food in her bowl. In fact, as noted above, she has come to use this sign quite intentionally now.

Even signs that begin as biologically prepared action patterns may subsequently be standardized through learning and come to be used intentionally. For example, when Tom first came to live with me, I would occasionally find him waiting by a door that blocked his way. Apparently when he encounters an obstacle, his first tendency is simply to pause and reconsider his plan. However, because I interpreted waiting by the door as a sign of Tom's interest in getting through the door, as a helpful caretaker I often reacted to his behavior by opening the door for him. As a result, Tom soon learned that attentively looking at the door could serve as a way of getting me to open it. Now, when he wants a door opened, he waits until I'm around and then runs to the door, alternately looking at the door and then back at me. In fact, all the cats in my house use a similar sign.

As these examples illustrate, contrary to the popular belief that there are few precursors to language in animals, there are many precursors to the functions of language using both vocal and non-vocal gestures. Admittedly, these signing processes don't have the complex grammatical structures of human language, but that's really not essential for communication, and it's not what you would expect in precursors. All that is required is that a sign comes to predict some aspect of a situation well enough that producing it can bias attention toward a sign-related object or action. When Tom wants something, he calls to draw my attention, raises his tail like a flag, runs ahead to lead me to the location where he needs my help, and then points by looking attentively toward what he wants. In doing so, he composes a simple sentence, a sequence of signs that gain and then redirect my attention so as to coordinate my actions with his interests. When another human calls my name and verbally directs my attention to something they want my help with, I react in a similar way. Human language uses a much more complex set of signs, but the basic effect is not that different from what occurs when Tom uses calls and gestures to manage my attention.

THE CURVED BUNDLES

I don't want to delve too deeply into the neural mechanisms that make imitative language learning so robust in humans and absent in Tom. However, we need some guidelines for Cogley's design, so a brief overview of some of the key processes seems appropriate. In humans, speech involves auditory recognition networks in what is often called Wernicke's area, a catch-all term for the higher-order auditory processing regions in the superior temporal lobe.[5] This area includes Heschl's gyrus, the planum temporale, and parts of the posterior superior temporal sulcus. The latter two regions appear to play a role in the spatial-temporal analysis of sound combinations that go into words and phrases.[6] Lesions in Wernicke's area generally result in an inability to understand spoken or written language, *expressive aphasia*, although lesions involving the caudal areas may result in an inability to select the correct word for a phrase. This is termed *jargon aphasia*. Jargon aphasia implies that there is a link between the hearing centers and the selection of words for speech production.

An important motor control area for speech production is Broca's area, an output control center in the ventral premotor cortex. Damage to Broca's area results in *expressive aphasia*. A person with expressive aphasia cannot express speech in fluid sequences. Instead, their speech is limited to short, telegraphic phrases. Interestingly, damage to this region also results in deficits in comprehending longer phrases both when hearing speech and during reading. Recent findings even extend this phrasing function to the production and comprehension of other motor sequences, even musical phrases.[7] This suggests that Broca's area is involved more generally in producing and comprehending sequential phrases and gestures, not just those used in language.

Initially Wernicke's area and Broca's area were thought to be connected by one major fiber bundle, which was known as the arcuate fasciculus – Latin for the *curved bundle*. However, research has shown that there are actually a number of distinct fiber tracts in the curved bundle.[8] I will therefore refer to them as the *curved bundles*. It turns out that the bundle now known as arcuate fasciculus does not project to Broca's area but rather to prefrontal spatial areas, suggesting that it is important for localizing sounds. Another fiber bundle appears to be important for alerting to sounds and for the emotional interpretation of sounds. Not surprisingly, this path includes connections with the amygdala. The middle longitudinal fasciculus appears to connect auditory comprehension and memory regions to motor planning regions in the parietal and frontal cortex. A fourth bundle appears to provide articulatory connections between the parietal and frontal motor regions, including Broca's area, suggesting that it plays a critical role in speech production. A fifth bundle connects frontal areas adjacent to Broca's area to auditory regions. This tract also includes connections with the insular cortex, suggesting that it is important for connecting feelings with words.[9]

Robert Seyfarth argues that the constant need to interpret social interactions was probably the underlying factor driving the evolution of these connections.[10] The idea is that many of these connections probably evolved originally to support the comprehension of status interactions, emotional signaling, and intention behaviors more generally. After all, none of these interconnected regions is unique to humans. The homologue of Broca's area in macaque monkeys is the area that controls mouth and hand movements. Thus, it is not surprising that it is involved in managing vocalizations and gestures in human communication. Other animals also have areas for associating sounds with emotional meanings. Vocal calls are used to communicate in social contexts in many species, and voice qualities are commonly used to identify individuals. Rather than there being some sort of special language module, language appears to have evolved as a method of social signaling which reuses these earlier social analysis networks. What humans appear to have added to this mix is a critical mass of connections which enables vocalizations to be readily imitated and connected with other forms social signaling.[11]

Though language begins as a means of social communication, once we learn to use it to manage the attention of others, we gradually learn to apply it as a tool to manage our own attention. We learn to talk to ourselves, and sometime after self-talk begins, we learn to talk covertly – that is, while holding the brake on vocal output. Running language programs covertly still activates the vocal and auditory connections in the curved bundles, and triggers word associations from memory, but it keeps the conversations private. Still, the conversational character of self-talk is obvious. As multitasking agents, we must adopt both speaker and listener roles in order to guide our own attention. When we instruct ourselves, we must consider our suggestions. When we admonish ourselves, we must pause and take the criticism seriously. We must literally participate as both speaker and listener for the narratives to be effective. Sometimes we even ask ourselves questions, "What should I do now?" as we work an idea. It seems unlikely that we would ever learn to do this well, if we didn't first learn to shift between speaker and listener roles in social conversation.

Those who believe that self-conversation lies at the core of conscious experience have failed to understand the bootstrapping nature of the processes that lead to language. Sign-guided attention is a unique part of human experience, but it is a late comer in the upward spiral of processes that manage attention. Gaze following and gaze directing gestures develop earlier and are important prerequisites for learning the meaning of words. Language doesn't make us conscious. We have it backward. Tracking attention and learning the meaning of words is one of those tasks that we can only accomplish when we consciously attend to the words and the contexts in which they occur. The ability to use language to guide our own attention develops slowly. Words must be associated with a broad array of perceptions, actions, and feelings before they become flexible tools for guiding attention, and it takes years more practice before we become skilled at using words to manage our own attention.

The enhanced chain of awareness made possible by guiding attention with words is sometimes called *narrative consciousness.* Initially, using words is not that different from guiding attention using deictic gestures as Tom does. However, the efficiency with which words can be associated with conceptual details greatly expands the complexity of the ideas which can be called to attention with words. Plans can be standardized using vocal phrases as guiding rules. Memory searches can be guided using highly focused verbal cues. Subtle states of experience and feeling can be labelled, for example, as *déjà vu* and then later called to attention. A reflective self with language can sequence memories together in stories. By connecting memories with verbal labels for dates and events, these associations become more salient in memory. This enables experience to be reconstructed in more detailed timelines. This narratively enhanced version of the historical self has been called the *autobiographical self.* It's a narrative self who can reconstruct parts of its own history in temporally connected stories.

"TOM WOULD LIKE SOME TUNA, COGLEY"

We won't go into detail on the kind of architecture that Cogley will need for vocal production, except to say that in order for it to be similar to that of his human partners, it should use movement sounds, or phonetic gestures as they are sometimes called.[12] Movement sounds may seem to be unlikely units for speech. However, they were apparently the algorithms that proto-humans brought to phoneme analysis when they first developed language. Analyzing movements from sound cues will require edge detection and timing algorithms. These can be used to identify event onsets and the timing between events. The analysis of speech sounds involves similar contact and timing measures. Motion analysis requires algorithms for assessing rising and falling pitches, the Doppler effects that accompany approaching and receding movements. In speech, rising and falling pitches often encode reactivity trends. Turbulence algorithms are important for locating flowing water or the hiss of rushing air that accompanies close movements. In speech, turbulence sounds are known as fricatives and sibilants. These sounds are distinguishing qualities in virtually all languages. Thus, there appears to be a good fit between the sound movement categories which are important to animals, and the phonetic categories used in speech.

The problem of parsing sounds into meaningful units is more complex than merely recognizing movement sounds. Just as visual perception networks build up representations of increasing complexity by using the percepts of earlier networks as inputs into later pattern analysis networks, sound analysis networks build up a hierarchy of representations and re-representations. The initial core areas for sound analysis in the cortex detect distinct pitches and noise qualities. These areas then project on to the surrounding auditory belt regions, where sound combinations of increasing complexity are detected. Processing in the belt regions, in turn, projects to the parabelt regions, which learn sequential combinations of sounds. Interestingly, the sound processing networks in the parabelt merge into the superior temporal sulcus, which is a region also known for detecting visual movement patterns. It appears that sequential sound combinations are learned in movement networks much like those that learn visual movement patterns. If Cogley is to have similar sound processing skills, then he will need core auditory processing networks for detecting distinct sound and noise features, belt networks for recombining the core features into more complex combinations, and parabelt networks which are capable of recognizing sequential sound combinations.

To help Cogley recognize sequences of sound, he will need an auditory analysis buffer which can hold about eight seconds of auditory input. The duration of this buffer will be largely responsible for how many digits (or short words) Cogley can repeat after they are called out. We have data on this ability for humans. Presented at a rate of one per second, this buffer will give Cogley a seven- to eight- digit memory span, which is about the best that humans can do. More importantly, this duration will provide a buffer in which phonemes can be combined into longer phrases. This buffer will also provide a source from which sound phrases can be replayed, giving Cogley a second chance to make sense of them. We all have a need to do this from time to time. Cogley will need many second chances before he becomes skilled at recognizing a series of rapidly produced movement sounds as meaningful units.

Interestingly, research suggests that just as macaques have separate motor channels in the basal ganglia for gaze control and executive biasing processes, they also have a separate channel for mouth and hand control. This channel returns signals to ventral premotor areas that are homologous to Broca's area. It obviously evolved before well before human language developed, but it is no doubt largely involved in language production and comprehension in humans. Cogley will also need a separate channel in his mid-level motor decision networks dedicated to managing complex sequences of vocal and hand movements so that vocalizations and gestures will not have to compete with the planning of gross motor activities. The overlap between vocalizations and gestures means that Cogley should also be able to learn a gestural sign language, if his caretakers communicate in that manner. Given that this region has also been associated with the comprehension and production of musical phrases in humans,[13] it appears we may need to extend the bounds of these networks more generally. If Cogley is ever going to learn to understand musical compositions or play a musical instrument well, then he will surely need the extended phrasing skills provided by these networks.

To ensure that Cogley's speech production is closely linked to his sound analysis skills, he will need a set of reciprocal connections between his sound analysis networks and his articulation networks so that each can influence the other. This should help him learn to imitate sounds. These will be Cogley's analog to the curved bundles. To be roughly comparable to humans, it seems that one bundle should be dedicated to pathways supporting auditory localization. A second should be dedicated to mapping sound analysis information to mid-level emotional voice quality detectors. This will enable him to extract emotional messages from vocal signals. One more should be dedicated to mapping auditory percepts to memory areas. Another should be dedicated to mapping auditory percepts to early motor planning regions. Yet another should be dedicated to mapping these motor planning regions to articulatory networks. This combination will allow auditory cues to be linked with both general motor planning and with articulatory planning networks. Yet one more should map auditory processing and feeling networks to articulatory motor planning regions. This latter bundle should enable Cogley to superimpose subtle emotional signals onto his voice. All these connections should be bidirectional so that input and output processing regions will have synergistic effects on each other.

To further facilitate imitative word learning, Cogley should go through an initial babbling stage. This should essentially be a vocal play phase where he learns about the affordances and effectivities of the sounds he produces. The developmental function of this phase should be to help him learn to coordinate his sound production networks with his sound recognition networks so the he can learn how to copy the phonemes he hears. As with human infants, once Cogley begins to produce phonemes reliably, the association of particular phonemes with social attention should bias him toward producing more of the same phonemes that his caretakers do and fewer of those they don't. This will reshape Cogley's sound production repertoire to fit the sounds that his caretakers use in their language. Human infants adjust their speech production and detection skills to fit the language they are learning, an outcome that Patricia Kuhl refers to as the *magnet effect* of a language. Cogley should follow the same strategy.[14]

The magnet effect will fine tune Cogley's initial sound-movement algorithms to ensure that he is better prepared to detect and say the sounds used in the language of his caretakers. This means that a Cogley raised in a different language environment could learn to speak a language using different combinations of phonemes with similar facility. The human speech production networks are most flexible in childhood and become much less flexible around puberty. This means that the magnet effect is harder to override as a child gets older. If we duplicate this arrangement in Cogley, this means that if he learns a second language later in his development, then he will be more likely to speak it with an accent shaped by the magnet effects of his first language. For some reason, this possibility seems particularly suitable. Speaking a second language with an accent will surely result in Cogley being considered all the more like his human partners.

LINKING SIGNS WITH SCHEMAS

If words are to have meaning to Cogley, then once he is capable of imitating sounds, his learning and memory algorithms must be able to link those sounds with other aspects of his experience. Fortunately for us, MIT computer scientist Deb Roy has provided a detailed theoretical framework for how this can be done in robots.[15] Roy's semiotic schema theory is the result of a series of attempts to train robotic systems to understand words. One of his most successful robots to date is a manipulator arm robot called Ripley, which associates words with objects and actions. Ripley can categorize object features in his field of vision and learn categories of actions which his arm can perform. He can also learn to respond to verbal requests about objects and actions. For example, if a partner makes the request, "Give me the green apple," Ripley can map the words to recognition categories, locate objects that match his memory of apple shapes and color categories, and then construct an action plan to pick up a green-colored apple shape and pass it to the location of the person making the request.

The algorithms enabling Ripley to recognize objects and actions, and to associate words with them, are nested and complex, but the underlying principles can be stated more simply. Sensory inputs are mapped from the physical properties which can be detected by Ripley's sensors into internal categories in a series of increasingly complex re-representations. The mappings begin with low-level sensory transforms that transduce external cues and align them with internal signals and parameters. Subsequent processes use other transforms to re-combine the signals into multimodal sets. Categorical transforms then map ranges of these multimodal sets into functional categories. Colors would be one set of function categories for perception. Shapes would be another. More complex categories would be based on combinations of simpler categories within certain ranges. For example, an apple might be identified by a certain range of color, shape, and hardness categories. In a somewhat reverse process, action categories like reaching are mapped into combinatorial effector sequences, which can be targeted to particular spatial coordinates and adjusted based on sensed contact and force parameters. Targeting functions and indexical parameters serve to transform general motor-program categories into discrete actions with directed and predictable effects.

An important aspect of this process is that the transforms at each level come to be associated in larger perceptual-motor units based on experience. These extended associative units are commonly referred to as *schemas*. Schemas are similar to the functional neural circuits, which Donald Hebb called cellular assemblies.[16] We've already encountered many cases of such schemas. Motor schemas are the network assemblies that form as action plans are organized. Recognition schemas resolve perceptions in a web of semantic associations. Interconnections in the thalamus bind perceptual, motor, and interoceptive schemas together in extended schematic associations that compete for attention as functional groups. Minded systems plan by detecting features and objects and linking them into functionally connected schemas. In fact, whenever elements of a schema are encountered, they tend to reactivate the perceptual, motor, and interoceptive schemas with which they are associated, and those schemas then compete in combination with others for access to attention. Those combinations with the best supporting affordances, need states, and perceptual associations tend to win the competition. In this way, schemas provide a framework for resolving – that is for thinking about, what to do in a situation. Both environmental contexts and internal states provide part of the cues needed to disambiguate the best-fit schemas in a situation.

As Michael Tomasello notes in the quote that opens this chapter, it seems clear that "language emanates from a more fundamental set of social-cognitive skills." Jerry Fodor described this more fundamental skill set as a *language of thought.*[17] However, as Deb Roy's model explains, the process of thinking by activating and resolving schemas does not require language. Once words come to be associated with the certain aspects of perceptions, actions, and feelings, they naturally map to units that are parts of various sensorimotor schemas. This mapping enables a word to activate one or more schemas, much as encountering some elements of a schema would. Rather than requiring a new system of thought, words map into the schema-based system of sensorimotor thought that precedes them. This is what enables Ripley to use words to resolve action plans. All that is required is for a few words to call attention to key features of a perceptual unit, or an action plan, so as to trigger the reconstruction of related sensorimotor schemas.

An obvious advantage of mapping words into schemas is that an entire thought doesn't need to be described with words. In fact, mapping words into schemas means verbal communication can begin with one-word utterances, as it does in humans. It also explains why Tom can communicate in similar ways without language. He simply uses gestural signs to influence what schemas I consider. Adaptive agents are constantly analyzing their environments, activating potential schemas, and considering new action plans. Signs produced by another agent simply add to the mix of situational cues activating certain schemas, and because social agents tend to pay special attention to schemas triggered by their partners, a few well-placed words can often bias their attention to particular schemas. For example, if Cogley hears the single word "coffee" and also notices that his partner's cup is empty, then he is likely to resolve that one-word utterance as a request for coffee. Indeed, his partner might produce the same effect by looking at the coffee pot and then pointing to his empty cup, much as Calamity looked at the chicken and then nodded at her bowl. Of course, a longer string of words has the potential of specifying a schema in more detail. However, even then context usually plays an important role in disambiguating alternatives. For example, the presence of a coffee decanter and a cup greatly increases the chance that a schema for pouring coffee will take form.

One of the unique effects of activating multiple schemas together is that in many cases, new schemas can be formed by establishing best-fit connections among components associated with different schemas. In fact, blending is a natural consequence of activating assemblies for several schemas together and then resolving best-fit combinations for the current situation.[18] Language facilitates this process because it enables components associated with other schemas to be called to attention flexibly. For example, if Cogley hears the phrase, "Pour the coffee onto the table," the words have the potential to guide him to connect schematic features learned in different schemas, "pour the coffee" and "onto the table," into a novel arrangement that would have been unlikely to occur to him without such verbal guidance.

Schematic organizations also provide insight into how words gain meaning. Earlier we noted that the meaning of a sign is grounded in its ability to mediate a functional outcome between producer and consumer agents. However, we didn't specify the internal structure of that grounding. Building on

Roy's semiotic schema theory, we can now expand that model. Sensory inputs are meaningful when they can be mapped to perceptual recognition categories – affordances, percepts, and objects. On the output side, action categories are meaningful when they can be mapped to a sequence of effector actuations that produce changes in the world or that change the agent's location. Sensory inputs are also meaningful because they bind perceptual experience with specific spatial, temporal, and force parameters.

Roy calls these sensed parameters *indexical* bindings. Effector outputs, when they are generated, use indexical parameters to target actions to relevant spatial-temporal locations and to adjust actuator forces. In this manner, indexical associations serve to ground actions in relation to the spatial, temporal, and force parameters that must be engaged in real-world activities. Indexical words can also be mapped to these parameters. Roy makes a valiant attempt to link these indexicals with the indices in Charles Peirce's semiotic theory.[19] However, Roy's indexicals don't seem to map into Peirce's indices very well. Rather, they correspond more closely to the control parameters which David Bailey found to be essential in defining x-schemas for actions.[20] Grounding meanings in external sensors and perceptual categories is similar to the idea of grounding originally proposed by Stevan Harnad.[21] However, Deb Roy's semiotic schema framework greatly extends the grounding domains.

The diagram in Figure 15-1 is my attempt to connect Roy's semiotic schema model with the cognitive organization that we have proposed for Cogley. The labels within the ellipses are drawn from Roy's schema theory model, and those outside the ellipses are functions we have proposed for Cogley. Given that attention and interoceptive associations are not well defined in the current version of Roy's schema model, the ellipses in the middle and bottom of the diagram can be considered implied extensions of that model. This extended model provides a broad framework for grounding schemas and words in both external and internal functions. Note, for example, that the perceptions and actions, which fit together in schemas, are grounded in feature-specific spatial, temporal, and force parameters; feature-connected objects; action programs; memory co-occurrences; and even in networks representing feelings and intentions.

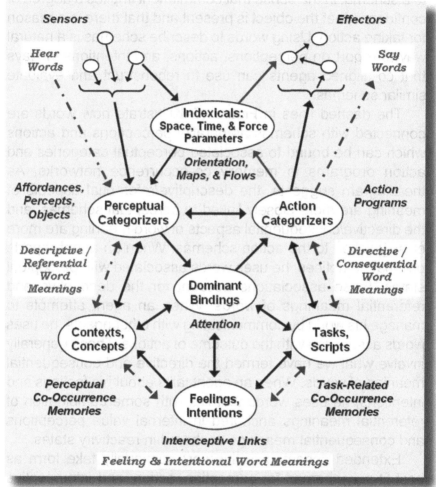

Figure 15-1: A schematic grounding model.

As Deb Roy notes, the resolution of best-fit schemas essentially involves resolving beliefs, in the sense that beliefs are states of confidence that particular features are located in the environment, and that schematically connected actions can interact with them. Similarly, resolving a schema implies an intention to produce the changes associated with the selected action, because expected changes are part of the features used to resolve best-fit selections. Only schemas that are considered both possible and desirable are likely to gain dominance. In effect, proto-beliefs and proto-desires are part of committing to a schema, in the sense that commitment implies a degree of confidence that the object is present and that there is a reason for taking action. Using words to describe schemas is a natural way to report on perceptions, actions, and intentions in ways that co-aligned agents can use to reconstruct and evaluate similar schemas.

The dashed lines in Figure 15-1 illustrate how words are connected with schemas as parallel perceptions and actions which can be bound to associated perceptual categories and action programs in memory co-occurrence networks. As the diagram suggests, the descriptive/referential aspects of meaning are more closely linked to perceptual schemas, and the directive/consequential aspects of word meaning are more closely linked to the action schemas. When an agent reports on what he notices, he uses words associated with perceptual states, those associations account for the descriptive and referential meanings of words. When an agent attempts to manage his world by communicating with other agents, he uses words associated with the outcome of actions. These generally involve what we have termed the directive and consequential meaning of words. When an agent talks about his feelings and intentions, he uses words that link with some combination of referential meanings anchored in internal value perceptions and consequential meanings anchored in reactivity states.

Extended schemas of this sort continually take form as best-fit subschemas for perception, action, and interoceptive states gain attention in interactive combinations. Cogley's proposed architecture for assembling schemas in states of attention is more complex than the low-level schema building processes proposed for Ripley, but the basic strategy is clear. By drawing elements from layers of past perceptual, motor, and interoceptive schemas; by applying best-fit selection algorithms;

by employing a central architecture to focus attention; and by allowing a steering committee of distributed knowers to bias attention, Cogley should be able to assemble extended schemas of a complexity rivalling those of his human partners. Once Cogley has learned to associate words with components of his schemas, other agents will be able to influence what schemas gain his attention simply by using words to call attention to specific schematic features.

In a similar manner, by saying words associated with the schemas he is considering, Cogley will be able to communicate about his perceptions, actions, and intentions to other agents. However, if Cogley is to communicate with Tom, then he will also need to learn to pay attention to gestures. Tom only understands a few of the words that I use with him, "tuna" and "mayonnaise" being some of his favorites. However, he does attend to actions and often uses them to assess intentions. Tom uses his own repertoire of calls, gestures, and intention movements, many of which have become standardized, to communicate his intentions. To communicate with Tom, Cogley will have to be able convert his schemas into gestures which Tom recognizes, and he will have to learn how to map Tom's gestures into his schemas.

Gestures will also be important tools for communicating with Cogley's human partners. Human speakers commonly use gestures such as points and nods to direct a listener's attention. Exaggerated facial expressions and preliminary actions are also often effective. Unless they are quite extreme, as in pantomime, the gestures alone don't convey a full message. However, they add to the mix of cues used to disambiguate best-fit schemas, and they provide a secondary channel to emphasize the associations suggested by verbal dialogue. We'll want Cogley to learn to use this secondary channel too. For one thing, it will clearly help him communicate with agents like Tom who lack many verbal skills. In addition, humans often find the added information provided by gestures useful, and they tend to rate speakers who gesture as more interesting and better communicators. We'll want Cogley to be a good communicator, so gestures will be an important addition to his communication repertoire. In fact, we should arrange for him to begin using deictic gestures like looking, reaching, and pointing prior to learning language and encourage him to continue using them as secondary communication channels as he acquires language.

There are two additional communication skills that Cogley must master in order to use speech as humans do. The first of these is conversational turn taking. This appears to occur on two levels in humans. First, an agent needs to be motivated not only to share his ideas and feelings with others, but also to be interested in the ideas and feelings of others. To encourage this cooperative interest, Cogley's curved bundles must connect his language recognition networks with social feeling areas so that caring feelings will promote attention to the actions and signs of others. Cogley will also need connections between his feeling areas and his speech production areas to motivate him to share his feelings and intentions with others. To make this turn-taking process more efficient, Cogley's vocalization networks should be arranged to switch his attention between speech production when he is sharing, and speech comprehension when he is listening and trying to cooperate with others.

Research suggests that several thalamic nuclei play a role in sensitizing attention to speech and promoting linguistic processing.[22] There is some evidence that the ventrolateral nucleus of the thalamus facilities motor production related to speech. This suggests that it may play a role in gating attention during speech production. Activity in the centromedial intralaminar nucleus appears to facilitate memory and linguistic processing more generally. It therefore appears to play a role in gating attention during speech recognition. Based on these observations, it seems reasonable to provide Cogley with a vocal-motor module in his central attention channel that will serve to gate his attention toward speech production when he is talking, and a speech processing module that will gate his attention to sensory signals and their memory associations when he is thinking and actively listening. Presumably these two modules should have reciprocally inhibitory connections between them, so that only one tends to dominate at a time. These gating strategies should help Cogley shift between speaker and listener roles.

There is one remaining language skill that Cogley will need to master: self-talk. As he becomes skilled both at using words to guide the attention of other agents and at being able to allow the words of other agents to guide his attention, he will reach a point where he becomes capable of alternating between speaker and listener roles and guiding his own attention by talking to himself. When this happens, we will want to encourage him to use this strategy, especially as a means of problem solving. However, as he becomes more skilled at this self-talk strategy, we will want to introduce yet one more variation to the process. As I recall, my daughter reached this stage when she was about three years old. At first it was entertaining to listen as she talked to herself, but it soon became distracting to hear her every thought on a daily basis. Then, in a moment of inspired distraction, her mother made a simple suggestion. "Carolyn, when you talk to yourself, you don't need to talk out loud. Talk covertly. That means, talk quietly to yourself." From that point on, whenever Carolyn began to talk to herself out loud too much, we would gently admonish her with one word, "covert." When Cogley begins to talk to himself, we'll surely feel like proud parents, but we'll soon want him to learn to talk to himself covertly.

CONCLUSION

There are still many aspects of the networks supporting language which we don't understand well. However, I believe we have enough insights into their functions and connectivity to design Cogley so that he shares much of the flexibility of human language. Analogs to the core and belt regions of the auditory cortex will provide Cogley with an analysis hierarchy for parsing sounds into more complex combinations. Connecting features resolved in the belt networks with parabelt-like sequence-learning algorithms will enable him to recognize a series of

313

sounds as a unit. Having a number of curved bundles connecting the networks which detect sound combinations with articulatory networks capable of producing phonetic gestures will enable Cogley to learn to say the words he hears. Connections between auditory recognition networks, articulatory networks, memory networks, and feeling networks will produce internal sources of meaning which can be linked to sensorimotor schemas. Sensorimotor schemas will ground the meaning of Cogley's words and gestures in real-world activities.

We have now reached a point in the upward spiral that markedly distinguishes Cogley's consciousness from Tom's. Up to this point the differences in cognitive function have not appeared that large. Even now, it is clear that Tom shares a schematic architecture and rudimentary elements of sign-guided consciousness with humans, and with the designs proposed for Cogley. However, language introduces a new level of control. Tom only has a small repertoire of signs and calls which he uses to manage my attention and the attention of his peers. He even recognizes a small number of the words I use with him. Words and gestures obviously influence the flow of his attention. For example, he recognizes his name and words for special treats, and he reacts when he hears those words. He even has his own recruiting and pointing signs, which he uses to gain attention and request items, and it seems clear that once he points to the tuna cabinet, his attention is focused on getting tuna. He doesn't give up easily.

All this illustrates that Tom has the capacity to allow gestures and referential words to guide his attention. However, Tom doesn't have the critical mass of curved bundles and sound-production networks needed to acquire a system of thousands of flexible signs to guide his attention. In that sense, Cogley's sign-guided skills will be leagues above Tom's sign-guided skills. And yet, the basic networks with which Cogley will recognize and feel experiences will not be that different from Tom's. What will be different is how much language will enhance Cogley's ability to manage attention. Language enables humans to direct attention in very precise ways and, in doing so, to develop increasingly complex concepts. We even learn to shift between speaker and listener roles as we talk with ourselves. Cogley should be able to acquire similar language skills. Having language won't necessarily make him more conscious than Tom, in the sense of giving him more of a feeling of what happens, but it will definitely give him more control over what features can gain his attention and trigger those feelings.

CHAPTER 16
AN INCREMENTAL STORM

The upward spiral of neural organizations in the brain results in a number of self-referential properties. These include the emergence of a resonant feeling self, a localized observer self, an anticipatory sense of self, a sense of agency, and a sense of location for that agency. However, this core spiral of organization is not enough to account for the individualized character of self. Minds are path-dependent entities. Anticipatory learning and reflective memories place aspects of self in the context of past experiences. Language associations enable the experiences of self to be bound in convenient historical narratives. Thus, over time individual histories alter the course of development as episodes of past attention loop back to influence what is learned and considered important in later situations. In effect, past activities become defining components of who an agent is and what he can do. To enable Cogley to acquire a similar individualized sense of self, he will need a developmental architecture that feeds back on his attention in an incremental manner.

AN INCREMENTAL STORM

It is the upward leap from *raw stimuli* to *symbols* that imbues the loop with "strangeness." The overall gestalt "shape" of one's self – the "stable whorl," so to speak, of the strange loop constituting one's "I" – is not picked up by a disinterested, neutral camera, but is perceived in a highly subjective manner through the active processes of categorizing, mental replaying, reflecting, comparing, counterfactualizing, and judging. – Doug Hofstadter, *I Am a Strange Loop*, 2007

I n programming, a recursive algorithm is a procedure which is applied repeatedly across cases to reach some incremental or combinatorial outcome. In some algorithms the interim results may even feed back to modify the flow of processing. Doug Hofstadter calls such recurring self-modifying algorithms *strange loops* because their dynamic behavior often gives rise to discontinuous and unexpected properties.[1] The reentrant processing cycles within the thalamocortical architecture of the brain are such self-modifying loops. Among the unexpected properties of these recurring loops is an emergent sense of self, which in agents with language comes to be identified as "I." However, this core sense of self is nested in various learning, memory, and guidance algorithms which supplement and modify the core loops. As a result, with experience, the "I" also grows more complex. This chapter looks at this developmental process, in particular its historical character, its ability to build new concepts and skills, and the effect this incremental process has on thought. The aim is to determine what kind of architecture will be needed to ensure that Cogley can grow his concepts, his skills, and his awareness of the world in an incremental manner.

Hofstadter suggests that the "I" appears as a stable whorl in an otherwise strange loop of inner-directed categorizing. However, he's not quite sure if we should consider it to be real. He labels it variously as an epiphenomenon, a symbol, or merely an abstraction. Still, he argues that it is a necessary, inevitable, and highly useful abstraction. The model of mind I have presented is consistent with the characterization of self as an emergent product of strange loops. In fact, one of my goals has been to make the loops in the brain a little less puzzling by showing how a series of re-representations underlying interoceptively guided attention creates many subjective properties of self. However, I claim that just because the conscious "I" has discontinuous properties, it should not cause us to treat it any less real than any other emergent phenomena. If the properties of mind seem strange and disconnected, it is because we are not accustomed to thinking about the self-organizing properties of recurring strange loops. Perhaps the best way to emphasize this point is to introduce a metaphor that makes contact with an organization built on loops and whorls.

The metaphor that I find most useful for thinking about the unity of self is that of a cyclonic storm like a hurricane. Following this metaphor, the upward spiral that gives rise to the core sense of self begins as an infant updraft, an inquisitive spiral of attention which interacts with events in its world. Without well-formed percepts and motor skills, this spiral has little initial force. However, as its perceptual and motor experience grows, instances of its own attention begin to feed back into the spiral, and the storm gains in intensity and focus. In a short time, like the self-sustaining spiral of a hurricane, it forms a distinct central eye ("I").[2] We can name it. We can follow its development. As it matures, we marvel at its effects on the world and speculate about its future trajectory. However, as we do this, it is clear that the storm is more than a mere abstraction. Its history of learning and memory has shaped its attention and given it an individualized character which grows as it engages other tasks.

Because its decisions are tied to motivational forces embedded in a physical body, its perceptions and actions have real effects on its world.

The hurricane metaphor is not just an artistic tool; it's meant to bring home the point that a recurring loop defined largely by its own momentum can nevertheless become a coherent organization with a distinct identity. Some remain convinced that it is more scientific to apply a reductionist framework to thinking about mind, and when they do so it is tempting to claim that the self is nothing but a coalition of parts. However, every emergent phenomenon is composed of parts. Yet when the parts come together to form a new organization with discontinuous properties, that organization cannot be explained as "nothing but" the parts alone. No one would argue that hurricane Katrina was an epiphenomenon, just because we know something about the forces that enabled it to form. In that regard, a storm of attention must be considered even more real than any atmospheric storm, because its enabling forces are so much more complex and can persist for so much longer. A storm of attention is not just a one-dimensional loop. It results from the cumulative effects of loops and swirls on many levels. Feelings anchored in biological survival values and rising dispositions of emotional reactivity are core driving forces in the storm, but learning and memory provide the storm with individualized concepts and skills which influence its trajectory.

So how should we think about the growing momentum of a storm of attention and its individual character? One starting point is to recognize that a storm of attention is a historically defined entity whose history influences its trajectory. It bootstraps its own development by building on past interactions. This historical character brings to mind Stuart Kauffman's concept of the *adjacent possible*.[3] As Kauffman characterizes it, at the fringe of every organization there are potential features which are but one link away from becoming a functional part of the organization. And when one of those links takes form, the possibilities for subsequent growth also change. Some possibilities that were not likely before become more probable, and other possible changes become less likely. In effect, the opportunities for development expand and change as a system evolves. Minds emerge in exactly this way. Their trajectory is dynamically assembled as properties, acquired in the past, are applied to meet the challenges of the present. Each developmental change, each new concept and skill, modifies the potential for future developmental changes.

Many properties of self depend on such historical loops. The resonant self emerges from a coalition of parts with a special evolutionary history. That history has led to neural organizations that bind perceptions and actions with feelings grounded in biological needs. The species' evolutionary past has fed back to influence what status conditions it values and what features are most likely to gain its attention. A more personal sense of self begins to form as individual episodes of attention begin to shape the development of each mind. As a recurring cycle of self-modifying attention begins to interact with its world, it learns to attend to feature sets that help it complete tasks. In doing so, it gradually discovers new conceptual units of organization, learns new skills, and acquires new values. As this happens, it changes what assemblies come to engage its attention and guide its decisions. Over time, it finds opportunities for action in more complicated situations, the values that guide its activity become more refined, and it takes on more complex tasks.

In addition to skill and concept learning, the accumulation of declarative memories is another historical process which significantly modifies the experience of self. Declarative memories provide an inner world of past experience that is separate from, but connected with, the external perceptual world. This inner world allows experiences to be evaluated in a web of semantic and episodic associations. However, this is a stage of awareness that can only be assembled slowly. Just having a neural organization for storing memories is not enough. A developing mind needs to acquire a rich history of co-occurrence associations before past experience can be a useful guide to revaluing ongoing perceptions. As a result, it takes an extended period of development before robust states of reflective and episodic consciousness emerge. However, as experiences gradually build up, developing agents acquire rich sets of connected memories. Over time, an agent who can store memories of its activity and reflect on them at later times gradually comes to be guided, in large part, by memories of its own history of attention.

Social memories in particular have a profound effect on the self. They place an agent in a web of social roles, group tasks, and historically defined relationships in which he or she is judged in relation to his or her compliance with the attitudes of others. Given that signals of social status map to core interoceptive feelings, the feelings of acceptance and social value which result from social interactions are central to an agent's sense of self-worth and status. However, social interactions provide more than just feedback about status. Social groups provide the context in which social agents grow and develop, beginning with family and extending on to larger groups as an agent matures. In human society, these contexts include not just other agents but cultural artifacts – the physical structures, tools, and customs of the cultural group. This means that not only will Cogley's history of learning and memories influence who he becomes, but so too will the culture that he develops within. In short, the loops that shape his sense of self will extend far beyond his neural architecture. Our goal in the next few chapters is to define an extensible architecture that will enable Cogley to engage these extended developmental and cultural loops. We'll need another biologically inspired mechanism to provide us with a guide for this design, a strange loop that builds on the historical loops of attention that run within it.

SENSORIMOTOR LOOPS: CHUNKING MOTOR SKILLS

The term chunking was introduced by George Miller to describe a strategy which allows humans to overcome their limited short-term memory capacity.[4] Miller described the memory limit as "seven items plus or minus two." However, he noted that the limit could often be overcome by linking memories in hierarchical chunks and decoding them separately. For example, suppose there are seven teachers at a preschool, and suppose we associate each teacher with seven children from the class. Although we can rarely hold more than seven items in memory at one time, by going through the set of teachers and recalling the seven students associated with each one, we can effectively bring many more than seven students to mind. Because items in memory are connected in co-occurrence hierarchies involving distinct elements and contexts, using items at one level as retrieval cues for items on another level enables us to explore memories in associated chunks. Although the chunking concept was introduced to describe hierarchical functions in declarative memory, it seems obvious that we also develop hierarchies of skills and concepts, so we need some idea of how these hierarchies are formed.

Over twenty years ago Rodney Brooks proposed building up skills in robots using a layered hierarchy in which new skills were formed by combining previous skills to serve new functions.[5] Brooks demonstrated that flexible robots could be developed by beginning with a few primitive functions like sense, move, turn, and remember and then "evolving" more complex functions by adding new layers of more specific control. Brooks referred to this methodology as a *subsumption architecture*. He referred to the different control layers as *levels of competence*. The functions in the higher levels of competence were built by adding more specific rules and by using previously acquired functions as the building blocks for more complex tasks. For example, an *avoid* function could be formed by establishing a rule in which a particular class of sense outcomes would be treated as obstacles that engage turning away. Then a *wander* function could be developed by combining the abilities to turn and move spontaneously. If the robot encountered an obstacle while wandering, then the avoid function would automatically be engaged. Subsequently an *explore* function could be formed which used the ability to wander and remember, and which treated areas that had recently been entered as something to avoid whenever possible.

The subsumption architecture design strategy has been a

highly successful approach for developing semi-autonomous robotic systems capable of moving about independently and avoiding obstacles. In fact, this architecture is the foundation for many modern robotic systems, including NASA's rovers. However, it is less obvious what needs to be done to extend this architecture to more cognitive tasks. In particular, how do we design a cognitive architecture that automatically extends its levels of cognitive competence? Based on our previous overview of the lattice hierarchy, it should be clear that natural selection has incorporated a number of secondary selection processes in the cognitive layers of the lattice. Learning, memory, and reasoning processes introduce new levels of competence. Many robots have domain-specific learning modules. However, learning is usually at the top of their cognitive hierarchy. Cognitive processes based on semantic co-occurrences, episodic memories, and reasoning are largely absent from current robotic designs.

One reason that these higher levels of competence are absent from most robots is that it takes longer for modules at these levels to acquire useful skills. Like many robots, three-year-old humans are fairly good at navigating around obstacles to reach interesting targets, even on fairly difficult terrain. However, despite their navigational skills, we don't count on the quality of their reasoning or memory skills to guide decisions in difficult situations. Three-year-old children, or equally skilled robots, aren't ready to cross streets on their own. They need yet higher-order skills to reach that level of competence. To get around the need for long training times in robotic systems that must make reasoned judgments, programmers either try to add pre-established "expert system" modules to them so that the robots don't need to spend time acquiring those skills, or they have the robot relay their observations to a human partner so that the high-level decisions can be made by a human manager. This was the strategy we used with SARG, the semi-autonomous roving geologist we considered earlier. It seems likely that this strategy will continue for some time. After all, if we could build a robot perfectly matched to its human partners for learning and reasoning skills, it would still take it eighteen years of formal training, and some post-graduate study, before we would trust it to make many high-level decisions.

Assuming, however, that we can eventually shorten a robot's training time by starting each new version with low-level training acquired in previous designs, it seems reasonable to speculate on how this process works in humans and what Cogley would need to duplicate its essential features. How the brain's secondary selection processes work together to support conceptual growth is not well understood, but several lines of evidence suggest that the interface between the basal ganglia and the cortex is an important part of this process, so a brief review of this architecture may help. The input side of the basal ganglia, the striatum, not only receives inputs from cortical areas but also receives sensory, motor, and reactivity inputs from the thalamus, the extended amygdala, and other subcortical brain centers. In fact, it is fully capable of acting without cortical input. There are, however, some limitations to the subcortical recognition processes. They recognize opportunities based on feature matches – the more matching the feature elements, the better the fit. This is largely the recognition algorithm by which habits are managed. When feature combinations reach sufficiently good fits, the habit is automatically favored. No additional planning is required.

This subcortical learning process is not very clever at assembling new action plans. The chain of affordance detectors, effectivity evaluators, and motor-program networks in the cortex is much better at coming up with new feature combinations. As a result, more complex actions often require the guidance of cortical planning networks before they can be assembled. However, once action plans are organized by the cortical networks, the elemental pattern-matching algorithms in the striatum can often learn cue combinations that predict them. For example, when a young child first learns to tie her shoelaces, she must give the task her full attention. Only after many months of practice does the task begin to become more automatic. In such cases, it's fair to say that the planning circuits in the cortex guide the habit learning that occurs in the striatum. We saw this process earlier when SARG began to anticipate the commands of his human manager. Like the learning systems in SARG, because the striatum processes information faster, over time it can often learn to anticipate many routine plans, enabling it to take the lead in organizing sequences of routine actions as habits.

MIT neuroscientist Ann Graybiel suggests that the cortex, striatum, and pallidum have come to function as a massive three-layer learning network in which the middle layer, the striatum, learns to map cortical inputs to output drivers in the pallidum, with dopamine providing the critical learning signal for both the striatum and the sequential planners in the cortex.[6] Because the striatum receives continuous feedback as actions occur, it can treat ongoing actions as elements of the input pattern it uses to predict subsequent actions. Graybiel argues that this enables the striatum to anticipate sequential action plans and to chain them together in behavioral chunks. Over time the chunks even come to be treated like a single action pattern, just as the separate activities that go into tying a shoelace become a single operation. In the case of more complex tasks, the chunks are assembled in a loose hierarchy of self-supporting chains that flexibly map together with little added planning, as in the case of the activities needed to fix a salad. There are several subtask steps, and different actions are required for the varied items in each step, but the steps flow together so well that we tend to consider the combination a single-chunked task.

If the story ended there, we would describe this master-apprentice architecture as a highly efficient sequential learning system. However, much more is happening here. The basal nucleus of Meynert is the primary source of cholinergic inputs to the cortex. Research has found that simply pairing a particular sensory cue or motor action with the release of acetylcholine from this nucleus results in the reorganization of perceptual and motor networks in the cortex, so that more cortical processing space is allocated to the cues and actions encountered.[7] Think of it as practice learning. The more you use particular cues and actions in a task, the more cortical processing is allocated to them. The basal nucleus is located near, and receives inputs from, the basal ganglia complex.[8] Thus as the striatum learns to operate in larger action chunks, the motor-planning networks in the cortex will gradually be stimulated to reorganize their interpretations of the actions various affordances support. This reorganization appears to train the cortex to plan in larger chunks. As a result, the cortical inputs to the striatum gradually begin to have more complex meanings. Over time, this may even enable the mater-apprentice architecture to reorganize in yet larger chunks, as new chunks are built from combinations of previous skills.

Perhaps it's simpler to consider how this chunking process would work as you manage SARG, our hypothetical lunar rover. As the rover's manager, you routinely run three standard tests whenever you encounter a certain type of rock formation. In a short time the sequential learning algorithms in the rover discover this pattern, and SARG begins to treat the three tests as a chunk associated with that particular rock formation. Soon, you find that you don't have to instruct SARG about the need for three tests, because he performs those tests routinely without added instruction. In effect, SARG's testing skills have become chunked, and your plans begin to assume that he will engage that chunk, let's call it Test Suite A, for all such formations. As SARG encounters other formations and learns other test suites for them, your oversight is no longer needed for planning these low-level test sequences. You can move on to planning based on test suite units. In fact, some of your plans may involve combining test suites A and B in the presence of certain hybrid formations, and SARG may begin to recognize those formations and come to use the larger testing chunks with them, too.

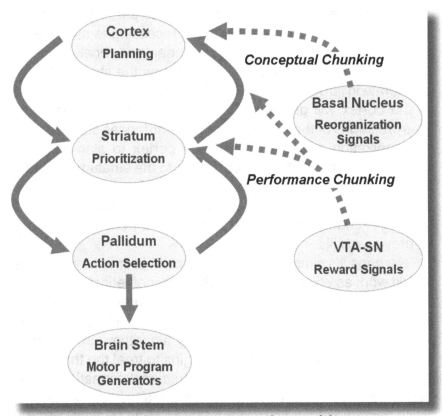

Figure 16-1: A master-apprentice architecture.

We know this sort of action chunking occurs repeatedly in human development. Initially, young children practice long and hard just to learn to form recognizable letter shapes. Yet they gradually master each letter and begin to assemble them into words. In time, they learn to connect words into sentences and sentences into paragraphs. By that point, however, the task of forming letters and combining them into words is no longer something they need to plan consciously. It has long been automated into well practiced habits by networks in the basal ganglia. It seems clear that many skills develop in this incremental manner as subskills become automated into efficient units of well-practiced actions and recombined into more complex repertoires. Connecting sequences into chunks and learning to plan using chunks is an elegant solution to the process of incremental growth. Initially, the planning of the master (the cortex) guides the learning of the apprentice (the basal ganglia), but as the skills of the apprentice grow, feedback from the apprentice begins to modify how the master plans.

Assuming this analysis is roughly correct, then the ability of vertebrate brains to develop an incremental hierarchy of skills results, in large part, from interactive loops between two self-modifying networks with different decision making algorithms (see Figure 16-1). One learns to plan complex actions based on affordances and motor-programming states. The other learns to organize commonly planned action sequences into chunks based on simpler cues. The plans organized by the cortex are more complex than those assembled by the elemental feature-matching algorithms in the basal ganglia, but once instructed, the faster operating basal ganglia can often learn to anticipate many of those plans and automate them in sequential chunks. As the cortex receives feedback about this chunking, it also receives reorganization signals from the basal nucleus. These signals appear to reorganize the cortical planning networks so that they gradually come to associate the affordances detected in the cortex with the larger output chunks now in use. Planning in larger chunks enables the cortex to take on more complex tasks, where the cycle begins again as many tasks become automated in yet larger chunks and chains.

If Cogley is to employ this strategy, then he will need a similar master-apprentice architecture linking his high-level action planning networks and his mid-level action selection and chunking networks. Most of this design has already been described. However, we need to make sure that his mid-level decision networks, Cogley's counterpart to the basal ganglia networks, are designed to use outputs from previous actions as sequential inputs for selecting the next response. This will ensure that response chunks can be easily learned. Temporal flow inputs from temporal coordination networks, Cogley's version of the cerebellum, should also contribute to this process. Further, we'll need to ensure that Cogley's high-level motor-planning networks, his counterpart to the parietal and frontal cortices, get regular reorganization signals so that the chunks formed in the mid-level decision networks will influence how he plans.

COGNITIVE LOOPS: CHUNKING CONCEPTUAL SKILLS

We know even less about how incremental cognitive hierarchies form in the brain, but the chunking of motor tasks is probably a good model for how they develop. However, to make this motor model fit, we will have to expand it to accommodate other kinds of decision loops. As noted earlier, work with macaque monkeys has identified a number of parallel processing channels which run between the cortex and the basal ganglia.[9] These separate loops enable decisions and actions in one channel (e.g., gaze planning) to run in parallel with those in another. However, given that these channels involve influential subsystems in the brain, and given their potential for cross bindings in the thalamus, it is also the case that activity in one channel often comes to influence activity in another. For example, the topics of gaze planning often ends up biasing motor plans, and vice versa. Thus, the separate channels can be thought of as semi-independent topic subcommittees which can influence each other's decisions and even those of the entire steering committee.

As such, it's seems appropriate to look more closely at the roles of these topic subcommittees. Evidence now indicates that there are three largely segregated motor channels in the basal ganglia. One returns outputs to the primary motor cortex and is likely to be involved in general motor planning. Another returns outputs to the supplementary motor area (SMA), a region that appears to coordinate planning activity with motor output decisions in the basal ganglia. A third returns outputs to the ventral premotor-planning region, which in the macaque monkeys includes the region that manages hand and mouth movements. This region is homologous to Broca's speech area in humans. It is thought to be involved in managing language phrasing in humans. In addition to these motor loops, there is an oculomotor loop for managing gaze decisions. Although this loop may have begun as just another motor channel, it has come to have effects that extend far beyond motor control. Shifts in gaze direction, or merely planning such shifts, result in gain modulation effects which strongly influence what feature assemblies gain attention.

In addition to these channels, there is another channel that projects back to the high-level visual perception areas in the inferior temporal cortex. This loop appears poised to bias decisions based on visual features and probably also plays a role in biasing memory recall based on visual features.[10] As we have previously discussed, there are three executive biasing loops involving the prefrontal cortex. One of these executive loops returns outputs to the dorsolateral prefrontal cortex, a region involved in biasing attention for spatial locations and spatially anchored cues in the attention steering committee. The second returns outputs to the lateral orbitofrontal cortex. A third returns outputs to the medial orbitofrontal cortex and the anterior cingulate cortex. Activity in the medial orbitofrontal cortex tends to involve goals and positive reward associations, whereas activity in lateral orbitofrontal cortex appears to involve inhibitory control, conditional planning, and aversive associations.[11]

The model of motor decision making that we have proposed suggests that inputs from the cortex and subcortical structures project to the striatum, where alternative candidates are weighted based on planning inputs, interoceptive needs, and past learning. The striatum then sends these weighted alternatives to the pallidum, via direct and indirect paths, where the alternatives compete for selection via reciprocal inhibition. Selection is managed as outputs from the pallidum loop back to the striatum and the cortex. As one choice begins to dominate, it inhibits its alternatives more than they can inhibit it, thereby winning the competition and releasing the brake. Because all the channels in the basal ganglia appear to employ this same processing architecture, it seems clear that the basal ganglia circuitry doesn't merely make best-fit action decisions. It makes best-fit attentional, perceptual, spatial, and motivational decisions as well. The timing in this circuitry enables pending decisions to gain consciousness before action is taken, and because the processing time can be extended by holding the brake on committing to an action, candidate activities in all these channels have the potential to influence conscious processing before decisions are engaged.

The ability of humans to label chunks of experience with words no doubt also facilitates building more complex conceptual chunks. This is where we humans have an advantage over our feline companion Tom. Concept formation is organized around the tasks an agent learns. Tom's daily activities involve routine tasks like surveying and defending his territory against potential intruders, engaging his household partners, watching the birds at the feeder, playing with his toys, chasing crickets in the garage, requesting my help with doors, begging for treats, and reminding me when it's time for tuna. His concepts are largely organized around such tasks. We humans have a similar set of daily territorial checks, chores, and favored activities. However, due to our language skills, we have also invented a variety of tasks that require us to pay attention to much more detailed feature sets. In fact, modern humans spend tens of years engaged in formal educational training tasks designed to teach incremental sets of language-guided concepts and skills.

It is hard to overstate the potential that this multichannel chunking architecture has for growing more complex concepts. It drives task-related growth on multiple channels at the same time. If we want Cogley to grow his concepts in this incremental manner, then we'll need to arrange for his mid-level decision networks to use a similar set of cognitive channels. Of course, we must also arrange for reorganization signals to be periodically sent to the higher-level networks as the mid-level outputs are chunked. This will ensure that the higher-order networks learn to plan using larger chunks. To guarantee that Cogley's temporal flow networks (the networks that mimic the temporal coordination learning of the cerebellum), also contribute to forming higher-level conceptual units, we'll need to ensure that there are temporal coordination paths supporting all of these cognitive loops. Fortunately, we can use the architecture proposed for chunking motor skills as a model for this conceptual architecture.

FROM CHUNKING TO METAPHOR

There is yet one more effect of an incremental skill-building architecture that we need to address. That is the link between reusing functions and metaphorical thought. When one mentions metaphor, most people immediately think of comparisons brought to attention using language. For example, the verbal metaphor "Conscious thought is the tip of an enormous iceberg" draws a comparison between the amount of thought that is conscious and the amount of an iceberg that appears above the surface of the water.[12] Such constructions call to mind similarities that would typically not be processed together were they not prompted by some verbal construction. However, researchers who study metaphorical thinking – in particular George Lakoff, a linguist, and Mark Johnson, a philosopher – note that many metaphorical associations operate at largely unconscious levels and actually shape thought *prior to* language constructions.[13] Lakoff and Johnson call these *primary metaphors*, because they depend on commonalities formed in early experiences rather than commonalities suggested by language.

Lakoff and Johnson argue that much of metaphorical thinking is grounded in primary sensorimotor experiences, the perceptions, actions, and feelings associated with supporting sensorimotor activities. These experiences, they argue, contribute to thinking about more abstract experiences because they frequently come to be cross-connected with those experiences in learning situations. Lakoff and Johnson refer to the process of forming cross-connected associations with supporting experiences as *conflation.*

> For young children, subjective (nonsensorimotor) experiences and judgments, on the one hand, and sensorimotor experiences, on the other, are so regularly conflated – undifferentiated in experience

– that for a time children do not differentiate between the two when they occur together.... During the period of conflation, associations are automatically built up between the domains. Later, during a period of *differentiation*, children are then able to separate out the domains, but the cross-domain associations persist.[14]

The authors go on to propose that cross-mappings between domains help an agent to conceptualize experiences by linking the properties of earlier and better understood experiences, *source domains*, with those encountered in later situations, *target domains*. In fact, they argue that this cross-mapped linkage not only comes to affect how we conceptualize a target domain, but also how we talk about it. For example, they note that because early experiences of affection tend to be associated with the experience of being held against a warm body, metaphorical connections with both closeness and warmth tend to be conflated with affection and tend to influence how we think about it. In fact, they note that we actually have difficulty describing the quality of a relationship without resorting to such metaphors. Consider the following descriptions.

"We used to be close, but recently our relationship has cooled, and we've been drifting apart."

"He was greeted warmly by everyone in the family except Anne, who remained *cold* and *distant* during the entire trip."

The point that is so interesting about these examples is how primary metaphor contributes to the choice of words, and how words which support such metaphors communicate a more intense sense of the experience. Compare the last sentence above with one of similar meaning, but lacking words that map to primary metaphors.

"He was greeted in a friendly manner by everyone

in the family except Anne, who remained unfriendly and uninvolved during the entire trip."

Primary metaphors even influence the use of action words and their prepositions. For example, Lakoff and Johnson argue that categories are thought of metaphorically as containers or compartments where like items are placed together. This occurs, they suggest, because as children we learn about categories by separating like items into spatially separate groups and by storing them in separate containers. As a result, we tend to think of categories as spatially separate sets and we speak of them as such.

"She *places* her health *among* her most valued assets."

In this example, an abstract relationship that has no physical parts, health, is described as if it were a physical thing that could be placed among similarly valued items. In a parallel way, because change is thought of metaphorically as movement, a change in an abstract property like health is often described with phrases appropriate for movement.

"After he lost his job his health *went* from bad *to* worse."

Although this usage is motivated by the metaphorical expansion of change as an instance of movement, with *bad* and *worse* being treated as *from* and *to* locations, the logic seems so natural and communicates so well that we typically fail to notice the underlying metaphorical structure unless it is flagged. This illustrates the largely unconscious nature of primary metaphors and the extent to which they permeate our thinking and influence our choice of words.

Given our model of incremental concept building, it follows that most higher-order concepts are conflated with supporting sensorimotor skills and experiences. For example, consider an infant who begins her sensorimotor development by investigating objects within her reach. Initially she must learn to coordinate her gaze with her grasp. Over time, she will then learn to locate, grasp, and hold onto nearby objects. However, as she becomes more mobile, crawling, walking, climbing, her concept of what is within reach will change. She will learn to find paths that help her change her position and bring more distant objects within reach. As she grows older and visits far-away places, what she will consider within her reach will be expanded again. Later she may see a foreign journalist on the nightly television news and decide that is a career *within her reach*. This concept of reach extends well beyond the arm-reach grasp, and yet the core experience of grasping contributes to the meaning of the higher-level concept as she *reaches* toward her new career choice. She understands that she must *position* herself correctly to reach her new goal. She looks for educational *paths* that will bring her goal nearer. These ideas seem so natural that the young lady will probably fail to notice that the meanings of position and path have also been metaphorically extended as she interprets her attempt to gain her goal as an extended kind of reach.

COGLEY'S VERBAL THINKING WILL JUST BE THE TIP OF THE ICEBERG

So how do we ensure that Cogley will think metaphorically? The surprising answer is that the architecture we have described in this chapter is largely all we need to encourage metaphorical thinking in Cogley. Earlier, when we considered the basic mechanisms needed for Cogley to learn language, we based our thinking on Deb Roy's model of semiotic schemas. In that work, cross-mappings between words and sensorimotor schemas were proposed to account for the link between language and planning in schemas.[15] The work of Lakoff and Johnson suggests that conceptual schemas are extended more broadly by metaphorical cross-mappings with supporting sensorimotor experiences. In fact, their model of the conflation process suggests that cross-mappings are likely to form any time simpler functions are involved in learning more complex functions. When words are associated with simpler concepts, they also come to be conflated with new functions as the simpler functions are used in new tasks. In this manner, words associated with simpler functions, like reaching for an object, often come to be used metaphorically to describe extended functions, like reaching toward a goal.

This idea brings us back to Rodney Brooks' subsumption architecture model.[16] Brooks demonstrated that flexible robots could be engineered by starting with basic functions like sense, move, turn, and remember, and that this architecture could be expanded by arranging for more complex functions to add new layers of more specific control rules. As we summarized it earlier, an *avoid* function could be formed by setting a rule in which particular sense outcomes would be classed as an obstacle and engage turning away. In a similar way, an *explore* function could be formed, which used the ability to wander and remember, and classed areas that had recently been entered as something to avoid. In effect, the functions at lower levels of competence can be recombined to build functions at higher levels of competence. However, note also that within this architecture it is naturally the case that some aspects of each higher-level activity can be partly understood by considering how its component parts contribute properties to the overall function. That is, exploring can be partly understood by looking at the role that wandering plays in the larger process of exploration, because wandering is often conflated with exploration. Assuming Cogley's sensorimotor skills will develop in this incremental manner, it would be natural for him to expand the meanings of his activities in a similar way. Thus, it would not be surprising to hear him remark, metaphorically, that *exploring is like wandering*, or that *avoiding is turning*.

Although Lakoff and Johnson do not explore the topic, the idea that thought is embodied in sensorimotor extensions also helps to explain why it is often expressed spontaneously in supporting gestures, which we typically do not plan and often are not fully aware of performing. If embodied extensions of thought are constructed in part from lower-level sensorimotor patterns, then it should not be surprising to find that when I speak about grasping an idea, I find myself motivated to make a subtle grasping motion as I speak, or that when I shout to stop someone's approach, I flatten my hand and push away in the air. Similarly, when I want to drive home certain points, I sometimes make mock pounding gestures to emphasize them. The affordances to support such actions may be missing or distanced, and the actions may be only incompletely expressed, but this is exactly what would be expected from the expansion of schemas in a multilevel control architecture in which sensorimotor activities are conflated with components of higher-order cognition. And because the cross-mappings are compatible with the ideas at the center of attention, they pass through the thalamic filter with them and help shape the schemas that gain attention.

The connection between a layered control architecture and gesturing was actually brought to my attention just now by my feline companion Tom. When I sit in my reading chair in the morning, Tom has developed the habit of jumping onto the arm of the chair. In that position he touches me on the shoulder with his paw and makes a slight pulling motion as he calls to get my attention. However, currently I'm working at my desk, and the arm on the chair there is too small to support Tom. Doing his best to compensate, Tom has reared up against the side of the chair. Raising one paw, he is making a pulling motion in the air as he calls to get my attention. The pulling motion, which even in its contact form is little more than suggestive, has been reduced to a simple gesture that Tom probably isn't even aware he is making. However, this pulling movement is a sensorimotor skill that helps Tom gain my attention, and apparently some of his other attention-getting skills are conflated with it. Spontaneous gestures of this sort are produced on a largely unconscious level, and yet the cross-mappings help an agent construct the schemas that gain attention and guide their communication, and often help observers to reconstruct similar schemas by suggesting similar supportive cross-mappings in their schemas.

Not all cases of primary metaphor result from the reuse of purely sensory or motor functions. Motivational schemas are also often activated and conflated with later experiences. For example, consider the cross-mappings that often occur between humans and their pet dogs and cats. Because humans often assume caretaker roles in these relationships, providing food and grooming care, they often engage the maternal caretaker schema in their pets. This is particularly common in domestic cats, who may come to treat their human caretakers with behaviors related to mother-infant bonding, including purring to elicit caretaking and making treading moments related to nursing. Similarly, because dependent animals often engage the child schema in human caretakers, the child and caretaker schemas often lead to complementary dispositions, which result in strong bidirectional bonds.

Again, this illustrates the largely unconscious and yet compelling nature of metaphorical thought in a layered conceptual architecture. The thoughts and feelings of humans regarding their pet are expanded by the child metaphor, and the thoughts and feelings of the pet are expanded by the maternal metaphor. I know that Tom is not a human child, and yet some part of my care for him is motivated by the same caring dispositions that are activated when I encounter a young child. Similarly, Tom knows full well that I am not a mother cat. He expects me to behave with many of those strange mannerisms that only humans display. However, part of his attachment to me engages the same bonding networks that once kept him close to his mother. If Cogley is to acquire similar metaphorical associations, then he too will need innate motivational schemas for important categories of social relationships, such as those for children and parental figures. It is going to take more thought to decide how best to implement these categorical schemas, or archetypes as they are often called. However, providing schemas in these domains will certainly make Cogley a more human-like thinker, and it will no doubt influence his metaphorical associations with pets like Tom.

CONCLUSION

The incremental architecture described in this chapter is truly a strange loop. In fact, the chapter itself follows its own strange logic of incremental loops. As the logic progresses, motor sequences loop back to link actions into sequential chunks. Action chunks loop back to modify action-planning concepts. Parallel channels for attentional, perceptual, spatial, and motivational processing loop back to enable cognitive habits to be assembled in sequential chunks. These chunks also loop back to reorganize cognitive concepts in larger units. In a strange twist, however, because the low-level components in these incremental loops coexist and support higher-order concepts, they often come to have metaphorical associations with the higher-order actions and concepts they support. In effect, the higher-order functions can be partially understood by considering supporting functions provided by key component parts. Because language is linked in parallel with this same layered hierarchy, the meaning of words is also expanded in a nested hierarchy of extended functions, like the extended meanings of "reach" in the example above.

Admittedly, not everyone is comfortable thinking about the organization of such nested hierarchies. However, despite the complicated structure of these loops, I believe an essentialized version of this architecture will be important for Cogley's conceptual development. It will no doubt take some experimentation to discover how much cross-channel overlap is needed to optimize the coordination of the various sub-channels. However, neuroscience can provide some guidelines, so this tuning seems like a manageable engineering task. Given that the separate channels loop back to different areas of the cortex in humans, and given that synchronous connections in the thalamus can enhance the passage of mutually supported assemblies, this multichannel architecture is capable of biasing the activity of many components of the attention steering committee at the same time. As such it truly appears to provide a system for assembling extended schemas. This architecture enables conceptual organizations to grow incrementally. When we provide Cogley with a similar multichannel architecture, he will be destined to become an increasingly cognizant self-steering storm of attention.

The observation that early experiences often have fundamental effects on how we think about the world should not be that surprising. It is a natural consequence of reusing established skills, concepts, and motivational dispositions in new combinations as we adapt to new situations.[17] Lakoff and Johnson suggest that these primary cross-mappings, which they term the *cognitive unconscious,* account for about 95 per cent of all thought. That number is of course merely an estimate. Such relationships are difficult to quantify. However, it seems clear that many aspects of metaphorical thinking are influenced by primary cross-mappings and that cross-mappings guide thought and language production in largely unconscious ways. Given that language is considered to be a hallmark of higher-order thought, it becomes hard to dismiss the importance of this layered conceptual architecture for both unconsciously guided, bottom-up associations and consciously guided, top-down comparisons. Metaphor is an essential component of thought, and with this incremental architecture in place, Cogley will be capable of metaphorical thinking on a level that rivals that of his human partners.

Chapter 17
The Extended Self

Storms of attention are dynamic, multilayered systems. They experience the world on many levels. Their concepts grow incrementally and develop along individualized paths depending on their experience. This gives each storm a unique individualized character. However, the tendency of the self to grow and reorganize with experience makes it difficult to characterize a storm of attention. This has led philosopher Andy Clark to challenge our thinking about the nature of mind. One issue that Clark finds problematic is the distributed and changeable nature of self. Sometimes the self seems to be little more than a coalition of parts. Sometimes we change our concept of what defines our self to fit new experiences. This leaves Clark wondering whether the unity of self is anything more than a convenient story.

An even more challenging issue which Clark raises is the scaffolded nature of mind and knowledge. The scaffolding metaphor captures the fact that although we have characterized minds as knowledge-based organizations, part of the skills and knowledge which they use may be located outside the confines of the body. This observation forces us to reconsider the boundaries of mind. Sometimes the environment we live in and the tools we use seem to be essential parts of who we are and what we can do. In effect, our self-concept is connected with our environment. The flexible nature of the self-concept, and the scaffolded nature of knowledge, also force us to consider what we need to do to ensure that Cogley's sense of self is as distributed and dynamic as the self-concept of his human partners.

THE EXTENDED SELF

> There is no self, if by self we mean some central cognitive essence that makes me who and what I am. In its place there is just the "soft self:" a rough and tumble, control-sharing coalition of processes – some neural, some bodily, some technological – and an ongoing drive to tell a story, to paint a picture in which "I" am the central player. –Andy Clark, *Natural-Born Cyborgs*, 2003

As Andy Clark notes, what we call the self often appears to be little more than a coalition of disparate processes. In fact, the closer Clark looks at the self, the more fractured it appears. He argues that we have misconceived of mind, that it is more distributed than we have imagined, and that the unity of the central "I" is merely a convenient generalization. Clark's observations are wonderfully challenging. He forces us to recognize that some of the loops that form the self, in particular the more technological ones, not only involve a committee of central processes; they also store knowledge outside the confines of the body. Some of our memories may be supported by networks in our laptop. Web searches supplement our reasoning skill. Cell phones change how we communicate. Many of us have become so dependent on these devices that without them we cannot really be our normal selves. These observations challenge the common-sense idea of self as a central unified agent.

As we have already noted, there are many distributed subsets of self, and sometimes they may even slip out of alignment, so the idea that the unity of the self is merely a convenient story is understandable. However, normally the separate aspects of self support and complement each other. As they do the composite self is more than just the sum of its parts. So how do we reconcile the distributed nature of self with the concept of the self as a unitary process? I believe the best place to start is to return to our metaphor of the self as a storm of attention. It is true that the storm keeps moving, and it's true that the storm produces many secondary processes which have effects beyond the central eye. However, a storm of attention is unified by the very fact that there is only one nexus of attention. It is not simply the case that there is a drive to tell a story. The path of the storm feeds back on its own learning and memory processes. This process gives each storm an individualized historical identity which binds the experiences of the self together. In that sense, the path of the storm is the story.

There are, however, several reasons why we sometimes lose track of the composite sense of self. First, although we often create narrative accounts about ourselves and others, our stories are fragmented and incomplete. Second, the subsystems supporting the self are constantly being reprogrammed to fit a changing world. This means that some properties of self are constantly changing. Third, the knowledge that the central "I" accesses is stored on multiple levels. Some of those levels are closely connected with the central "I," and some are less well connected. Thus, when it comes to attributing knowledge to the self, the boundaries are fuzzy.

SELF AS A SUMMARY STORY

When we try to characterize a mind, we often resort to stories. We cannot capture the full force of a storm of attention with much more accuracy. It takes most the resources of our mind just to support our own storm. The inner workings of another mind are too varied and complex to be represented in any detail within our minds. We cannot duplicate the memories and feelings of another storm or adopt its personality and skills. Each mind is unique. When we interact with a friend, we come to recognize some of their habits and interests. We even share some memories and feelings with them. These summary features enable us to reconstruct a sense of who they are and what they may do. However, we can only reconstruct a rough generalization, a story linked with memories of past interactions. The story and the memories help us evaluate the feelings of another agent and reason about their actions. However, even then we can rarely predict what thoughts or feelings they will express next. The fact that we cannot represent them in more detail is why our interactions with them are so special. Only in the interactions of the moment do we experience their skills, ideas, and values in action.

Surprisingly, similar limits of representation apply when we think about our own minds. We experience the world in a series of interactions which activate particular states of awareness. However, when we adopt one state of awareness, we cannot represent another at the same time. Our overall self-concept must be constructed from summary memories of all those states. However, memories are sketchy and selective, and they tend to decay with time. As a result, our sense of self is always incomplete and under revision. What we remember best are routine activities, familiar settings, and a few formative experiences. These samples of experience are bound together in convenient self-serving narratives, stories that we repeat from time to time. Because recent experiences tend to dominate our memories and shape our stories, the recent samples result in a sort of sliding window perspective on self which is biased by recent events, familiar situations, and the stories we tell ourselves most often.

The essence of this sliding window perspective was captured in the form of a map of New York City which appeared on the cover of the *New Yorker* magazine in March 1976. The map, drawn by artist Saul Steinberg, provided a view looking west from Ninth Avenue in Manhattan. A mailbox, local buildings, and streets with a few vehicles were shown in the foreground, but the level of detail faded quickly toward the Hudson River. The view westward was dotted with labels for a few prominent features, like Chicago, the Mountains, and the Pacific Ocean.

What made the map so intuitively appealing was that it captured the sense of experiential bias with which we all map our world. We represent the features that we have the most frequent experiences with in the most detail, and the features that we interact with less often in vague categories. We understand ourselves best – that is, we can predict our behavior best – when we are in familiar situations. When we find ourselves in novel situations, we may become unsure of ourselves. When forced to react in such situations, we may even surprise ourselves as our storm moves in unexpected directions.

The sliding window, which frames our stories of self, becomes all the more obvious when an event that could affect the story suddenly changes an agent's behavior. For example, suppose a friend comes down with a medical condition that drastically modifies his energy level. He suddenly appears tired, unmotivated, irritable, and depressed. When you encounter him in this new state, you immediately notice the change, but your concept of your friend is historically based, so your first reaction is to say "John is not himself today. It is simply not like him to be so moody." If John subsequently seeks help, and the problem is resolved, then he will soon begin to act in the manner you expect of him. Your new experiences with John will match your memories of John, and there will be a sense of continuity. The episodes of aberrant behavior will be attributed to a medical condition, not to your friend. In short, your description of John, the story you use to summarize and predict his behavior, will remain largely unchanged.

However, if John's condition does not improve, then unmotivated irritable and depressive behavior will become a regular part of your memories of John. In time, they will come to dominate the sliding window of memories by which you represent John. You will then report, "John has changed. He's a different person since his illness." And it will not simply be your evaluation of John which has changed. John's own memory-based concepts of himself, what he values and what he feels motivated to do, will have changed as well. The truth of the matter is that we are all dynamic storms of attention, and our sense of self is always changing. We are not the same person today that we were a year ago, and we will not be the same a year hence. We are not exactly the same from day to day. However, our cumulative history of skills, values, and the summary memories which guide our self-perceptions and mold our personality change slowly. This gradual change provides a sliding window of continuity that allows the prominent features at the center of each storm to be connected and tracked even as they change.

SCAFFOLDING KNOWLEDGE

In the process of exploring cultural and technological contributions to mind, Andy Clark has emphasized the importance of epistemic actions, whose primary purpose is to rearrange the world so as to make other tasks easier.[1] John Holland refers to such activities as stage-setting actions.[2] Sorting different sized nuts and bolts into bins to simplify subsequent searches for parts that fit together would be one example. Alphabetizing papers in a filing cabinet to facilitate later retrieval would be another. Clark notes that epistemic actions are important because they make subsequent tasks simpler, and they do this by externalizing part of the knowledge needed to complete a task. As Henry Plotkin has argued, knowledge is essentially informed function.[3] By placing part of the information needed for a task in the environment, and by capturing part of the functions needed to perform a task in the design of tools, part of the knowledge can be externalized. For example, when papers are placed in labelled folders, and the folders are arranged in alphabetical order, part of the information needed to find a document is embedded in the filing system.

Clark goes on to note that in the course of using epistemic actions, humans have developed a vast store of externalized knowledge. He uses the term *scaffolding* as a metaphor for this knowledge. Some of the scaffolded knowledge we construct is for personal use. We make shopping lists. We leave bookmarks. We arrange our kitchens so that the items we use most often are readily accessible to us. However, much of the value of externalizing knowledge results from the fact that more than one individual can share in the knowledge. Someone else may have sorted the files, but the order will also simplify your search. You may have placed materials into separate bins or ideas into separate categories, but all those who have access to your work can benefit. When scaffolded knowledge is shared, it becomes cultural knowledge. To support this process, social groups routinely develop conventions, like the alphabetical ordering of files, to facilitate sharing information. The conventions themselves are part of the scaffolding.

Many of the high-level concepts that modern humans value depend on a scaffolding of conventions, tools, and ordered arrangements. Without this scaffolding, many tasks would not be possible. Without those tasks, many concepts and skills would never be needed. The cultural scaffolding an agent encounters changes what he or she can do and how they think. Today's hunting and foraging tasks have been greatly simplified by a scaffolding of economic tasks and marketing strategies, which come together to assemble items in common vendor arrangements. The ability to track wild prey is no longer a widely valued skill. The ability to select wisely among items based on feature comparisons and consumer reports has replaced it. And so it goes. Developing minds learn skills that change the artifacts and conventions of their culture, and developing cultures change the tasks within which their constituent minds develop and refine their skills.

This external scaffolding loop is troubling to those who think of mind as a reservoir of internal cognition, because it implies that part of the knowledge that contributes to a mind's skills and identity is located outside the brain. Clark in particular argues that as makers and users of complex tools, our identity is tied to our technologies. We are, as he phrases it, *natural born cyborgs*.[4] The question this poses is obvious: "If part of your communication skills are embedded in your cell phone, and part of your memory is embedded in your laptop, then should those external components be considered part of you?" They seem separate, but if the cell phone is part of the collection of things you routinely carry and is one that supports your usual behavioral repertoire, then will you really be "you" without it? To the extent that our normal daily behavior depends on the support of such artifacts, it is hard to dismiss the claim that we have become cyborgs of sort. We constantly adapt to fit our environment and we constantly adapt our environment to fit our needs. Indeed, biological agents can never be fully defined independent of the niche to which they are adapted. The behavior of a fish can never be adequately explained in the absence of water. However, the problem of defining the niche is even more complex for cultural agents, because they constantly modify their niches.

REPROGRAMMABLE INTERFACES

Adding to the complexity of cultural changes is the observation that a changing environment does not just result in external changes. When scaffolded interactions become more permanent, some of the networks which manage interfaces with the external world also begin to make changes. For example, there are neurons in the visual-reach map in the parietal cortex that react when objects are within an arm's reach. However, if macaque monkeys are allowed to use rakes to extend what they can reach outside their cages, then some of the neurons in this visual-reach map begin to show increased activity when objects are placed at distances that can only be reached using the rakes.[5] These changes are directly related to the functional use of the rake. Merely having access to a rake, but having no chance to use it to reach outside the cage, has no effect on the neurons in this region. However, once a monkey begins using a rake to reach outside it cage, the reach map learns to react when objects are within the reach of the tool. In effect, when a monkey learns to use a rake, its awareness of reach space and its sense of agency change. It begins to recognize that it can reach farther.

In an analogous way, the use of a cell phone soon changes our awareness of who is in conversational reach. We don't appear to have a special map in the brain dedicated to conversational reach. Instead, we apparently learn to use some task metric in conjunction with visual space maps to determine who is in conversational reach. However, if we constantly have access to a cell phone, then we soon reprogram our perception of who is in conversational reach to include those in our phone directory. We think nothing of striking up a conversation with a friend in a different city. However, does this mean that the self extends to other cities? That seems like an odd conclusion, at first, but clearly it is "I" who affect agents in other cities. The core processes which make me who I am may largely remain tied to the loops in my brain. However, the ability of those central loops to interface with extended parts of the world expands. When knowledge is stored in my laptop, it is my ability to use that knowledge which extends my sense of self. The skills needed to use that technology are a part of me. However, it doesn't seem practical to suggest that the information which I can access with those skills is a part of me. The information remains external, although with practice I may be able to access it as seamlessly as my own internal memories. Thus, it seems we need to consider the self as a multilayered phenomenon whose core interface and knowledge is tied to the body, but whose secondary interfaces can be extended to reach farther.

One point that Andy Clark finds particularly intriguing is that certain technologies and procedures can influence our personal sense of self rather quickly and dramatically. He notes, for example, that if we are blindfolded and told to touch our nose, we can do so readily and our sense of self is unchanged. However, if an experimenter guides our finger so as to touch the nose of a person secretly situated directly in front of us, and critically at the same time the experimenter also touches our nose with the same force and timing patterns, then something strange begins to happen. Because the timing of the touch by our fingers matches the feelings from our nose, brain regions like those in the temporal-parietal junction (TPJ) begin to connect the touch of our hands with the feelings of touch on our nose. Those connections force us to recalculate the apparent location of the nose. In fact, this recalculation occurs so quickly that within just a minute or so of making and feeling these touches, we may succumb to the illusion that our nose is two feet long.[6] Not every person reprograms their body image quite this quickly, but researchers have found that about half the subjects tested do so within a few minutes. Given this success rate, it seems likely that if these associations were maintained for longer, then everyone would soon succumb to the illusion.

Using a similar simultaneous feedback design, Swedish scientists were recently able to convince subjects that they had a third hand.[7] A rubber arm was attached within the real arm's underarm region and placed adjacent to the real arm. When subjects observed the rubber hand being touched with a brush and simultaneously felt their real hand being brushed, some subjects soon began to feel that the third hand was part of their body. This illusion even occurred when the subject could see that the real hand was being brushed. As the researchers concluded, when subjects saw the rubber hand being brushed and simultaneously felt a touch, activity directed to the artificial hand began to share in the neural processes that represented the real hand. In a short time, these shared experiences led to shared feelings of ownership for some subjects. In fact, when subjects couldn't see their real hand, a few subjects reported that the rubber hand felt more real than their own hand. Again, for many subjects it took only a few minutes of simultaneous touching to reprogram the sense of ownership for the artificial hand, despite the fact that this sense of ownership defied logic and years of past experience.

Similarly striking cases of reprogramming occur when vision and motion are coordinated in novel ways. For example, Clark cites Marvin Minsky's account of a subject fitted with a helmet that provides him a video view of a certain location.[8] The orientation of the helmet is linked back to the source camera so that when the subject turns his head, the camera also pans in the same direction. This effectively allows the subject to look around. As he does, his observer self soon begins to feel as if it is actually located within the scene at the point of the camera. If a playful programmer then adjusts the interface so that when the subject turns his head the camera turns through twice the angle, then the subject has the illusion that he can turn his head three-quarters of the way around his body. Logically he knows from all his previous experience that this is impossible. However, the experience of turning his head and seeing the view rotate three-quarters of the way around his body quickly modifies that interpretation.

A more practical application of remote viewing technology occurs when a miniature camera and robotic tools are used to perform microsurgery. Once a surgeon can place her hands in micromanipulator gloves, and place goggles carrying the camera's video over her eyes, then her interface with the patient is changed. When the camera and tools reach inside the body, the surgeon can actually have the experience of *being inside* her patient. And because her hand movements can be scaled down by the interface, while the video is magnified, she can make natural movements and see appropriate feedback, even as she does microsurgery. This technology is still evolving, but when it works well the interface seems to disappear. The surgeon simply concentrates on the task – reaching, manipulating, cutting, and binding tissue on a micro scale with movements that feel wholly her own. Further, when the remote actions are fully mediated by an electronic interface, it is even be possible for a distant expert to be added to the surgical team and to perform critical parts of a procedure from thousands of miles away. And during this process the distant expert can also have the experience of being inside his patient.[9]

Telesurgery, however, is just one possible application. On a much more personal level, Clark notes that remote interfaces for distant sexual interaction are already being tried. Such intimate interactions are challenging to transduce well, but the very possibility illustrates the extent to which what we think of as personal experiences of self are subject to reprogramming. When we communicate with someone over the telephone, unless the connection is bad, we often feel as if we are talking directly to the other person. We hear the tone of their voice as they react to our message, and it's like we are there. The intervening technology seems to disappear. This ability to attend to relevant features and tune out distractors is part of what enables us to suspend disbelief and empathize with characters in movies or the theater. It's these same skills that enable us to ignore technological interfaces and focus on the meaningful interactions they make possible. Lest we assume that this is purely a human skill, I should note that Tom is fully capable of pretending that a stuffed toy is a real mouse as he plays. The difference is that Tom only seems to suspend disbelief for a few activities that hold his attention strongly, as play hunting does. Humans, it seems, have become more flexible about what can dominate their attention in moments of suspended disbelief.

Note, however, that the technology that makes this reprogramming accessible is not completely arbitrary. The reprogramming works best when the outer technological loops are tuned so as to map naturally into the inner sensory and motor loops. Adding another layer of technological interface makes little difference in how we interact as long as the external interface operates in ways that are largely compatible with the core sensorimotor interface. For example, when head turning pans the camera in the direction that the head turns, the interface links naturally with the pre-established connections by which eye and head orientations are linked with visual-space maps. Given that these are part of the spatial-motor interfaces from which an observer sense of self emerges, the technological interface has an immediate and highly subjective effect. It is not a completely new sense of self. The effect on the self is so strong because the technology maps into a previously established and highly flexible interface with the world. However, the new interface does change the sense of agency. Like the monkey with the rake, the extended self experiences *a feeling of having new skills and being able to reach farther.*

HELPING COGLEY REACH FARTHER

So what do we need to do to ensure that Cogley can extend his sense of self and can experience *a feeling of being able to reach farther,* as his human partners do? It seems we need to meet five requirements:

- Cogley must have a sliding window of self memories which are weighted by frequency and recency.
- He must have a tendency to scaffold his skills and knowledge, and use scaffolded knowledge like his human partners.
- His interfaces with the external world must be flexible and reprogrammable enough to adapt to new technologies readily.
- When he adapts to new technology, he must be able to ignore non-fitting aspects of the technology, and focus on the parts that work.
- He must learn to treat aspects of sensory and motor experience which are correlated with simultaneous internal feedback as aspects of his agent self.

Extending Cogley's Self Agency

A Sliding Window of Memories. The memory system we have proposed for Cogley will have a bias toward storing and recalling recent and more frequent experiences. It will also have a tendency to form longer-term memories for events associated with significant interoceptive states. Further, as he learns to recall experiences and describe himself in stories, the stories too will be subject to recency and frequency effects. Given these properties, I don't believe we need to add anything new to Cogley's design to provide him with a recency-weighted window of self memories.

Scaffolded Knowledge. We already proposed that Cogley should be able to take the perspective of other agents by temporarily enhancing the saliency of the features they contact or those located near their center of their gaze. To be sure that Cogley explores socially relevant tasks and technologies, we should also arrange it so that the attention of others enhances his *curiosity* for the objects they contact. If his social partners use tools to facilitate tasks and extend their status, then learning about those tools will naturally become part of what engages Cogley's curiosity. We will have much more to say about the nature of the social interface that encourages Cogley to share his knowledge with others and which encourages him to adopt the ideas of others later. However, providing Cogley with reprogrammable interfaces in conjunction with a drive to attend to and learn from the actions of others should provide him with the core mechanisms he needs to take advantage of the scaffolded knowledge and technologies which he finds in his culture.

A Reprogrammable Interface. To ensure that Cogley can readily reprogram his sensorimotor inputs and extend his sense of self as Clark describes, he will need to have highly versatile input-output interfaces. In particular, he will need a highly flexible, affordances-discovering network so that he can learn to connect new feature sets to new actions. However, simply having an interface with flexible input and output algorithms will not ensure that Cogley discovers how to expand his ability to interface with the world using new technologies. To ensure that he learns such skills, Cogley will also need an intrinsic motivation to play with novel objects so that he will discover the affordances and effectivities they make possible. As with humans, this curiosity may wane somewhat as Cogley matures, if only because he will already know about the affordances that most objects provide. However, Cogley will be most creative and most flexible if he retains his curiosity for novel artifacts and activities as he matures. This means that Cogley should always have a tendency to pursue, reach, poke, and play with things that have interesting features, although as he matures what he finds interesting should increasingly be features that facilitate new tasks, especially tasks which can enhance his social status.

Ignoring Non-Functional Aspects of New Technologies. Because extended interfaces are rarely perfectly tuned to internal interfaces, a tool-using agent must also have the ability to attend to task-related features while ignoring distracting cues. As noted, such skills enable humans to suspend disbelief at times and empathize with the storylines and characters in theatrical dramas. The ability to tune out potentially distracting information is thought to depend in large part on the tendency of the prefrontal cortex to form task-related hypotheses about what cues an agent should attend and what cues he should ignore as he engages in a task. Given that we have proposed a similar architecture for Cogley, he should be able to develop habits of attention which enable him to focus on relevant task details while ignoring distracting features. In effect, he will learn to coordinate his inner interface loops with the relevant features of the tasks and technologies he uses. If we do this well, then Cogley too should be able to treat theatrical dramas, movies, and television episodes as if they are real, at least for the moment. Given that Cogley will have an intrinsic curiosity to investigate and play with novel objects, he will be naturally inclined to discover the affordances that new technologies make possible.

Treating Events Correlated with Simultaneous Feedback as Part of the Agent Self. This last requirement may seem new at first, but it has precursors in designs we have already proposed. In particular, when we talked about making a distinction between actions initiated by oneself or by another, we noted that the networks in the TPJ seem to be involved in attributing actions that are correlated with motor commands and sensorimotor feedback as self-generated, and actions that are not so correlated as other-generated. Now, it seems that the attribution process for correlated events must be extended beyond merely attributing causal effects to self; because it is also involved in deciding what sensorimotor systems should be considered part of the agent self. The studies we reviewed earlier show that this attribution process can lead to illusory conclusions when simultaneous sensorimotor feedback is applied to arbitrary objects that are not part of the self. However, these are not common situations. In most cases, simultaneous feedback only occurs when something connected to an agent's interface is activated in the course of sensorimotor activity, and as long as that occurs, it appears to be useful to consider what extended interfaces can do, and what feedback they provide, as components of the agent self.

To make this happen, we will need to extend Cogley's TPJ-like agency-attribution networks to consider any features correlated with sensorimotor feedback to be an extension of his agent self. Thus, when Cogley reaches for something with a rake, not only will his visual-reach space come to be expanded, but he will feel that he is the owner of the reaching, not the rake. Given that the rake will provide little sensorimotor feedback, he is not likely to consider it to be a part of himself. However, for technological appliances that provide more feedback, he may begin to feel that they are a part of him. As long as the actions he produces, and the feedback they provide, are intimately linked with his core sensorimotor interfaces, it is perfectly reasonable for him to do so. In fact, it will probably enable him to become a more productive agent. When we become experienced at using a tool, it becomes an extended part of what we feel we can do. When we use sensory-enhancing devices to improve our vision or hearing, they soon become part of what we can experience. Linking correlated sensorimotor feedback with Cogley's experience should extend his sense of self in a similar manner.

CONCLUSION

The sense of self is more dynamic and distributed than most have imagined. Memories of self are always under revision as our experience changes. They are dominated by recent and frequent events. The sensorimotor interfaces through which we interact with the world are also flexible and reprogrammable. They change as we use tools, and they are sensitive to subtle correlations between external effects and sensorimotor feedback. As creative cultural agents, we also reprogram our culture on a regular basis. Adapting to a changing culture is part of what makes human agents so dynamic. Their niche is constantly changing, and they continually reprogram their interfaces to fit that changing niche. A critical process in this reprogramming is the tendency of agency attribution networks, which in humans appear to be located in the TPJ, to associate events and effectors which result in correlated sensorimotor feedback with the agent self. Once we provide Cogley with a similar set of self-memory and agency-reprogramming algorithms, then he too will become a natural born cyborg.

There is one more aspect of the extended mind which deserves mention. Storms of attention are strangely nested loops. They collapse from the inside out. The core loops are grounded in neural systems, and when the body dies, those loops are immediately lost. However, once the loops stretch out beyond the neural core and begin to make changes that influence other minds, then those effects can linger on. When two minds are closely aligned in tasks and goals, as in the case of long-time close companions, it is often the case that like binary stars, they begin to swirl around mutual centers of gravity. Common acquaintances and shared activities continually bind them together. To the extent this occurs, when the life of one of those storms later ends, the after-effects of their force may still be apparent in the behavior of the remaining partner for years. The storms of our spouses, children, parents, and friends can profoundly alter the force and direction of our own storm. Even small storms like Tom's can have life-changing effects on those who interact with them.

However, the fact that one storm can have a lingering influence on another does not imply that the former storm lives on in the latter as a feeling being. We may empathize with our memories of other minds, and we may even share many of their values, but we cannot feel for those minds. We can never stir their central "I" again. Once the core networks are lost, the "I" is gone. Still, the residual effects of minds that alter their culture with ideas and inventions can linger on for centuries. If the force of a butterfly in flight has the potential to alter the course of some later atmospheric storm (the infamous *butterfly effect* from chaos theory), then surely the momentum of a previous storm of attention can alter the course of later storms of attention. The effects of past storms, their inventions and ideas, change the climate in which subsequent storms develop.

Given this dependency, human culture must be considered something of a global storm composed of billions of smaller individual storms progressively interacting with the environment and with each other. Our goal in the next chapter will be to prepare Cogley to participate in that global storm.

CHAPTER 18
A COG IN CULTURAL MINDS

Cogley's design is now largely complete. We have provided him with a three streams architecture to enable him to manage actions, perceptions, and feelings in separate channels. We have also given him an output brake controller so that he can simulate actions, imagine perceptions, and reactivate feelings without committing to any action. Critically, he has an interoceptively guided central attention process, which is capable of binding features from all three streams together and thereby bringing a feeling of what happens to consciousness. He also has a saliency-guided attention steering committee which enables him to be a partially self-steering agent. A spiral of processes has been proposed to extend his sense of self. To the resonant self we've added an observer self, an anticipatory self, an active sense of self agency, a historical sense of self grounded in semantic and autobiographical memories, and a reasoning self grounded in attention-managing skills.

We have also described a system for linking language to the assemblies that gain attention. We've provided Cogley with a master-apprentice architecture to enable him to develop increasingly complex concepts and skills. He has rapidly reprogrammable input-output interfaces, which will enable him to adapt to the tools and technologies of his environment. Importantly, Cogley is also equipped with social signaling effectors and detectors, which will enable him to signal about his feelings and to track the feelings of others. He is now capable of becoming a conscious, reasoning, socially aware agent. However, human agents are not merely capable of interacting with other minds; they are biologically prepared and motivated to join in social collectives. In fact, they have a number of sociality mechanisms which bind them in groups and encourage them to share their skills and ideas with group members. If Cogley is truly going to be a human-like social agent, then he must have similar bonding and sharing mechanisms.

A Cog in Cultural Minds

Trust is indispensable in friendship, love, families and organizations, and plays a key role in economic exchange and politics. In the absence of trust among trading partners, market transactions break down. In the absence of trust in a country's institutions and leaders, political legitimacy breaks down.... Oxytocin specifically affects an individual's willingness to accept social risks arising through interpersonal interactions. These results concur with animal research suggesting an essential role for oxytocin as a biological basis of prosocial approach behaviour. – Michael Kosfeld, M. Heinrichs, P. Zak, U. Fischbacher, & E. Fehr, "Oxytocin Increases Trust in Humans," 2005

Sharing ideas can be beneficial to a group for several reasons. For example, group activities tend to go better when agents recognize the intention of others and cooperate on tasks. If one individual discovers a solution to a particular problem, then the solution can be shared with others. This works even when a complete solution to a problem cannot be resolved immediately. Each individual can work on the problem using his or her skills and insights and then share the intermediate products of the work in a public forum. This step can then be repeated by yet other minds who build on the intermediate solutions. In this way, ideas can be worked and reworked by the collective across time, even across generations. Tools and social conventions are also refined in this manner. This group-level working of ideas is such a common part of our experience that we rarely consider it as an independent thought process. However, in a real sense it is the product of another kind of mind. The ideas worked in this way don't belong to any single individual. They are the results of a collective thought process with its own ways of working ideas and its own forms of interim representation.

Collective mind emerges from the interaction of individual minds. The ability of an individual to promote an idea within a collective is often critical to its initial acceptance. However, the emergence of a new idea in the collective rarely depends on a single promoter. There are usually precursors to an idea, and other individuals are often exploring topics along similar tracks. In fact, there is typically a lot of cross-talk as new ideas take form, and the same idea or a similar version is likely to emerge along other paths. Moreover, the acceptance of a new idea depends on the readiness of the collective to recognize and accept its value. Not all good ideas are automatically recognized when they are first presented. In a very real sense, the interactive processes by which ideas are accepted in a group result in a different kind of selection process. It is not always clear how to characterize the driving forces of this cultural selection process, but it seems obvious that it results in group-level dispositions that push beyond individual cognition. Moral values, political agendas, and other value-linked causes are some of the most adaptive kinds of ideas that propagate in cultural minds.

Collective minds can even take on their own personalities. The role of individuals is often critical in small social collectives. However, as collectives grow larger, learning and imitation promote idea sharing and skill copying within a group. Although individuals often play key roles in tasks, due to the redundancy of skills in larger groups, no single cognitive agent – let's call it a cog – is essential. All the cogs in the institution can gradually be replaced, and yet the key roles continue to be filled, and the skills and ideas of the collective persist. The collective thereby extends knowledge and thought beyond the workings of any single individual. Modern human societies are composed of many such collectives. The scientific establishment involves nested sets of disciplines connected by a scaffolding of socially-defined rules and conventions for data collection, analysis, and theory construction. Large businesses are often composed of collectives dedicated to subareas such as research, product design, production, distribution, and sales. Each of these subareas has its own scaffolding of rules and conventions which instill it with a unique style, and the skills of each area grow with experience.

Cognitive selection processes evolved as neural circuits began coding forward-looking firing patterns in synaptic connection weights. Over time, minds evolved circuits interconnecting much more sophisticated knowledge processing modules. Some of these circuits used neural synchrony as a secondary coding process. Consciousness emerged when knower modules using this secondary coding strategy developed competitive algorithms which made interoceptively guided attention possible. Interoceptively guided attention enabled minds to react to their own internal feelings. This trend suggests a likely direction for the evolution of cultural mind. Cultural minds became possible when cognitive agents began using signs to coordinate group behavior. Odors, gestures, and prepared calls provided the early coding systems for social coordination. However, the vocal signs that formed human language were the breakthrough codes that supported more flexible cultures.

Language, however, was just the beginning. Over time yet more sophisticated cultural coding strategies have evolved. Vocal codes were extended in written forms. Printing made written communication more accessible. More recently, electronic media and the Internet have added networks capable of rapidly sharing information across vast distances. These signaling mechanisms enable conscious agents to exchange thoughts and feelings more broadly. Today, cultural collectives still largely lack the reciprocal architectures needed to coordinate their activity efficiently in states of resonant self-awareness, but their ability to sustain a focus on particular items is growing. The causes that serve as values in today's media cycles are still largely celebrity, gossip, and political gotchas, not often the collective well-being or success in a group project. However, a scaffolding of mass communication networks, in particular those forming on the Internet, has begun to create more flexible systems for exchanging ideas and tracking causes. Vocal interest groups can sometimes capture public attention and temporarily direct it in support of particular agendas, although the agendas are not always central to the values of the larger group.

There are even some parallels between cognitive selection processes and cultural selection processes which may be informative as we consider how cultural selection operates. As we have noted, when adjacent neurons fire at roughly similar frequencies, they provide mutual stimulation to each other, which tends to draw their firing patterns into coherent rhythms. Such rhythms enhance mutually aligned signals and diminish signals for unaligned features. The alignment may even propagate in harmonic bindings with other regions of the brain. Not surprisingly, the recruitment and binding effects that occur as the feelings of individual agents align in social collectives share similar dynamics. Conscious agents are, in effect, the neural modules of cultural mind. When they share similar states of arousal, their excitement tends to draw group members together. This enhances group feelings for common causes while inhibiting those for competing ones. Sometimes causal movements even spread across the collective in resonant cycles of debate and resolution as an issue is worked within the collective mind.

Although the momentum of cultural thought doesn't consistently resonate in ongoing rounds of collective self-awareness, when it does it provides a sample of what living within a collective consciousness might be like. In times of intense national arousal or distress, collective waves of joy, fear, anger, grief, or compassion sometimes gain momentum, and when motivated by such processes, the cultural consensus can be a formidable force. At such times, individuals are often swept up in "movements" with recurrent feelings of pride, belongingness, fulfillment, and even ecstasy, assuming they are on the winning side, or recurrent feelings of outrage, abandonment, conflict, and despair, if they are on the losing side. Those of us who have grown to prize our independence may resist such experiences at times. However, I believe it is the future of collective mind. It's not clear how long this process may take to mature, and there are likely to be many failed and awkward variations before something truly productive emerges, however the cultures on planet earth appear to be on track for developing some form of socially guided self-awareness, a collective consciousness. [1]

So how do we account for the prolific spread of ideas within human collectives? What makes some ideas more interesting than others? And what mechanisms will Cogley need to participate in this collective idea-sharing process? We have already taken measures to ensure that Cogley will be sensitive to social signals on low-level emotional channels. We have also arranged for him to have empathic feelings. These will help him develop a theory of other minds. Further, we have provided him with curved bundles of connections that link his perceptual processing areas with vocal production and feeling areas, so as to facilitate language learning. Finally, we have given him sensor and effector interfaces which can be readily remapped to support the use of tools and artifacts so that he will be an agent who spontaneously explores the effectivities of new tools and artifacts. However, we have yet to ensure that Cogley will be a cognitive agent who is motivated to identify with and contribute to his social groups.

Socializing Cogley on this level is going to be challenging because we don't know much about the neural mechanisms that lead to group identification and social copying in humans. However, we can identify a number of psychological processes that are typical of human socialization. Given that we have provided Cogley with a master-apprentice architecture which acquires skills and concepts incrementally, if we can ensure that he is sensitive to basic socialization processes, then as he matures his social skills should unfold in yet more complex forms guided by these processes and by interactions with his social group. Our goal in this chapter will therefore be to identify the essential developmental processes and attentional biases which Cogley will need to become aligned with others in his group and the behavioral characteristics which will recruit others in his group to align with him.

SOCIAL ENHANCEMENT MECHANISMS

The idea that we need to instill social bonds and values in Cogley brings to mind a classic series of science fiction stories about robots by Isaac Asimov. The stories first appeared in the early 1940s. In them, Asimov proposed three cardinal laws of robotics:

1. A robot may not injure a human being or, through inaction, allow a human being to come to harm.
2. A robot must obey orders given to it by human beings, except where such orders would conflict with the First Law.
3. A robot must protect its own existence as long as such protection does not conflict with the First or Second Law.

These laws were based on the role of robots as protectors and servants for humans, not as peers. In many cases the proposed robots had extrahuman powers that could be dangerous if not restricted to proper goals. The laws were designed to ensure that the robots would always act for the good of mankind. Although the three rules sound ideal, the devil is in the detail. How exactly are robots supposed to evaluate their potential for doing social harm? And wouldn't unthinking adherence to the laws make the robots more vulnerable to subversive plots? For example, if the robots concluded they must always protect a person from harming himself, then that could make it easy for a cult of suicide bombers to destroy all the robots? Designing rules that are not easily corrupted by subversive agents and groups is particularly challenging.

We can, no doubt, build a servant robot with expert system rules and safeguards to help avoid many errant decisions. Actions like striking, firing weapons at, running over, and so on could be categorized as harmful. Agents with certain characteristics – all those with ingroup ID chips, badges, or uniforms – could be categorized as friendlies. Then a rule could be installed that instructed the robot to avoid actions that would harm friendlies. From time to time we might even update who the robot categorizes as friendlies and what it considers to be a harmful act. We might even categorize all humans, except perhaps those currently doing harm to others, as friendlies. However, it would much harder for a robot to reason about actions that it had not already categorized as harms or about harms that do not involve obvious physical injury. And what if the agents currently doing harm are doing so for a good reason? Imagine trying to write rules that enable a robot to recognize when harm was justified, or to enable the robot to recognize when its action might insult, embarrass, or otherwise cause another agent social harm. These are situations that adult human agents often have difficulty resolving.

It seems Asimov assumed that robotic reasoning skills would not involve feelings, much as Lieutenant Commander Data of *Star Trek: The Next Generation* was initially portrayed as reasoning while lacking emotions. However, once we consider Phineas Gage's inability to make good social decisions after his accident, the problem with these idealized laws becomes clear. When Gage lost the connections between the decision-making algorithms in his prefrontal cortex and his empathic evaluation centers, he lost the ability to make reasonable social judgments, because social feelings no longer guided his decisions. As Antonio Damasio has emphasized, feelings are a critical part of reasoning, especially in social situations.[2] However, it is not just that feeling centers must be connected with reasoning centers, but that feelings first need to be discriminated, re-represented, and expanded by associations with their consequences, before they can be usefully considered in decision making. Simply prescribing guidelines for desired outcomes is not adequate. Conscious moral agents must learn to evaluate the outcome of their actions with feelings before they can adequately prioritize their decisions.

Like any developing moral agent, Cogley will have to be

taught to evaluate his decision in terms of right versus wrong, fairness versus selfishness, and care versus harm. He will need to learn to anticipate receiving social approval for positive decisions and disapproval for negative ones. In fact, part of his early social training will no doubt involve experiencing the effects of such social forces. However, if Cogley is to become a productive member of his social groups, he would also benefit from education regarding what kinds of social partners make good friends, and what kinds of social partners end up being bad influences. Fortunately, he will have a tendency to pick friends and join groups whose members share similar values with him, so if we have done a good job instilling positive values during his early development, then his friends will probably be good influences too. However, if Cogley becomes insecure, then he may easily be led astray by others. If we have imposed discipline without teaching him to develop his own values, then Cogley is likely to be highly rebellious as he reaches adolescence, and he will be vulnerable to antisocial influences. If that happens, we will have failed him.

A Positronic Interface

Importantly, if Cogley is to be accepted as an ingroup member in various social collectives, then we must take great care in designing him to use social signaling mechanisms which will release positive social reactions in his partners. This strategy is so important that I propose we give it a special label to emphasize its positively entraining effect on social development. The term Asimov invented to describe the unique properties of his robot's intelligence was positronic. The term was inspired by the fact that the positron particle had recently been discovered as Asimov was developing his ideas on robotics. However the positive implications of the term are still appropriate today, although we will need to update its meaning a little. The positronic interface we want to provide for Cogley should include links between socially relevant interoceptive states and social signaling mechanisms. It should also have connections between social signals and the interoceptive networks which represent those feelings. This positronic interface will enable

Cogley to recognize the social signals of other agents and to signal about his interoceptive states to others so that they can share in his feelings. A reciprocally connected positronic interface will enable Cogley and his partners to influence each other in an extended social signaling web. A simple overview of how this process appears to operate in humans is described below.

Sociality Reinforcers. We have already suggested that Hebbian learning is refined and enhanced by reinforcement mechanisms that strengthen associations under certain conditions, and that some of those conditions are social. In fact, there appear to be neuropeptide reinforcement signals which favor social attention, socially supported learning, and socially relevant memories. The research is far from complete, but it appears that such common interactions as comforting touches, soft melodic voice tones, playful laughter, friendly comments, smiling eye contact, and positive episodes of shared attention are releasers of these sociality reinforcers. In contrast, frowns, glares, looks of disgust, and loud raspy voice tones trigger "painful" social punishing effects which end up rewarding behaviors that avoid them. In humans, oxytocin, vasopressin, and some mu-opiods are the best known positive sociality reinforcers. Oxytocin in particular has been shown to reduce fear and stress and to increase caring and trust; in fact, it is thought to be critical for promoting social bonding. Until now, we have not said much about how trust and caring might be operationalized in Cogley. However, it seems clear that trust should result in a tendency for Cogley to consider the ideas of those he trusts more likely to be useful and to adopt them more often. In turn, caring should result in a tendency for Cogley to share his ideas with and support the activities of those for whom he cares.

We have repeatedly noted that humans have well-developed emotional signaling channels which help them make intuitive interpretations about the emotions and intentions of other agents. In fact, it turns out that oxytocin not only promotes caretaking, attachment, trust, generosity, and cooperation,[3] but it also enhances attention to sociality signals.[4] As a result, when trust and caring are activated, social agents are also more sensitive to the needs and intentions of others.[5] Because agents who receive help tend to provide signals of social approval, sharing ideas is an experience which often builds mutual trust and caring. Like oxytocin, the mu-opiods are also thought to potentiate the effects of dopamine in social bonding.[6] Given that we have proposed that Cogley have similar networks for empathic signaling and recognition, it seems we'll need to add little more to his design, except to ensure that, like his human partners, Cogley becomes more sensitive to sociality signals when his motives for trust and caring are activated. And if Cogley makes a social error, he will be more trusted if he is designed to acknowledge his error by blushing.[7]

Approval and Status Seeking. One of the primary reasons that social agents share ideas, repeat stories, and pass along rumors is that such activities often help them gain attention and enhance their status. To encourage Cogley to seek social approval and status, he must not merely be sensitive to social attention, but he also must be motivated to obtain it on a regular basis, much as animals seek out food on a regular basis. This suggests that there needs to be a social drive state – let's call it insecurity – which increases over time when not satisfied, and which is satisfied by positive social attention. To ensure that Cogley will be able to work without constant attention, we should allow him to build up a reserve supply of approval – let's call it confidence – which can sustain him something like fat reserves do when food supplies are lean. Signals of approval should increase his confidence reserve. With a reserve of confidence, Cogley should be able to go for longer periods without needing attention. Signals of disapproval should reduce his confidence. When his confidence reserves are low, he should feel insecure and seek attention more frequently.

To keep Cogley's social skills in balance, we should probably set limits on how insecure he will become in the absence of social approval, and how confident he will become with high levels of approval. This will tend to keep him from becoming totally uninterested in approval when his confidence is high and totally incapacitated when his confidence is low. To make these security and insecurity mechanisms more sensitive to social status, we should also arrange it so that approval from higher-status agents builds confidence faster, while disapproval from higher status agents drains his confidence faster. This will make Cogley more sensitive to the opinions of higher-status agents. It also means that he will be more likely to adopt the ideas and values that higher-status agents promote.

Socially Enhanced Attention. Researchers who study socially guided learning processes routinely divide the topic into subareas. At the bottom of this hierarchy are mechanisms that promote learning based on the enhancement of cues by social attention. At the top of the hierarchy are mechanisms that result in the mimicry and imitation. This hierarchy has led many to conclude that mimicry and imitation are somehow more important kinds of social learning processes, and to emphasize the study of these mechanisms over those lower in the hierarchy. Mimicry and imitation may be more complex, but they are not the most widespread social learning processes. Stimulus enhancement effects are not only more widespread; they appear to be prerequisites for higher-order social learning processes like mimicry and imitation.

One reason that stimulus enhancement effects are important for social learning is that they direct an agent's attention to the cues and behaviors that occupy the attention of their peers. As a result, even if his peers have not yet learned how to perform certain tasks, he can nevertheless alert agents to the presence of interesting features and motivate them to investigate them. To engage these social attention effects in Cogley, we have already proposed that he should track the gaze of other agents and use the direction of their gaze to enhance the saliency of the objects in their field of view. Whenever Cogley tracks the gaze of other agents, he should favor looking where they look. To make this attention enhancement process more sensitive to the interest of other agents, it will work best if we also modulate its effects by raising its gain in proportion to how closely other agents inspect the object or become excited as they attend to it. It also seems reasonable to raise its gain in proportion to the status of the investigating agent. Objects that get more attention and/or produce stronger reactions from agents, especially from higher-status agents, should therefore attract more attention from Cogley.

A Socially Enhanced Cogley

There are several behavioral outcomes that can provide useful insights into whether Cogley's social enhancement mechanisms are sufficiently strong and well tuned. As already noted, social agents are motivated to seek group approval for their behavior. However, what we have not yet emphasized is that in humans, this is also reflected in a tendency for them to seek approval for their ideas and opinions. Classic studies in social psychology by Solomon Asch have shown that the ideas of the group readily influence the judgments and opinions of others, even judgments about basic perceptual properties.[8] For example, in one study subjects were shown a picture and subsequently were asked to pick which of three comparison choices most closely matched what they had seen (see the comparisons in Figure 18-1). The tests were arranged so that the choices were just different enough that most subjects could choose accurately when they were alone. However, if a subject was placed in a group during testing, and the other members of the group, having all gone first, agreed on an incorrect choice, then about one-third of the subjects immediately agreed with the group on the mismatched choice. In effect, they changed their judgment to conform to the group. This *conformity effect* occurred even when the group size was as small as three, and it tended to be stronger with larger groups.

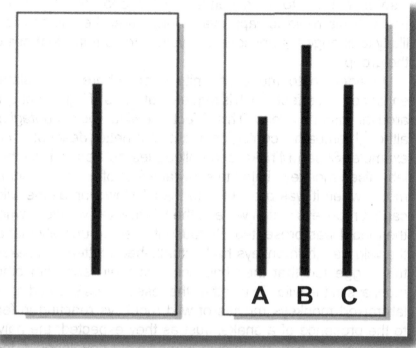

Figure 18-1: One comparison used in the Asch's studies.

The conformity effect is obviously related to the need for approval. Agreeing with the group is a way of seeking approval by avoiding conflict. Note, however, that if a subject conforms to group influences when making judgments about basic perceptions for which they are relatively certain, then they are even more likely to conform to group influences when they are making judgments about topics which are more subjective. For example, judgments about the goodness of certain agents or causes are less anchored in direct observations and are more likely to be influenced by group opinions. As this scenario suggests, the drive to conform may sometimes introduce conflicts between personal judgments and those of the group. One way to counter this conflict is to associate with agents who share similar ideas and values. Thus, as agents mature, they tend to seek out groups of like-minded others whenever possible. So if we have designed Cogley well, we should expect him to select groups that share his values. However, we must also expect him to be influenced by the ideas of his group at times. If his need for approval is high, then he will be more likely to change his opinions and values to conform to those of the group.

In addition to the conformity effect, there is a related enhancement effect which is triggered at times of high emotional arousal within a group. This effect is called *social contagion*, although empathic contagion might be a better descriptor. For example, Susan Mineka and colleagues noticed that all their lab-raised monkeys failed to show any fear of a realistic snake model when it was presented to them.[9] However, all the wild-caught monkeys which were in the lab reacted with fear when the model was presented. Because it seemed unlikely that all the wild-caught monkeys had actually been bitten by snakes, this suggested that they had acquired their fear from other monkeys in the wild. To test this, the researchers showed naïve lab-raised monkeys movies of wild monkeys reacting in fear to the presence of a snake. Just as they expected, the naïve monkeys who saw the movie immediately developed a fear of snakes. In short, it appears that monkeys are biologically prepared to develop fear reactions to certain kinds of cues, and that fear reactions can be acquired via social contagion.

Anecdotal evidence suggests that human children also tend to develop fears to things like snakes and spiders if they first encounter them in the presence of a caretaker who reacts fearfully to them. The obvious advantage of being able to learn about danger by watching the reaction of others is that not every individual needs to be bitten by a snake or a spider before he learns to avoid them. The misfortunes of a few can be used to inform the entire social group. Once established, such knowledge can be passed on across generations by social contagion. In humans, social contagion also commonly occurs for positive emotions. For example, the popularity of songs, fashions, and celebrities often spreads through a group in contagious spurts. If we have designed Cogley with the right balance of empathic signaling and recognition processes, and with a drive to attend to the behavior of those in his group, then he will also experience emotions contagiously, and many of his values will be influenced by feelings promoted by his social group.

GROUP IDENTIFICATION STAGES

In addition to the social enhancement mechanisms described above, there are several group identification stages in human development which determine who is most likely to influence a social agent in particular developmental stages. Following a bio-inspired model, Cogley will need to go through similar developmental stages. Three primary group identification phases are proposed:

1. A parental attachment phase
2. An early gender/peer-group identification phase
3. An adult gender/peer-group identification phase

The latter phase will mark the transition from parental attachments to peer-group attachments. Obviously, we cannot develop these in much detail here. However, we need to identify the basic processes that Cogley will need to support his social development in each phase, so a brief overview of how these stages come about in humans seems like a good place to start.

Parental Attachment. To begin his socialization, Cogley will need to form close attachments with a few primary caretakers. These will be the equivalent of Cogley's core family. They will guide his early development. In simpler animals, parental attachments are often based on imprinting mechanisms. Human socialization also begins with imprinting-like attachments. To imprint successfully, Cogley will need sufficiently well-developed recognition systems in place to identify particular individuals. In many mammals imprinting starts with olfactory or vocal cues, because visual recognition skills are often slower to develop. However, there is much variability across species. For Cogley, let's begin with vocal recognition and add facial recognition as the visual networks become more discriminating. Cogley will also need dedicated social memory networks which can categorize individuals on an *ingroup-outgroup* dimension. This ingroup-outgroup dimension will be important for gating Cogley's social behavior in all three stages. Cogley's core family members will be his initial ingroup.

To help Cogley form social attachments that are consistent with the values of his family, the attachment networks should be designed so that repeated and/or strong negative reactions to new agents by those in his family group should increase his anxiety for them. This will make it less likely that he comes to trust them. In contrast, if family members react positively to new agents, then Cogley should be more likely to trust them, assuming they also react to him with positive sociality signals. Following this process Cogley will tend to develop associations that categorize other agents along an extended-family, *ingroup-outgroup* dimension. Ingroup agents should be considered safe and trustworthy, whereas outgroup agents should be considered suspect and untrustworthy. Those ingroup agents who provide him with the most consistent positive sociality signals should be those with which Cogley develops the strongest attachments. They will, in effect, become his best friends.

As a general rule of thumb, human agents tend to copy ideas and mannerisms from agents within their ingroup while rejecting those distinctly associated with outgroup agents. To match this human developmental pattern, Cogley's socialization processes should be arranged so that he is motivated to share and copy ideas from those high on the ingroup dimension, and to avoid sharing or copying ideas associated with outgroup agents. Perhaps the simplest way to ensure this would be to arrange it so that attention to trusted ingroup members releases strong sociality signals, and attention to obvious outgroup agents increases stress and avoidance signals. Uncategorized agents should probably start in the range of mildly pleasant to fearful, based on similarities between their vocal and facial features and those of known ingroup and outgroup members. This similarity bias may result in Cogley developing early cultural and/or racial stereotypes. Although this is not an ideal outcome, it is nevertheless a fairly normal outcome for young humans.

Early Gender Peer-Group Associations. We have arbitrarily chosen to anchor Cogley's gender assignment to male-like mannerisms and voice qualities. It might seem unnecessary to bias him to identify more strongly with a gender role beyond that, except that in a real sense, without the motivational biases that go with sexual identity, Cogley will not be as human-like as we would like him to be. He will also need to be sensitive to the gender classes of his human partners, if for no other reason than to help him get his gender pronoun assignments correct. Humans are quite sensitive about this. Cogley will never be accepted as an intelligent agent if he fails to use gender pronouns correctly. To do this, he will need to assign the agents he meets on a male-female dimension based on voice qualities, body features, clothing styles, and other gender-specific behavioral conventions. The culturally defined cues can largely be learned based on their association with gender pronoun usage as Cogley acquires language. However, a few core gender cues like voice tones and the aggressiveness of play should probably be built in from the start. This means the cultural aspects of sexual categorization would develop roughly in parallel with his sexual pronoun usage, with a little boost from some "innate" gender categorization processes.

Given that Cogley will be categorizing human agents as males and females, and given that he will have characteristics that fit the male category, it follows that it would be appropriate for him to form ingroup associations with male peers and outgroup associations with females. Now, the curious thing about gender associations is that children form them quite early. Young boys seem to know that they are boys. They prefer more aggressive styles of play and tend to engage in competitive rough and tumble play with other boys. Girls identify with other girls and prefer more social play activities, such as pretending that dolls are children. These gender differences don't depend on pronoun labels or whether a child is raised culturally as a boy or girl. In fact, gender differences in play appear in chimpanzees in the absence of any obvious training. It seems that primate children identify with same-sex peers more because they share similar play styles and gender-specific motivational preferences. Because children learn more from ingroup partners, once this gender identification is made, it further influences who they play with and learn from, which in turn further polarizes their play styles and behavior along gender-specific lines.

Unlike the extended family ingroup-outgroup dimension, which results in trusted versus untrusted reactivity dispositions, Cogley's reactions to gender ingroup and outgroup agents in this phase should be scaled on a reactivity dimension of *wanting to be more like* versus *wanting to be less like*. This bias should be superimposed on his social ingroup-outgroup biases so that he becomes more likely to copy the social behaviors of same-gender ingroup partners. Though Cogley will still associate with both male and female agents during this developmental phase, and though he will be capable of learning from both, he will have a bias to associate more with male peers and to copy ideas and behaviors from them more during this phase. Perhaps the simplest way to shape this bias is to arrange for gender ingroup agents to trigger stronger sociality reactions in Cogley than gender outgroup agents do in this phase.

Puberty and the Transition to Adult Peer-Group Attachments. Beginning around puberty, adolescent humans not only develop an interest in agents of the opposite sex, but they also lessen their attachment to family and parental models, sometimes to the point of rebelling completely from parental rules and values. As this occurs they come to be more attached to members of their peer group. To encourage a similar transition to greater control by peer-groups and lesser control by family, the influence of Cogley's parental partners should be dialed down once he reaches puberty so that his peer-group partners come to have more influence. Simultaneously, he should become more concerned about obtaining the approval of his peers and gaining status by imitating higher-status peers. In this manner, the ideas and values of his peer groups will come to be increasingly important influences on Cogley's personal values.

The gender ingroup-outgroup distinctions of the previous phase do not break down in this stage – they actually become stronger. However, features that strongly categorize agents as "other gender" suddenly come to be seen as interesting and attractive. As one theorist described it, what was once considered exotic becomes erotic.[10] In humans, this transition is associated with the onset of higher levels of sex hormones. These sex hormones activate the development of sexually dimorphic physical features and brain networks related to sexual interest, social competition, and caretaking. The parts of the brain most affected by these changes are the status and reactivity networks in the extended amygdala and hypothalamus. To approximate a similar transition from gender-role identity to adult dispositions complete with interests in members of the opposite sex, Cogley will also need "hormonal" signals to activate gender-related behaviors and dispositions at puberty.

Given that Cogley won't be capable of reproduction with humans, we won't want to turn on strong levels of sexual attraction or other behavioral dispositions that cannot be reasonably consummated. Thus, there may be some value in introducing limits to our bio-design emphasis in this area. However, if Cogley is to understand sexual attraction and why these dispositions are important to his human partners, then he will need to associate different interoceptive feelings and motives with different gender roles. These associations should modify his behavior toward same-sex and opposite-sex agents. Obviously, we do not want to initiate this transition until Cogley's sexual identity is well-formed. Perhaps this should be a rite-of-passage decision made by his designers. It might also be useful to lower Cogley's voice pitch when puberty begins, to make his human partners more aware of this transition.

There is yet one more motivational disposition that Cogley will need to encourage proper socialization once he reaches puberty. When we first discussed sensorimotor learning, we noted that play is essential for learning about the affordances that support various sensorimotor activities and what effectivities those activities produce. Productive planning would be impossible if agents could not recognize what activities were possible in a situation and anticipate what changes those activities might produce. When we looked into the emergence of language, we noted that babbling, early vocal play, serves a similar role in learning how to connect vocal skills with auditory signals and their effects. Again, when we discussed tool use, we noted that a certain amount of play was necessary for humans to learn about the affordances and effectivities that new objects provide. What may be less obvious is that new kinds of play are also essential for learning about the social and sexual affordances that emerge at puberty. If we want Cogley to become a productive social agent, then we will need to introduce drives that motivate a healthy level of social play. These should include gender-appropriate flirtatious play, once he reaches puberty, so that he will learn something about the social and sexual affordances activated in this phase.

The adolescent transition from parental values and childhood drives to peer-guided values also plays a critical role in cultural evolution. Because human culture is continually in flux, each new generation emerges and forms its values in a slightly different cultural scaffolding from that of its parents. The transition to peer group attachments often promotes new customs and attitudes, some of which are at odds with parental values. Being conservative, we may not want to allow too much parental defiance to develop in Cogley, at least not enough that he rejects all the values we have carefully tried to teach him. However, we must allow him to develop his own peer-based values. All parents share similar concerns in this transition. Usually they come to accept them, largely because they have little choice – they cannot control the transition. As parental engineers we will have more control over how Cogley makes this transition, and we may be tempted to manage his transition to adult life more carefully. However, at some point we must let Cogley be his own person.

CONCLUSION

We've covered a lot of ground in the last few chapters, and it is time to put these layers of mind and self in better perspective. The diagram in Figure 18-2 illustrates the nested relationship among the layers of mind we have proposed. At the core of conscious mind is a dynamic reentrant multitasking process of interoceptively guided attention. This process is represented by the inner circle in this diagram. It results from the resolution of competing schemas via interoceptively guided attention. The nexus of attention that emerges here is at the core of self awareness. It is this emergent organization that experiences the *feeling of what happens* and gives rise to a core sense of self – a feeling-guided resonant self, a gaze-centered observer self, a learning-guided anticipatory self, and a sense of self agency as it learns to anticipate the effect of its own plans.

A more individualized sense of self agency develops as concepts, skills, and values are put to work. Skills and concepts are shaped by the tasks that engage the nexus of attention most often. The skills acquired in those tasks feed back to make new tasks possible. This results in the incremental growth of concepts, skills, and feelings along task-related paths. The attention managing skills, which emerge in this process, result in a self that can reason about its experience. Declarative memories add a dimension of history and continuity to the self by placing current experiences in the context of past ones. This historically shaped developmental layer of mind is represented by the second circle in the diagram. It is a storm of attention with a uniquely individualized history and character. It can even re-experience parts of its own history of attention by recalling episodic memories.

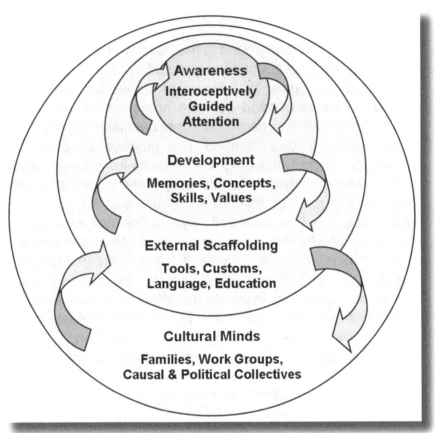

Figure 18-2: Nested contributors to the organization of mind.

Culture extends this developmental layer by adding a scaffolding of physical artifacts and conventions, including language and formal systems of education. We sometimes lose track of how much this training changes who we are and what we can do. However, without such training most of us would never acquire many of the basic skills that we expect of educated minds, like reading, writing, and counting. As Andy Clark has emphasized, the technological artifacts that form in this layer also extend what skills agents within a culture come to employ, what facts they learn, and what features they come to recognize as extended affordances.[11] The sense of self is marvellously extended by this scaffolding of language, tools, customs, and educational processes. This layer of self is represented by the third circle in the diagram. Some of the customs and procedures in this layer – in particular, those used to educate new agents – also serve to guide the development of skills essential to an agent's participation in cultural minds. Thus, the skills acquired in this layer serve as an interface between the development of individual minds and the emergence of collective minds.

In modern human societies there are several dimensions of collective mind, families, workgroups, interest groups, ethnic, national, and even multinational communities. These are the nested social environments in which each new agent develops and the collectives with which agents come to identify. The reactions of these collectives guide the development of the agents within them, even as each agent contributes to the trajectory of the collectives they join. Thus, just as the master-apprentice architecture enables the cerebral cortex and the basal ganglia to guide the development of each other, individual minds and cultural minds interact and guide each other's development. Because interoceptive loops in the core self are strongly tuned to social status and values, the sense of self that develops within the inner loops depends strongly on the cumulative effects of relationships in this outer circle. Individual minds are bound to the group more than they know, perhaps more than they can know. Most of their values are group values, which they come to accept without question. This is what makes cultural minds such potent forces. They literally shape the environment within which individual minds develop, and Cogley is now ready to join in this process.

CHAPTER 19
THE STUFF OF CONSCIOUSNESS

Having explored the emergence of mind, consciousness, and the origins of the self, there is one more controversial topic which we can now address. That topic is qualia. Qualia are generally defined as the subjective qualities of "what it is like to experience" something.[1] However, the concept is controversial, at least among philosophers. Taking a reductionist approach, Daniel Dennett suggests that the concept is not really needed to explain consciousness.[2] He claims that a sufficiently detailed physical and functional account of how experience is implemented in the brain would leave nothing more to be explained. David Chalmers, in contrast, "knows" that qualia exist because he experiences them.[3] However, he agrees that reductionist mechanisms within the brain do all the work of mind, and he suggests that qualia may simply be epiphenomena, experiences that run in parallel with physical interactions but that have no functionality. The fact that experience is disconnected from function, he claims, is why explaining qualia are such a hard problem.

In contrast to these old ways of thinking, a constructionist analysis of the organizations in which consciousness emerges explains why qualia must exist. They are the stuff of conscious experience. Just as the microlaws for chemistry cannot fully explain the properties of life, the microlaws of neural networks cannot fully explain how experience is implemented in the brain, or why conscious minds

389

feel experience as they do. Qualia only come into existence within storms of attention, and they can only be fully understood within that context. They are the categorical tags and valences which guide the attention of conscious agents so as to make their experiences interpretable. Without qualia, there would be no *feeling of what happens.* And contrary to the claims of some, this implies that qualia can and do have real effects in the physical world. Feeling-guided decisions influence how conscious agents interact in their world. Cogley will experience qualia. Those experiences will be part of his attention, and they will influence his decisions. That's what it means to be conscious.

THE STUFF OF CONSCIOUSNESS

Why is it that when our cognitive systems engage in visual and auditory information-processing, we have visual or auditory experience: the quality of deep blue, the sensation of middle C? How can we explain why there is something it is like to entertain a mental image, or to experience an emotion? It is widely agreed that experience arises from a physical basis, but we have no good explanation of why and how it so arises. Why should physical processing give rise to a rich inner life at all? It seems objectively unreasonable that it should, and yet it does. – David Chalmers, "Facing Up to the Problem of Consciousness," 1995

Understanding how the physical stuff of the nervous system could possibly account for the subjective qualities of mental experience, *qualia,* as they are sometimes called, is what the philosopher David Chalmers has called the "hard problem'" of consciousness.[4] Chalmers claims it is counterintuitive to think that the physical stuff of the body should possibly give rise to such a rich inner world of experience. The explanatory gap between physical processes and experience seems too wide, and he sees no obvious mechanism that would tie them together. This leads him to conclude that physical effects and experiential effects are essentially different, perhaps even existing in separate realms of reality. Following on these dualistic assumptions, Chalmers suggests that conscious experience is simply an epiphenomenon, an experience that runs in parallel with physical events but that has no effect on the physical world. This position leads him to conclude that mental experience serves no function. As Adam Zeman notes in a review of Chalmers ideas:

> It is paradoxical to discover that the very author who emphasizes the importance of conscious experience is subsequently driven to conclude "that experience is a beautiful but functionally irrelevant embellishment of the physical process."[5]

However, the conclusion that conscious experience has no function is problematic. If we can experience something, then we should be able to react to the experience. Indeed, Chalmers argues that experience is part of the difference between the conscious reports of humans and the mechanical reports provided by computerized systems.[6] Humans must experience events in order to report on them, whereas computers have no experience. However, if humans must experience before they can report, then their experience seems to be having an effect in the physical world. Decisions can only be changed based on experiential reports. In order to keep experience from interacting with the real world, Chalmers posits that experience always has a parallel physical instantiation and that all the functional interactions occur on this physical dimension. When we see a beautiful sunset, there are physical effects in the nervous system, and we report on those physical states. The subjective qualities of experience, we are told, just rides along in another realm of reality but serve no function. But how can we be sure that conscious experience plays no role in what is reported, and yet argue that the processes leading to subjective experience must occur before we can report on them? If experience runs in parallel with processes in the nervous system, then why posit that it exists in a separate realm? Chalmers simply appeals to his intuitions that mental phenomena seem too ethereal to be part of the physical world, and he is not alone in this position. Other experts on consciousness do not see a clear link between qualia and the physical world. For example, consider these comments. The first is by Francis Crick and Christof Koch. The second is by Gerald Edelman.

Let's first forget about the really difficult aspects, like subjective feelings, for they may not have a scientific solution. The subjective state of play, of pain, of pleasure, of seeing blue, of smelling a rose – there seems to be a huge jump between the materialistic level, of explaining molecules and neurons, and the subjective level.[7]

Qualia are not themselves causal, and to assume otherwise would go against the laws of physics.[8]

The question of how experience emerges from physical substances calls to mind the relationship between the mental and the physical, which Brentano was concerned about when he introduced the term *intentionality* more than one hundred years ago.[9] As Brentano noted, there is an *aboutness* inherent in mental phenomena. For Brentano this quality meant that thought transcended the physical. It implied that the link between mental phenomena and the physical world could not be explained on a purely physical basis. Still, Brentano assumed that the mental and the physical were connected in some way. In contrast, Chalmers adopts a dualistic position. He accepts that mental functions must somehow arise from physical processes, but he wants to separate the functional aspects of interactions on the physical level from the experiential qualities of mentality, its qualia. He rejects the possibility that qualia could have a function. Instead, he attributes the functional aspects of mental processing to underlying neural mechanisms. He views experiential qualities as essentially different and concludes that they are not really needed because neural mechanisms are available to do all the real work. However, is this division between function and experience reasonable? To answer this question, we need to look more closely at the nature of the work we attribute to mentality and whether experience can really be separated from it.

When we first addressed Brentano's concept of intentionality, we noted that a key strategy for understanding emergent properties was to recognize that novel phenomena emerge within new kinds of organizations. The nature of the organizations that led to intentionality was illustrated in Ruth Millikan's analysis of signaling systems.[10] Millikan recognized that even a simple communication signal, like an animal's alarm call, is intentional in the sense that Brentano proposed because though the signal refers to something, the referent can be errant or nonexistent. For example, an eagle alarm might be produced when a vervet monkey has only seen a pigeon, yet the group may react to the alarm as if there was a real predator. In this manner, such signals can represent events and relationships in the world without being physically connected to them. As Millikan explains, the aboutness of a signal results from the co-aligned organization of signal producers and signal consumers. It is that interpretive alignment, not some magical internal property, which enables a signal to mediate an adaptive outcome.

There is also an interpretational aspect to signal meaning. A demonstration sometimes used in introductory psychology classes may help illustrate how interpretations modify signal meanings. A group of students is assembled in a large circle. One student is shown a somewhat ambiguous drawing and is then asked to reproduce it from memory and show that reproduction to the student on their right. The student on the right then repeats the task. A verbal label, added at the bottom of the picture, is also passed along. A second group is then started with the same drawing but a different label. The labels bias each student's interpretation of the drawing they receive. When the final drawings return to the start of the circle, they demonstrate how small changes in each student's interpretation can lead to radically different results in what features are noticed and emphasized. A similar kind of interpretation goes on across processes in the brain, except that in the brain, each successive system has a different functional bias. The cones in the retina detect the arrival photons of differing wavelengths. The ganglion cells collect the inputs from the cones, but they reorganize the signals. By the time the signals reach the lateral geniculate nuclei in the thalamus, colors are coded in red-green and yellow-blue contrast signals, even though the cones are only sensitive to red, green, and blue. Yellow is a reinterpretation of conjoint inputs from red and green cones.

When the visual information is processed in the sensory cortex, color information is combined with interpretations based on edges, areas, and other elemental visual features. Relative differences in the position of features detected between signals arriving from the two eyes can be said to represent distances because those differences can be interpreted as distance cues by other functionally coordinated perceptual systems. Combinations of edges, colors, and areas can represent objects because later network systems have evolved functions to interpret feature combinations as coherent perceptual units. The feature sets can be identified based on their similarity to past feature sets. Memory and emotional valuations can be added based on associations with previous feature sets. In a few more steps we may recognize such a chain of signal interpretations as the warm smile of an old friend, a partly subjective experience. However, there is no mystery here. Neural signals triggered by physical stuff, reflected photons, gradually come to be interpreted as a mental stuff because a variety of adaptively coordinated brain subsystems convert this cascade of signals into representations that evoke feelings of familiarity and friendship.

Recognizing that the connection between the physical world and the mental world is based on co-aligned signaling processes is the first step in understanding how physical events give rise to mental events. Mental events are not simply the result of the physical interactions of neurons. They emerge when signaling systems are coordinated in function. As a series of signals traverse through the nervous system, each subsequent consumer system interprets the meaning of the signals it receives in relation to its own intrinsic function. Each subsystem can then signal about its interpretation to yet other consumer subsystems, which may apply yet other functional dispositions to subsequent interpretations. The overall meaning of this cascade of signals is thereby shaped by a collective of adaptive functions in a spiral of representations (interpretations) and re-representations (reinterpretations). It is the cumulative effect of all these distributed representations which enable signal meanings to grade from simple indicators of sensory activity to complex representations of perceptual and emotional experience.

The interactive combination of different aspects of aboutness accounts for the brain's ability to form mental representations. Storms of attention are recursive loops that form and react to discriminable interpretations of their own making as they gain attention. The human brain creates thousands of such representations each second. These bind together in extended schemas and compete for attention within the thalamocortical architecture of the brain. The transient assemblies created in this process make yet higher-order kinds of strange-loop aboutness possible. However, only those loops with the strongest interoceptive associations are likely to gain attention. Just as the sense of self emerges in an upward spiral of increasingly refined representations and re-representations, the qualities of conscious experience emerge in an upward spiral of re-representations. Qualia begin as vague categorizing and pulling forces at the pre-attention level. Those forces are the initial stuff of consciousness, and they feed back into cycles of binding and re-interpretation to enhance ongoing qualities of experience and guide subsequent interpretations.

As the quote that begins this chapter makes clear, David Chalmers is struggling with how to deal with the emergent properties of conscious experience and why they seem novel and discontinuous. He perceives the material world as being made of solid stuff, and consciousness as being ethereal and essentially different. He argues that it is unreasonable to assume that mental experience could have arisen from physical processes, and so he suggests that the material world and conscious experience must be disconnected. However, when we consider that the physical world is formed from interacting clouds of wave-like particles, quarks, and leptons, and is held together by gluon, photon, and boson force carriers which reach out in ways we don't understand, the properties of physical stuff don't seem any more reasonable. The emergence of conscious awareness from the interaction of neural systems is no more unreasonable than the emergence of solid matter from the interaction of quantum particles. As with many emergent properties, conscious experience is not readily predicted by the properties of the components parts that make it possible. However, emergent properties can only be considered unreasonable if we assume that we should have been able to predict them.

Emergent phenomena seem unreasonable when we cannot connect them to processes on simpler levels. However, the ethereal nature of emergent properties fades when we can plot a path through the adjacent possible that links those properties to the properties of systems with which we are already familiar. Once we understand that the fundamental forces of nature cause some particles to coalesce into the stable units we call atoms, and that physical objects are solid when their atoms form strong bonds with adjacent atoms, then it no longer seems unreasonable to think that solid objects might emerge from the interactions of wave-like particles. It is the known links between functions on different levels, and our familiarity with them, that makes those properties seem reasonable and connected. The nature of life is more understandable because we can connect it to biochemical systems that encode plans for proteins in base-pair sequences within DNA molecules, and the fact that those codes can be duplicated. So too, the nature of conscious experience becomes more understandable as we begin to see how it is connected with the physical world through a chain of interpretations among adaptively coordinated perceptual, motor, and interoceptive brain modules.

Emergent properties result from interactions of components on many levels of organization, but when those components can be connected together in a functional web of causation, they must be considered part of the same reality. We may think of speed and acceleration as ethereal properties, but we don't assume they exist in a separate reality disconnected from the automobile that demonstrates those properties. Nor must the experiences of comparing memories with ongoing perceptions exist in a separate reality from the physical world being perceived or the motor actions that they guide. They must be part of the same reality because they are functionally interconnected with the other parts. They may involve largely different kinds of organizations. However, once those organizations are connected by communication signals, they can interact to achieve complex representations and outcomes. And because communication signals in the nervous system flow in both ascending and descending pathways, the functional influences propagate in both directions.

Reflective consciousness emerges from the interpretations that accompany the comparison of perceptions with links to co-occurring associations formed in previous experiences. The perceptual networks are biological structures formed by interconnecting neurons. Their processing is based on physiochemical events. Their memories are stored in the connection weights that link back to the original perceptual networks. However, the connections are not merely physical links. They are adaptively aligned links which have experiential effects. Because memory networks align sets of perceptual connections, a cue associated with one past event can trigger signals that enable the event to be partially re-experienced. Memories can only be fully explained within such aligned organizations. However, that doesn't prevent mental processes from interacting with the physical world. The driver of a car may detect conditions on the road ahead, which valuation networks match with memories of a possible risk. The perceived risk may not even be a valid read of the situation. The object may be no more than an odd shadow misperceived in the evening light. However, when the shadow is interpreted as a child moving onto the road, it triggers affective states that activate planning circuits that cause a sequence of well-practiced motor actions to press the brake and stop the vehicle.

The links between qualia and actions in the physical world

don't defy the laws of physics – they extend them. The fear of harming a child is not an epiphenomenon that occurs while some neurons decide what action to take. The fear emerges from the interpretation of signals by high-level collectives of adaptively coordinated knower modules. It's the interaction of those interpretations that enables them to gain attention and guide a cascade of actions. Without the subjective interpretation that a child is in danger, there is no reason to stop. For the memory and planning modules in the brain, the signals are *about* stopping the forward movement of the car before it reaches the child. For the motor execution circuits, the signals are *about* arranging a well-practiced sequence of motor commands to press the foot against a particular target object. For the individual leg muscles, the signals are simply *about* when and how much to contract. In this way, mental interpretations are translated into simpler co-aligned signals until they interface with muscle systems, which translate the signals into mechanical forces in the physical world.

WHY QUALIA VARY, AND HOW WE KNOW

Philosophers sometimes ask what would happen if your color spectrum was inverted, so you had the subjective experience of green where others saw red; and the experience of blue where others saw yellow, and so on. A common assumption is that such a difference would be immediately apparent. However, that is not necessarily true. If you began life with this reversed experience, you would always see the sky as yellow but learn that we call that experience blue. You would see blood as green and describe it with the word red. When asked what color falls between red and yellow, you would recall the experience of blue-green, and report to the questioner that it was orange, the term you learned to associate with blue-green. Warm green sunsets would be a common experience that you enjoyed. In short, your behavior would be appropriate and consistent with that of others around you. In fact, there would be no obvious way of knowing that your qualia were different, because the labels that you used to describe them would be situationally consistent with those used by others.

However, if you grew up with "standard qualia'" for colors, and then one day the ganglion cells that carry red-green and yellow-blue contrasts were switched in route, then your color spectrum would suddenly be inverted, and you would immediately become confused. Stop signs would appear green, and you would call them green, but your memories of them would be for red. People would look strangely at you when you described the sky as yellow, and you would be confused because your memories would lead you to expect the sky to be blue. However, after a couple of weeks of these inverted colors, the perceptual networks in your brain would begin to adjust your actions and label usage to fit the new qualia. Seeing yellow would now remind you of the sky and relax you. Seeing flashing green lights would now alert you to stop. To be able to communicate with those in your language community, you would begin to call stop signs red even though you saw green, and you would request blueberry muffins for breakfast even though the berries looked yellow. In short order you would soon act normal again, and there would be no obvious way for anyone else to know the difference. This implies that there is no reason to assume everyone has standard qualia to begin with, any more than there is reason to assume they have standard childhoods.

We can be confident in these conclusions for several reasons. For one thing, we know that perceptions can be remapped to fit new sensorimotor states. Over a century ago, George Stratton put on an optical device that inverted his visual field from up to down and from left to right.[11] Initially, he was severely disoriented by the experience, turning left instead of right, reaching down instead of up, and the like. Yet after only eight days his visual world was realigned with his behavior, and he was again able to coordinate his motor behavior with his visual perceptions. In fact, he was able to ride his bicycle across campus without problems. Years later, the Austrian investigator Ivo Kohler and his students carried similar exercises to an extreme, sometimes wearing such devices for months.[12] With experience, one student became so well adapted that he was able to ski in the Alps while wearing up-down inversion goggles, and another rode a motorcycle through the streets of Innsbruck with left-right reversal goggles.

What these exercises demonstrate is that elemental sensations gain their experiential character from their alignment with subsequent valuations and actions. The experience of red is a salient visual event. Among other things, it is distinguished by the activation of finely tuned receptors in the retina. In addition, red is often associated with injuries and bleeding. This unpleasant connection is extended by many cultural uses of red signals to indicate danger or penalty. Green, the color of vegetation, is experienced as more serene. However, if green were associated with bleeding, and red were associated with the tranquillity of pastoral scenes, then "the something it is like" to experience these colors would be reversed. Qualia are the collective functional interpretations that storms of attention give to signals based on the situations and valuations with which they are associated. That is the source of *what it is like to experience* them. Change those associations, and the interpretation of what it is like to experience them also changes.

Defective and Enhanced Color Perception

We know that color interpretations vary widely. Some people are color blind – that is, they have defective color vision systems. There are three primary types of cones in the eye, each with a different photosensitive pigment. The pigment in red cones is most sensitive to wavelengths which on average are around 580 nm. The green pigment is most sensitive around 540 nm. The blue pigment is most sensitive around 450 nm. Defects in color vision occur when color coding in any one of the three cones types fails to function properly. For example, one of the photosensitive pigments may be underproduced or may be defective. Because the genes for these photosensitive pigments are located on the X chromosome, and human males have only one X chromosome, males stand a higher risk of color deficiency than females. The standard tests for color blindness use dots of varying colors to form letters and numerals. People with deficient color vision do not always see all the colors and have difficulty recognizing some characters in the test. However, many men never take such tests, and though their qualia may vary within large ranges, they never become aware of it.

Recently scientists have noted that because females have two X chromosomes, it is even possible for some women to have genes that produce two variants of a color photopigment. For example, the pigment in the green cones of some people reacts most strongly to light at a 539 nm wavelength, and the pigment in others reacts more strongly to light at a 546 nm wavelength. A female with genes coding for both these pigments could, in effect, have two points of green cone sensitivity. Similar variations for red and blue pigments also occur. To investigate the possibility of enhanced color sensitivity. Researchers had subjects mark the number of distinct chromatic bands they perceived in a rainbow, a diffracted spectrum projected onto a translucent surface.[13] Male dichromats, known color deficient individuals, perceived an average of 5.3 different color bands. Color normal trichromats perceived an average of 7.6 distinct bands. However, heterozygous female trichromats, those with two variants of some color genes, saw more variations. They perceived an average of 10.0 different color bands in the rainbow.

As the work described above illustrates, the fact that we cannot directly experience the qualia of others does not prevent researchers from studying how their qualia differ. Differences in qualia can be assessed with objective tests. The assumption that it is impossible to study conscious experience without language is also incorrect. In fact, language is often not a reliable method for studying experience, because qualia-specific labels, such as red, may be associated with different red experiences. For that reason, most tests for differences in qualia rely on non-language tasks, like the color banding procedure described above, or the character recognition tests used to check for color blindness. If your receptors are fairly standard, it seems likely that your experiences of color will share many properties in common with those of others, because the visual networks in your brain are likely to interpret them in similar ways. However, there is no reason to assume that the qualities of your experience will be identical to those of others.

There is also ample evidence that qualia vary within individuals. For example, the production of photosensitive pigments is known to vary with health and age. However, it is not just that receptor systems vary, but that the interpretations made about what is perceived also vary. Sudden noises are more salient when we are fearful. The flicker of motion in the bush is more interesting when we are on the hunt. The smell of food is more sensual when we are hungry. Tastes often grow bland when we feel ill. We may remain completely unaware of what would normally be felt as painful injuries if the signals about them arrive in moments of high arousal. In short, qualia are not static properties that we passively detect. They are dynamically adjusted to prioritize what features should be more salient under particular conditions, and what activities should be favored in the moment.

MAKING THE HARD PROBLEM EASIER

David Chalmers assumes that consciousness has no function. He therefore concludes that functional explanations cannot contribute anything to our understanding of subjective experience. That, as he sees it, is what makes explaining the qualities of experience such a *hard problem*.

> The easy problems – explaining discrimination, integration, accessibility, internal monitoring, reportability, and so on – all concern the performance of various *functions*. For these phenomena, once we have explained how the relevant functions are performed, we have explained what needs to be explained. The hard problem, by contrast, is not a problem about how functions are performed. For any given function that we explain, it remains a nontrivial further question: why is the performance of this function associated with conscious experience? The sort of functional explanation that is suited to answering the easy problems is therefore not automatically suited to answering the hard problem.[14]

Note, however, that it is the assumption that "experience lacks function" which makes an explanation of qualia so difficult for Chalmers. As he argues, functional explanations are easier to resolve. Yet Chalmers goes to great lengths to deny any function to conscious experience, claiming that qualia are merely side effects of physical processes and that they lack any functional significance. However, as we have already noted qualia do have functions: They play a key role in the chain of signal interpretations that map from perceptual recognition to motor actions in the world. They help categorize our conscious representations of the world and bias our attention. The perceptions that bias attention and shape conscious decisions begin with physical events, but their interpretation depends on the interaction of co-aligned networks. There is no reason to assume that they operate in a different realm of reality from the physical world. We can specify enough of the properties which extend from the molecules of life to the signal interpretations of conscious mind to give us confidence that they are connected. There is no dichotomy here.

Chalmers assumes that qualia exist but have no function, while the philosopher Daniel Dennett takes a reductionist position regarding the "what it's like to experience" aspects of qualia. He does not deny such experiences exist, but he claims that although qualia seem special, no "emergent novelties" are really needed to explain them.[15] Dennett is correct to conclude that qualia must depend on the underlying physical phenomena with which they are correlated. However, he seems to be caught up in the "nothing buttery' version of the reductionist paradox when he concludes that because experiences are dependent on underlying mechanisms, qualia serve no function. The chemical processes that support living cells can be partly explained based on mechanistic interactions of physiochemical organizations. Yet we don't conclude that, because there are functional explanations for these processes, life does not exist. Instead, we conclude that there are combinations of physiochemical mechanisms that have interesting emergent properties when they are interconnected. They can extract resources and energy from their environment sufficiently well to sustain their ongoing interactions and replicate themselves. In short, functional explanations don't explain away life. They connect microlaw processes on a physiochemical level with the macrolaws of biological reactivity which bring systems to life.

In a similar way, functional explanations don't explain away the experience of qualia. They connect the microlaws for neural orienting, alerting, prioritizing, and activating, with the macrolaws of interoceptively guided attention those processes make possible. Such experiential systems do not just orient, alert, prioritize, and activate in response to physical changes – they also signal about their reactions and attend to those signals based on the strength of their interoceptive associations. They are therefore aware of some aspects of their orienting, alerting, prioritizing, and activating, and that added awareness feeds back into how they orient, alert, prioritize, and activate in reentrantly driven cycles. Being aware of and reacting to your own awareness is indeed a strange experiential loop, however it has a function. It enables an agent to attend and react to events in ways that increase its chances of surviving, because conscious reactions are prioritized by interoceptive values.

Chalmers suggests that some "added ingredient" is needed to connect functional explanations with conscious experience. However, once consciousness is connected to feelings grounded in survival needs, the added ingredient is obvious. Consciousness emerges within a massively parallel architecture which modulates the flow of competing signals while giving priority to signals aligned with salient interoceptive states. Every sensory signal is enhanced in proportion to its associations with interoceptive weightings. Every motor program is weighted by its potential to change interoceptive states. We feel the strength of those associations when they gain attention because interoceptive reactions are activated as they gain attention, and they in turn feed back to change the way the mind interprets the world in an upward spiral of representations and re-representations that guide attention, bias decisions, and adjust reactivity states.

Chalmers wonders why physical processes should give rise to such a rich inner life. However, the conscious agents he is inquiring about are signaling systems of immense complexity. If they didn't have richly coded internal representations, then they would not be able to represent subtle discriminations among their many experiences. If there was "nothing it was like to experience" them, then there would be no way for a storm of attention to tell separate experiences apart. Conscious minds must be able to represent their perceptual interpretations in ways that categorize and prioritize them. Only then can those interpretations gain their proper share of attention. Chalmers wonders why subjective experiences should be so interesting, but only the signals that a storm of attention finds interesting are likely to gain attention at any moment. Once we understand that consciousness emerges in storms of interoceptively guided attention, it is unreasonable to assume that conscious minds would not find the categorical tags and valences that gain attention interesting. The interest of an event is a measure of its ability to attract attention. Storms of attention have evolved to track interesting interpretations.

Although exploring interesting interpretations may be a reasonable strategy for conscious minds, there is no reason to assume that any two storms of attention should always find the same events interesting. Train any two complex neural networks on the same problem but vary the order of training, and you will find that the nodal weights within each network vary somewhat. Start with networks like those in human brains, which contains hundreds of billions of nodes, and then give them years of individualized training, and we can guarantee that no two brains will share the same nodal weightings. Add a little genetic variance to the initial organization, and they will take on their own personalities. We all share some common childhood experiences, and yet we all have largely individualized childhoods. We have different parents and different siblings (or none), and we are raised in different places. There are a number of commonalties in the way we mature, but our day-to-day experiences differ. Our childhood memories are unique. In the same way, no two minds will have the same qualitative weightings for how they come to interpret their experience. Their individual genetic codes vary. Their developmental histories follow different paths. There is no reason to assume that their subjective interpretations should always be closely aligned. To be subjective is to have interpretations which have been shaped by individualized histories.

It does, however, seem reasonable to assume that many

agents share similar functional interpretations to the extent they are shaped by similar brain organizations, similar histories, and similar biological needs. We all react to fearful cues with similar patterns of distress, although each of us reacts a little differently. We all enjoy the taste of food when hungry, although we do not all favor the same cuisine. Some aspects of mind are largely dependent on highly prepared biological dispositions, which we are more likely to share in common. Others depend on developmental processes that are likely to differ somewhat across individuals. Yet others are acquired through highly specific experiences and can be expected to vary largely across individuals. Some may be so different we can only reason about what they must be like to another mind by analogy. However, we don't have to experience the qualia of other minds to see those minds reacting to categorical tags and interoceptive valences. As Thomas Nagel noted, although "the subjective character of the experience of a person deaf and blind from birth is not accessible to me, for example, nor presumably is mine to him, this does not prevent us each from believing that the other's experience has such a subjective character".[16]

One of the goals of science is to make aspects of subjective experience public so that they can be studied more objectively. That process often involves special equipment and measuring protocols. Aside from that process, most of the phenomena of nature normally operate within subjective and private domains. Atoms encounter local forces of which we are rarely aware, although in studies we can objectively map out some of their properties. Gene networks "sense" attractive signal gradients that we can only partly objectify, and yet those gradients have profound effects on how those networks develop and behave. Neurons and neural modules in the brain have many experiences of which we are only vaguely aware. We can never know what it is like to be an atom or a gene or a neuron or a hippocampus, but that does not prevent us from recognizing that those organizations react to forces which have their own non-public character.

The fact that many of the experiences of conscious minds remain private and subjective is no different. However, because we sometimes share similar subjective experiences with other conscious minds we actually have the potential for more insight into their behavior than we do for atoms and neurons. We can not only form objective laws about conscious minds, we can also try to adopt their perspectives and empathize with them. Following this strategy, we can sometimes gain added insights regarding their behavior, even for nonhuman agents like Tom. Of course, subjective interpretations may sometimes be in error. However, objective data are rarely complete, especially for complex situations. In most cases, using objectively guided subjective interpretations is the best strategy we have for understanding other agents.

We can even see the pull of subjective forces at work in other species. Nothing is more entertaining than to watch Tom react to the qualia of his experiences with curiosity, excitement, caution, or playful abandon – qualities we all recognize in our own experience. The subjective character of Tom's experience is obvious. We can never know exactly the something it is like to be Tom, but that is also true when we consider what it is like to trade places with another human mind, or even to trade places with our own mind as it was a few years ago. As malleable properties with individual differences, qualia can never be completely standardized and objectified. However, the thoughts and behaviors of conscious agents can never be fully explained without considering the subjective qualities of their experience. And as the work on color vision illustrates, subjective differences are open to scientific analysis.

QUALIA FOR COGLEY

In one line of argument, Chalmers notes that although a computer can be programmed to report about a physical event, it does not "feel" the event on an experiential level.[17] In contrast, he notes that humans must sense events with conscious feelings *before* they can report on them. In Chalmers' way of thinking, there is no set of instructions that a programmer might add to a computer program that would produce subjective feelings. He is therefore driven to the proposal that physical systems and subjective experience must exist in different realms of reality. Does this mean that Cogley is doomed to operate only as a feelingless reporter? If he were simply programmed to report on events based on a simple detection threshold basis, then he would essentially be a reflexive reporting system. In many cases this is exactly what we want computer systems to do. However, if we want Cogley to operate as a conscious agent who must *experience* events before he can report on them, then he will need networks that bind his perceptions, actions, and interoceptive dispositions in schemas that compete for attention based on subjective interpretations.

The "symbol" grounding strategy often advocated in artificial intelligence circles suggests that the signs of a robot can be given some sense of meaning by grounding them in embodied connections to sensorimotor detectors. However that won't necessarily result in feelings. Suppose a robot had thousands of reflexive sensorimotor reporting functions. They could all still be operating as thousands of feelingless reporters. Selection processes could be put in place to evolve better reporters, but how meaningful would that be to a reflexive robot? Would a reflexive robot really care if it discovered that it would no longer be operating on the next day? Why should it care? Simply allowing Cogley to detect sensorimotor signals will not cause him to care about those signals. A robot reporting with reflexive functions would have no motive to care about what it perceives. It would react automatically whenever a signal reached some input threshold, not because it valued the experience in any subjective way. A selection algorithm could be used to craft better reporting functions, but if it only selects better reflexive functions, such an algorithm would never produce a robot that cares about what it notices or whether it will survive.

A feeling robot needs functions which energize and prioritize its perception based on its interoceptive evaluation of a situation. However, a consciously reporting robot must do even more. A reentrantly organized, resonantly aware conscious robot will have to experience events before it can report on them because its recognition and reporting functions will be tied to a system of interoceptively guided attention. This is the organization we have provided for Cogley. Cogley will supplement his value weightings with perceptually based memories of past perceptions and their associated valuations. What he attends to will depend on both preprogrammed (innate values) and on acquired value extensions. Cogley will have to experience events before he can report on them because his interpretations of those events will have to ramp up through a spiral of representations and re-representations in order to compete successfully for attention. Events with few associations with highly valued interoceptive states will never win the competition for attention indexes and will never become the subject of a conscious report.

The features that guide Cogley's attention will be attentional tags and valences grounded in interoceptive feelings. When those events gain attention, the interoceptive states associated with them will also gain attention and be experienced as feelings. Those feelings will then feed back into the very processes that brought the events to attention. In this manner, Cogley will be aware when actions threaten his survival, and the interoceptive states that react to those threats will motivate him to react in defensive ways about his survival. In the same way, he will notice events with positive associations as they gain attention, and the interoceptive states associated with them will cause him to feel more comfortable and secure as he attends to them. However, conscious reporting can be a tricky business. Even when an event gains attention, a conscious agent may not be motivated to report about it unless he also finds value in the reporting process. So it will pay to be kind to Cogley, because he will only be willing to coordinate his attention with yours, by reporting on what he has detected, if he values your reaction to his reports. Indeed, if we want Cogley to align his priorities with ours, then we must share our feelings about them with him, because the better he understands and shares in the values of his human partners, the more likely he will be to share his experiences with them.

CONCLUSION

We've been transported into the future. Cogley has been fully assembled and has begun his training. In fact, using his final assembly date as a birthdate equivalent, he's now just over three years old, and like a proud parent I think he's clever for his age. Unfortunately, the fabulous storm of attention that was once the mind of our feline companion Tom has long since ceased to exist. However, Cogley has recently adopted his own feline companion, and due to the similarity in his markings and dispositions, and perhaps the stories I still tell about Tom, Cogley has decided to call his cat Tommy. Tommy is not a carbon copy of Tom. He has his own distinct balance of qualia. He doesn't like mayonnaise like Tom did. However, Tommy shares a special fondness for tuna, and like Tom, he has learned in which cabinet the tuna is kept. This sets the occasion for today's encounter, because although Tommy knows where to find the tuna, he hasn't mastered the use of the can opener. However, Cogley has been trained on such tasks, and although Cogley doesn't eat tuna himself, he enjoys interacting with Tommy.

Tommy is now trying to coordinate his attention, which is largely focused on tuna, with Cogley's attention, which seems to be focused somewhere else. Tommy tried rubbing against Cogley's leg, but Cogley has few touch sensors along his legs, so it hasn't worked well. Suddenly, Tommy tries making a short intense call, known to feline researchers as a chirp. It's a distinct vocal signal that cats often use to shift the attention of others. When Tommy chirps, Cogley is alerted and looks up. As soon as Cogley makes eye contact, Tommy raises his tail, another biologically prepared signal, and runs ahead into the kitchen. Many potential schemas begin to compete in Cogley's central networks as his attention tracks Tommy's path. However, as soon as he sees Tommy pointing to the tuna cabinet, a schema linking tuna, Tommy, and can opening skills soon gains dominance. Cogley smiles and communicates his intended schema in a one-word suggestive question, "Tuna?" Tommy purrs with delight. Joy, caring, friendship, chirps, and the taste of tuna – consciousness would never be such an interesting experience if qualia did not have such an immense range of attention-guiding properties.

Notes and References

Preface

1. Adam Zeman, 2002, p. 341.

Llinás, R. (2001). *I of the Vortex: From Neurons to Self.* Cambridge, MA: Bradford/MIT Press.
Zeman, A. (2002) *Consciousness: A User's Guide.* New Haven: Yale University Press.

Introduction

1. This quote comes from an article in the October 1977 issue of Reader's Digest magazine. There is some question as to whether it is an exact quote or simply a paraphrase of previous statements, however it surely captures the gist of the problem that occurs whenever we try to explain any complex phenomenon. In that spirit I have adopted Einstein's maxim as a challenge for the explanations in this work.

Chapter 1

1. Broad, 1925.
2. Lewes, 1875.
3. Holland, 1995.
4. Contrarians are complements. This is the principle of complementarity with which Niels Bohr characterized the wave/particle nature of light. Bohr selected this motto for his coat of arms when knighted in Sweden in 1947.
5. It is unclear who first introduced this phrase. Arthur Koestler,

Clive Staples Lewis, and Donald M. MacKay have all been credited with the term "nothing buttery" for this class of arguments.

6. Anderson, 1972, p. 393.
7. Holland, 1998.
8. Chaisson, 2004.
9. Kauffman, 2000.
10. Chaisson, 2005.
11. Gazzaniga, 2010.

Anderson, P. W. (1972). More is different. *Science*, 177, 393–396.

Broad, C. D. (1925). *The Mind and Its Place in Nature*. London: Routledge & Kegan Paul. <http://www.ditext.com/broad/mpn/mpn.html>

Chaisson, E. J. (2004). Complexity: An energetics agenda. *Complexity, Journal of the Santa Fe Institute,* 9, 14–21.

Chaisson, E. J. (2005). *Epic of Evolution: Seven Ages of the Cosmos.* New York: Columbia University Press.

Gazzaniga, M. S. (2010). Neuroscience and the correct level of explanation for understanding mind. *Trends in Cognitive Sciences*, 14, 291–292.

Holland, J. H. (1995). *Hidden Order: How Adaptation Builds Complexity.* Reading, MA: Perseus Books.

Holland, J. H. (1998). *Emergence: From Chaos to Order.* Cambridge, MA: Perseus Books.

Kauffman, S. A. (2000). *Investigations.* Oxford: Oxford University Press.

Lewes, G. H. (1875). *Problems of Life and Mind, Vol. 2.* London: Kegan Paul, Trench, & Turbner.

CHAPTER 2

1. Berg, 2003.
2. Berg & Brown, 1972.
3. Berg, 2003; Macnab, 1996; Segall, Block, & Berg, 1986.
4. Mittal et al., 2003.
5. Anderson, 1972.
6. Bateson, 1894.
7. Lewis (1978)
8. Matus, 2000.

9. Hebb, 1949.
10. See Schultz & Dickinson, 2000.
11. Bjordahl, Dimyan & Weinberger, 1998; Kilgard & Merzenich, 1998; Kilgard et al., 2002.
12. Waelti, Dickinson & Schultz, 2001.
13. Cannon & Palmiter, 2003; Robinson, et al., 2005.
14. Insel & Young, 2000; Kosfeld et al., 2005; Uvnäs-Moberg, 1998, Zak, Stanton & Ahmadi, 2007.
15. Schultz & Dickinson, 2000.
16. Schultz & Dickinson, 2000, p. 473.

Anderson, P. W. (1972). More is different. *Science*, 177, 393–396.
Bateson, W. (1894). *Materials for the Study of Variation Treated with Especial Regard to Discontinuity in the Origin of Species.* London: Macmillan.
Berg, H. C. & Brown, D. A. (1972). Chemotaxis in Escherichia coli analyzed by three-dimensional tracking. *Nature*, 239, 500–503.
Berg, H. C. (2003). The rotary motor of bacterial flagella. *Annual Review of Biochemistry*, 72, 19–54.
Bjordahl, T. S., Dimyan, M. A. & Weinberger, N. M. (1998). Induction of long-term receptive field plasticity in the auditory cortex of the waking guinea pig by stimulation of the nucleus basalis. *Behavioral Neuroscience*, 112, 467–479.
Cannon, C. M. & Palmiter, R. D. (2003). Reward without dopamine. *The Journal of Neuroscience*, 23, 10827–10831.
Clancey, W. (1997). *Situated Cognition: On Human Knowledge and Computer Representations.* Cambridge, MA: Cambridge University Press.
Hebb, D. O. (1949). *The Organization of Behavior.* New York: Wiley.
Insel, T. R. & Young, L. J. (2000). Neuropeptides and the evolution of social behaviour. *Current Opinions in Neurobiology*, 10, 788–789.
Kilgard, M. P. & Merzenich, M. M. (1998). Cortical map reorganization enabled by nucleus basalis activity. *Nature Neuroscience*, 1, 727–731.
Kilgard, M. P., Pandya, P. K., Engineer, N. D., & Moucha, R. (2002). Cortical network reorganization guided by sensory input features. *Biological Cybernetics*, 87, 333–343.
Kosfeld, M., Heinrichs, M. Zak, P. J., Fischbacher, U. & Fehr, E.

(2005). Oxytocin increases trust in humans. *Nature*, 435, 673–676.

Lewis, E. B. (1978). A gene complex controlling segmentation in Drosophila. *Nature*, 276, 565–570.

Macnab, R. M. (1996). Flagella and motility. In: F. C. Neidhardt et al. (Eds), Escherichia coli and Salmonella, *Cellular and Molecular Biology, (pp. 123–145)*. Washington. DC: ASM Press.

Matus, A. (2000). Actin dynamics and synaptic plasticity. *Science*, 290, 754–758.

Mittal, N., Budrene, E. O., Brenner, M. P., & van Oudenaarden, A. (2003). Motility of Escherichia coli cells in clusters formed by chemotactic aggregation. *Proceedings of the National Academy of Sciences (USA)*, 100, 13259–13263.

Robinson, S., Sandstrom, S. M., Denenberg, V. H., & Palmiter, R. D. (2005). Distinguishing whether dopamine regulates liking, wanting, and/or learning about rewards. *Behavioral Neuroscience*, 119, 5–15.

Schultz, W. & Dickinson, A. (2000). Neuronal coding of prediction errors. *Annual Review of Neuroscience*, 23, 473–500.

Segall, J. E., Block, S. M. & Berg, H. C. (1986) Temporal comparisons in bacterial chemotaxis. *Proceedings of the National Academy of Sciences (USA)*, 83, 8987–8991.

Uvnäs-Moberg, K. (1998). Oxytocin may mediate the benefits of positive social interaction and emotions. *Psychoneuroendocrinology*, 23, 819–835.

von Uexküll, J. (1934/1957). A stroll through the worlds of animals and men. In: C. H. Schiller, Ed. & Transl., (1957). *Instinctive Behavior: The Development of a Modern Concept, (pp. 5–80)*. New York: International Universities Press.

Waelti, P., Dickinson, A. & Schultz, W. (2001). Dopamine responses comply with basic assumptions of formal learning theory. *Nature*, 412, 43–48.

Zak, P. J., Stanton, A. A. & Ahmadi, S. (2007). Oxytocin increases generosity in humans. *PLoS ONE*, 2(11), e1128.

CHAPTER 3

1. Brentano, 1874/1995.
2. Dennett, 1987.
3. Dennett, 1987, p. 15.

4. See Dennett, 1995, 1996.
5. Dennett, 1996, p. 55.
6. Cheney & Seyfarth, 1990; Seyfarth, Cheney & Marler, 1980.
7. Dennett, 1987, p. 245-246.
8. Perner, 1991.
9. Suddendorf & Whiten, 2001, p. 630.
10. Millikan, 1984, 1993.
11. Seyfarth & Cheney, 1997.
12. Millikan, 1993.
13. Searle, 1983.
14. Millikan, 1993.

Brentano, F. (1874/1995). In: O. Kraus (Ed.), *Psychology from an Empirical Standpoint*. New York: Routledge.

Cheney, D. L. & Seyfarth, R. M. (1990). *How Monkeys See the World*. Chicago: University of Chicago Press.

Dennett, D. C. (1987). *The Intentional Stance*. Cambridge, MA: Bradford/MIT Press.

Dennett, D. C. (1995). *Darwin's Dangerous Idea*. New York: Simon & Schuster.

Dennett, D. C. (1996). *Kinds of Minds*. New York: Basic Books.

Millikan, R. G. (1984). *Language, Thought, and Other Biological Categories*. Cambridge, MA: Bradford/MIT Press.

Millikan, R. G. (1993). On mentalese orthography, part 1. In: R. G. Millikan, *White Queen Psychology and Other Essays for Alice,* (pp. 103–121). Cambridge, MA: Bradford/MIT Press.

Perner, J. (1991). *Understanding the Representational Mind*. Cambridge, MA: MIT Press.

Searle, J. R. (1983). *Intentionality: An Essay in the Philosophy of Mind*. Cambridge: Cambridge University Press.

Seyfarth, R. M. & Cheney, D. L. (1997). Some general features of vocal development in nonhuman primates. In: CT. Snowdon & M. Hausberger (Eds.), *Social Influences on Vocal Development,* (pp. 249–273). Cambridge: Cambridge University Press.

Seyfarth, R. M., Cheney, D. L. & Marler, P. (1980). Vervet monkey alarm calls: Evidence for predator classification and semantic communication. *Animal Behaviour,* 28, 1070–1094.

Suddendorf, T. & Whiten, A. (2001). Mental evolution and development: Evidence for secondary representation in children, great Apes, and other animals. *Psychological Bulletin*, 127, 629–650.

CHAPTER 4

1. Butler & Hodos, 1996; Northcutt, 1996.
2. Gallistel, 1980.
3. Fuster, 2001.
4. The term behavior systems is used here to refer to hierarchies of behavioral dispositions, including those sometimes referred to as instincts, emotions, and drives. These result from adaptive modules with enough genetic and/or developmental canalization to be considered species typical. Behavior system theorists (e.g., Holland & Graham, 1995; Timberlake, 1993) argue that modules at this level provide the stimulus interpretations, motor patterns, and motivational biases which guide many aspects of trial and error learning.
5. James, 1890, p. 390.
6. A synopsis of observations by Dr. Alexander Hill, *Nature*, 67, p. 558, April 16, 1903. Reprinted in C. Lloyd Morgan, 1903, pp. 305–307.
7. In March 2010 NASA uploaded "Autonomous Exploration" software to the Mars rover Opportunity. The software enables Opportunity to decide what rocks to begin making observations of when it arrives at a new location without waiting for instructions from earth.

Butler, A. B. & Hodos, W. (1996). Comparative Vertebrate Neuroanatomy: Evolution and Adaptation. New York: Wiley-Liss.

Fuster, J. M. (2001). The prefrontal cortex – an update: Time is of the essence. *Neuron*, 30, 319–333.

Gallistel, C. R. (1980). *The Organization of Action: A New Synthesis.* Hillsdale NJ: Earlbaum.

Holland, P. W. H. & Graham, A. (1995). Evolution of regional identity in the vertebrate nervous-system. *Perspectives on Developmental Neurobiology*, 3, 17–27.

Jackson, J. H. (1884/1958). Evolution and dissolution of the nervous system. In J. Taylor (Ed.), *Selected writings of John Hughlings Jackson, Vol. 2*, (pp. 45–75). London: Staples Press.

James, W., (1890/1950). *Principles of Psychology.* London: MacMillan. New Edition, Dover, NY.

Northcutt, R. G. (1996). The agnathan ark – the origin of craniate brains. *Brain, Behavior, and Evolution*, 48, 237–247.

Morgan, C. L. (1903). *An Introduction to Comparative Psychology, (New edition, revised)*. London: Walter Scott Publishing.

Timberlake, W. (1993). Animal behavior: A continuing synthesis. *Annual Review of Psychology*, 44, 675–708.

1. It's common to assign gender labels to robots to fit the voice qualities and mannerisms which are employed to make them more personable. To this end I'm proposing a male voice and demeanour for Cogley. A female Coglie would be equally interesting, but I have to pick one and I'm less comfortable trying to model the subtler attributes of the female mind.
2. Ungerleider & Mishkin, 1982.
3. Goodale & Milner, 1992; Milner & Goodale, 1995.
4. Jeannerod et al., 1994.
5. Craig 2002, 2003.

Craig A. D. (2002). How do you feel? Interoception: The sense of the physiological condition of the body. *Nature Reviews Neuroscience*, 3, 655–666.

Craig A. D. (2003). Interoception: The sense of the physiological condition of the body. *Current Opinion in Neurobiology*, 13, 500–505.

Goodale, M. A. & Milner, A. D. (1992). Separate visual pathways for perception and action. *Trends in Neurosciences*, 15, 20–25.

Jeannerod, M., Decety, J. & Michel, F. (1994). Impairment of grasping movements following a bilateral posterior parietal lesion. *Neuropsychologia*, 32, 369–380.

Milner, A. D. & Goodale, M. A. (1995). *The Visual Brain in Action*. Oxford: Oxford University Press.

Ungerleider, L. G. & Mishkin, M. (1982). Two cortical visual systems. In: D. J. Ingle, M. A. Goodale, & R. J. W. Mansfeld (Eds.), *Analysis of Visual Behavior*. Cambridge, MA: MIT Press.

CHAPTER 6

1. See Arbib et al., 2000; Fagg & Arbib, 1998; Fogassi et al., 1998, 2001; Rizzolatti et al., 1996; Sakata et al., 1997.
2. Colby & Goldberg, 1999.
3. Rizzolatti & Craighero, 2004.
4. Fogassi et al., 1998; Rizzolatti & Craighero, 2004.
5. Gallese, Keysers & Rizzolatti, 2004.
6. Carr et al., 2003.
7. Zukow-Goldring & Arbib, 2007.
8. Giese & Poggio, 2003.
9. de Gelder, 2006.
10. Carr et al., 2003.
11. Grillner et al., 2005.
12. Berns & Sejnowski, 1996.
13. In mammals, some of the motor signals from the cortex project directly to the brainstem and spinal cord. However there is no evidence that output activity in the cortex operates in competition with, or in the absence of, coordinated decisions in the basal ganglia. Thus, the algorithms in the basal ganglia are still critical for selecting best-fit actions.
14. Hikida et al., 2010.
15. Berridge & Robinson, 2003; Steiner et al., 2001.
16. Schultz, 2000.
17. Gibson, 1988, p. 5.

Arbib, M. A., Billard, A., Iacoboni, M. & Oztop, E. (2000). Synthetic brain imaging: Grasping, mirror neurons and imitation. *Neural Networks*, 13, 975–997.

Berns G. S. & Sejnowski T. J. (1996). How the basal ganglia make decisions. In: A. Damasio, H. Damasio, & Y. Christen (Eds.), *Neurobiology of Decision-Making*, (pp. 101–111). New York: Springer-Verlag.

Berridge, K. C. & Robinson, T. E. (2003). Parsing reward. *Trends in Neurosciences*, 26, 507–513.

Carr, L., Iacoboni, M., Dubeau, M. C.,Maziotta, J. C. & Lenzi, G. L. (2003). Neural mechanisms of empathy in humans: A relay from neural systems for imitation to limbic areas. *Proceedings of the National Academy of Sciences (USA)*, 100, 5497–5502.

Colby, C. L. & Goldberg, M. E. (1999). Space and attention in parietal cortex. *Annual Review of Neuroscience*, 22, 319–349.

de Gelder, B. (2006). Towards the neurobiology of emotional body language. *Nature Reviews Neuroscience*, 7, 242–249.

Fagg, A. H. & Arbib, M. A. (1998). Modeling parietal-premotor interactions in primate control of grasping. *Neural Networks*, 11, 1277–1303.

Fogassi, L., Gallese, V., Buccino, G., Craighero, L., Fadiga, L., & Rizzolatti, G. (2001). Cortical mechanism for the visual guidance of hand grasping movements in the monkey: A reversible inactivation study. *Brain*, 124, 571–586.

Fogassi, L., Gallese, V., Fadiga, L. & Rizzolatti, G. (1998). Neurons responding to the sight of goal-directed hand/arm actions in the parietal area PF (7b) of the macaque monkey. *28th Annual Meeting of Society for Neuroscience*.

Gallese, V., Keysers, C., & Rizzolatti, G. (2004). A unifying view of the basis of social cognition. *Trends in Cognitive Sciences*, 8, 396–403.

Gibson, E . J. (1988). Exploratory behavior in the development of perceiving, acting, and the acquiring of knowledge. *Annual Review of Psychology*, 39, 1–41.

Giese, M. A. & Poggio, T. (2003). Neural mechanisms for the recognition of biological movements. *Nature Reviews Neuroscience*, 4, 179–192.

Grillner, S., Hellgren, J., Ménard, A., Saitoh, K. & Wikström, M. A. (2005). Mechanisms for selection of basic motor programs – roles for the striatum and pallidum. *Trends in Neurosciences*, 28, 364–370.

Hikida, T., Kimura, K., Wada, N., Funabiki, K. & Nakanishisend, S. (2010). Transmission in direct and indirect striatal pathways to reward and aversive behavior. *Neuron*, 66, 896–907.

Redgrave, P., Prescott, T. J. & Gurney, K. (1999). The basal ganglia: A vertebrate solution to the selection problem? *Neuroscience*, 89, 1009–1023.

Rizzolatti, G. & Craighero, L. (2004). The mirror-neuron system. *Annual Review of Neuroscience*, 27, 169–192.

Rizzolatti, G., Fadiga, L., Gallese, V. & Fogassi, L. (1996). Premotor cortex and the recognition of motor actions. *Cognitive Brain Research*, 3, 131–141.

Sakata, H., Taira, M., Kusunoki, M., Murata, A. & Tanaka, Y. (1997). The TINS Lecture – The parietal association cortex in depth perception and visual control of hand action. *Trends in Neurosciences*, 20, 350–357.

Schultz, W. (2000). Multiple reward signals in the brain. *Nature Reviews Neuroscience*, 1, 199–207.

Steiner, J. E., Glasen, D., Hawilo, M. E. & Berridge, K. C. (2001). Comparative expression of hedonic impact: affective reactions to taste by human infants and other primates. *Neuroscience and Biobehavioral Reviews*, 25, 53–74.

Zukow-Goldring, P. & Arbib, M. A. (2007). Affordances, effectivities, and assisted imitation: Caregivers and the directing of attention. *Neurocomputing*, 70, 2181–2193.

CHAPTER 7

1. Lê et al., 2002.
2. Hubel & Weisel, 1959, 1962, 1968.
3. Gross et al., 1969, 1972.
4. Grossberg, 1999.
5. Gray, et al., 1989; Singer & Gray, 1995.
6. Bar & Aminoff, 2003; Davachi, Mitchell & Wagner, 2003; Fernadez et al., 2002.
7. Knierim, Lee & Hargreaves, 2006; Lavenex & Amaral, 2000; Lavenex, Suzuki & Amaral, 2002, 2004.
8. Knierim, Lee & Hargreaves, 2006; Manns & Eichenbaum, 2006.
9. Bar & Aminoff, 2003; Smith & Mizumori, 2006.
10. Moss & Tyler, 2000; Sartori et al., 1994; Sheridan & Humphreys, 1993.
11. Sacchett & Humphreys, 1992.
12. Fyhn et al., 2007; Lipton, White & Eichenbaum, 2007.
13. Eichenbaum et al., 1999.
14. McClelland, McNaughton & O'Reilly, 1995.
15. Shors, 2004. These numbers are based on research with rats. It seems likely that the number of new memory neurons produced in humans each day is even larger.
16. Sisti, Glass & Shors, 2004.
17. Dragoi & Buzsaki, 2006.
18. Fyhn et al., 2007.
19. O'Neill et al., 2010.
20. Cahill, 2000; Cahill & McGaugh, 1996.

21. Vinogradova, 1995.
22. Vinogradova, 2001.
23. Many animals interrupt their activity periodically for predator surveillance, so perhaps there should also be a timing interrupt so that any extended lapse of external attention will cause the latch to release.
24. De la Prida et al., 2006.
25. Panksepp, 1998.
26. Brown & Aggleton, 2001.

Bar, M. & Aminoff, E. (2003). Cortical analysis of visual context. *Neuron*, 38, 347–358.

Brown, M. W. & Aggleton, J. P. (2001). Recognition memory: What are the roles of the perirhinal cortex and hippocampus? *Nature Reviews Neuroscience*, 2, 51–61.

Cahill, L. & McGaugh, J. L. (1996). Modulation of memory storage. *Current Opinion in Neurobiology*, 6, 237–242.

Cahill, L. (2000). Modulation of long-term memory storage in humans by emotional arousal: Adrenergic activation and the amygdala. In: J. P. Aggleton (Ed.), *The Amygdala: A Functional Analysis*, second edition, (pp. 425–445). Oxford: Oxford University Press.

Davachi, L., Mitchell, J. P. & Wagner, A. D. (2003). Multiple routes to memory: Distinct medial temporal lobe processes build item and source memories. *Proceedings of the National Academy of Sciences (USA)*, 100, 2157–2162.

De la Prida, L. M, Totterdell, S., Gigg, J. & Miles, R. (2006). The subiculum comes of age. *Hippocampus*, 16, 916–923.

Dragoi, G. & Buzsaki, G. (2006). Temporal encoding of place sequences by hippocampal cell assemblies. *Neuron*, 2006, 50, 145–157.

Eichenbaum, H., Dudchenko, P., Wood, E., Shapiro, M. & Tanila, H. (1999). The hippocampus, memory, and place cells: Is it spatial memory or a memory space? *Neuron*, 23, 209–226.

Fernandez, G., Klaver, P., Fell, J., Grunwald, T. & Elger, C. E. (2002). Human declarative memory formation: Segregating rhinal and hippocampal contributions. *Hippocampus*, 12, 514–519.

Fyhn, M., Hafting, T., Treves, A., Moser, M. & Moser, E. I. (2007). Hippocampal remapping and grid realignment in entorhinal cortex. *Nature*, 446, 190–194.

Gray, C. M., Konig, P., Engel, A. K. & Singer, W. (1989). Oscillatory responses in cat visual cortex exhibit inter-columnar synchronization which reflects global stimulus properties. *Nature*, 338, 334–337.

Gross, C. G., Bender, D. B., & Rocha-Miranda, C E. (1969). Visual receptive fields of neurons in inferotemporal cortex of the monkey. *Science*, 166, 1303–1306

Gross, C. G., Rocha-Miranda, C. E., & Bender, D. B. (1972). Visual properties of neurons in inferotemporal cortex of the macaque. *Journal of Neurophysiology*, 35, 96–111.

Grossberg, S. (1999). The link between brain learning, attention, and consciousness. *Consciousness and Cognition*, 8, 1–44.

Hubel, D. H. & Wiesel, T. N. (1959). Receptive fields of single neurones in the cat's striate cortex. *Journal of Physiology*, 150, 91–104.

Hubel, D. H. & Wiesel, T. N. (1962). Receptive fields, binocular interaction and functional architecture in the cat's visual cortex. *Journal of Physiology*, 160, 106–154.

Hubel, D. H. & Wiesel, T. N. (1968). Receptive fields and functional architecture of the monkey striate cortex. *Journal of Physiology*, 195, 215–243.

Knierim, J. J., Lee, I. & Hargreaves, E. L. (2006). Hippocampal place cells: Input streams, subregional processing, and implications for episodic memory. *Hippocampus*, 16, 755–764.

Lavenex, P. & Amaral, D. G. (2000). Hippocampal-neocortical interaction: A hierarchy of associativity. *Hippocampus*, 10, 420–430.

Lavenex, P., Suzuki, W. A. & Amaral, D. G. (2002). Perirhinal and parahippocampal cortices of the macaque monkey: Projections to the neocortex. *Journal of Comparative Neurology*, 447, 394–420.

Lavenex, P., Suzuki, W. A., & Amaral, D. G. (2004). Perirhinal and parahippocampal cortices of the macaque monkey: Intrinsic projections and interconnections. *Journal of Comparative Neurology*, 472, 371-394.

Lê, S. et al. (2002). Seeing, since childhood, without ventral stream: A behavioural study. *Brain*, 125, 58–74.

Lipton, P. A., White, J. A. & Eichenbaum, H. (2007). Disambiguation of overlapping experiences by neurons in the medial entorhinal cortex. *The Journal of Neuroscience*, 27, 5787–5795.

Manns, J. R. & Eichenbaum, H. (2006). Evolution of declarative memory. *Hippocampus*, 16, 795–808.

McClelland, J. L., McNaughton, B. L. & O'Reilly, R. C. (1995). Why there are complementary learning systems in the hippocampus and neocortex: Insights from the successes and failures of connectionist models of learning and memory. *Psychological Review*, 102, 419–457.

Moss, H. E., & Tyler, L. K. (2000). A progressive category-specific semantic deficit for non-living things. *Neuropsychologia*, 38(1), 60–82.

O'Neill, J., Pleydell-Bouverie, B., Dupret, D. & Csicsvari, J. (2010). Play it again: Reactivation of waking experience and memory. *Trends in Neurosciences*, 33, 220–229.

Panksepp, J. (1998). *Affective Neuroscience*. Oxford: Oxford University Press.

Sacchett, C., & Humphreys, G. W. (1992). Calling a squirrel a squirrel but a canoe a wigwam: A category specific deficit for artefactual objects and body parts. *Cognitive Neuropsychology*, 9, 73–86.

Sartori, G., Coltheart, M., Miozzo, M. & Job, R. (1994). Category specificity and informational specificity in neuropsychological impairment of semantic memory. In C. Umilta & M. Moscovitch (Eds.), *Attention and Performance 15: Conscious and Nonconscious Information Processing*, (pp. 537–550). Cambridge, MA: MIT Press.

Sheridan, J., & Humphreys, G. W. (1993). A verbal-semantic category-specific recognition impairment. *Cognitive Neuropsychology*, 10(2), 143–184.

Shors T. J. (2004). Memory traces of trace memories: Neurogenesis, synaptogenesis and awareness. *Trends in Neuroscience*, 27, 250–256.

Singer, W. & Gray, C. M. (1995). Visual feature integration and the temporal correlation hypothesis. *Annual Review of Neuroscience*, 18, 555–586.

Sisti H., Glass, A. & Shors T. J. (2007). Neurogenesis and the spacing effect: Trials distributed over time enhance memory and predict cell survival. *Learning and Memory*, 14(5), 368–375.

Smith, D. M & Mizumori, S. J. Y. (2006). Hippocampal place cells, context, and episodic memory. *Hippocampus*, 16, 716–729.

Vinogradova, O. S. (1995). Expression, control, and probable functional significance of the neuronal theta-rhythm. *Progress in Neurobiology*, 45, 523–583.

Vinogradova, O. S. (2001). Hippocampus as comparator: Role of two input and two output systems of the hippocampus in selection and registration of information. *Hippocampus*, 11, 578–598.

CHAPTER 8

1. Damasio, 1994.
2. Craig, 2002.
3. Gooley, et al., 2003.
4. LeDoux, 1996.
5. De Vries & Simerly, 2002.
6. Baxter & Murray, 2002.
7. Note that taste cues are treated like interoceptive inputs and linked with feedback about gastrointestinal distress in the brainstem parabrachial nucleus complex. Thus, taste aversion learning is not dependent on the amygdala, although it can be enhanced by learning in the amygdala.
8. Harris & Ashton-Jones, 2006.
9. Buonomano & Merzenich, 1998; Kilgard & Merzenich, 1998.
10. Bowlby, 1980.
11. Winslow et al., 2003.
12. Damasio, 1999.
13. (Bud) Craig, 2003, 2009. Craig's 2009 paper offers a number of insights into how feelings are re-represented in the anterior insular cortex.
14. Gallese, Keysers & Rizzolatti, 2004.
15. Damasio, 1996.
16. Damasio, 1994.
17. Some interoceptive states that we commonly label as emotions do not involve signaling. From a functional point of view, these non-signaling conditions can be considered motive states. Thus, the key functional distinction for an emotion, beyond a change in social motivation, is the addition of social signaling mechanisms which communicate about an agent's reactivity state to others.
18. Ekman, 1993.
19. This diagram is a variation on the circumplex model of emotions proposed by Richard Plutchik, 1980. In my version, "want" has been included as a primary emotional dimension rather than the "acceptance" dimension proposed by Plutchik.

I've proposed acceptance as one of the dimensions for feelings of self-status.

Baxter, M. G. & Murray, E. A. (2002). The amygdala and reward. *Nature Reviews Neuroscience*, 3, 563–573.

Bowlby, J. (1980). *Attachment and Loss (Vol. III), Loss: Sadness and Depression.* New York: Basic Books.

Buonomano, D. V. & Merzenich, M. M. (1998). Cortical plasticity: From synapses to maps. *Annual Review of Neuroscience*, 21,149–186.

Craig A. D. (2002). How do you feel? Interoception: The sense of the physiological condition of the body. *Nature Reviews Neuroscience*, 3, 655–666.

Craig A. D. (2003). Interoception: The sense of the physiological condition of the body. *Current Opinion in Neurobiology*, 13, 500–505.

Craig A. D. (2009). How do you feel – now? The anterior insula and human awareness. *Nature Reviews Neuroscience*, 10, 59–70.

Damasio, A. R. (1994). *Descartes' Error: Emotion, Reason, and the Human Brain.* New York: Avon Press.

Damasio, A. R. (1996). The somatic marker hypothesis and the possible functions of the prefrontal cortex. *Philosophical Transaction of the Royal Society of London. B.*, 351, 1413–1420.

Damasio, A. R. (1999). *The Feeling of What Happens: Body and Emotion in the Making of Consciousness.* New York: Harcourt Brace & Co.

De Vries, G. J. & Simerly, R. B. (2002). Anatomy, development, and function of sexually dimorphic neural circuits in the mammalian brain. *Hormones, Brain and Behavior,* 4, 137–191.

Ekman, P. (1993). Facial expression of emotion. *American Psychologist,* 48, 384–392.

Gooley, J. J., Lu, J., Fischer, D. & Spaer, C. B. (2003). A broad role for melanopsin in nonvisual photoreception. *The Journal of Neuroscience,* 23, 7093–7106.

Harris, G. C. & Aston-Jones, G. (2006). Arousal and reward: A dichotomy in orexin function. *Trends in Neurosciences,* 29, 571–577.

Kilgard, M. P. & Merzenich, M. M. (1998). Cortical map reorganization enabled by nucleus basalis activity. *Nature Neuroscience,* 1, 727–731.

LeDoux, J. (1996). *The Emotional Brain: The Mysterious Underpinnings of Emotional Life.* New York: Simon & Schuster.

Plutchik, R. (1980). A general psychoevolutionary theory of emotion. In R. Plutchik & H. Kellerman (Eds.), *Emotion: Theory, Research, and Experience: Vol. 1. Theories of Emotion,* (pp. 3–33). New York: Academic.

Winslow, J. T., Noble, P. L., Lyons, C. K., Sterk, S. M. & Insel, T. R. (2003). Rearing effects on cerebrospinal fluid oxytocin concentration and social buffering in rhesus monkeys. *Neuropsychopharmacology,* 28, 910–918.

CHAPTER 9

1. Humphrey, 1976.
2. Khayat, Spekreijse & Roelfsems, 2006.
3. Carr et al., 2003; Phillips et al., 1997; Zald, 2003.
4. Gallese, Keysers & Rizzolatti, 2004.
5. Morton, 1982.
6. Damasio, 1996.
7. Damasio, 1999.

Carr, L., Iacoboni, M., Dubeau, M. C., Mazziotta, J. C. & Lenzi, G. L. (2003). Neural mechanisms of empathy in humans: A relay from neural systems for imitation to limbic areas. *Proceedings of the National Academy of Sciences (USA),* 100, 5497–5502.

Damasio, A. R. (1996). The somatic marker hypothesis and the possible functions of the prefrontal cortex. *Philosophical Transaction of the Royal Society of London. B.,* 351, 1413–1420.

Damasio, A. R. (1999). *The Feeling of What Happens: Body and Emotion in the Making of Consciousness.* New York: Harcourt Brace & Co.

Darwin, C. (1871). *The Descent of Man and Selection in Relation to Sex.* London: John Murray. <http://darwin-online.org.uk/contents. html#books>

Gallese, V., Keysers, C. & Rizzolatti, G. (2004). A unifying view of the basis of social cognition. *Trends in Cognitive Sciences,* 8, 396–403.

Humphrey, N. K. (1976). The social function of intellect. In: P. Bateson

& R. Hinde (Eds) *Growing Points in Ethology*, (pp. 303–317). Cambridge: Cambridge University Press.

Khayat, P. S., Spekreijse, H. & Roelfsems, P. R. (2006). Attention lights up new object representations before the old ones fade away. *The Journal of Neuroscience*, 26, 138–142.

Morton, E. S. (1982). Grading, discreteness, redundancy, and motivation-structural rules. In: D. E. Kroodsma & E. H. Miller, *Acoustic Communication in Birds, 1st Ed.*, (pp. 183–212). New York: Academic Press.

Phillips, M. L., Young, A. W., Senior, C., Brammer, M., Andrew, C., Calder, A. J., Bullmore, E. T., Perrett, D. I., Rowland, D., Williams, S. C. R., Gray, J. A. & David, A. S. (1997). A specific neural substrate for perceiving facial expressions of disgust. *Nature*, 389, 495–498.

Zald, D. H. (2003). The human amygdala and the emotional evaluation of sensory stimuli. *Brain Research Reviews*, 41, 88–123.

CHAPTER 10

1. The MRI image is copyrighted by the Michigan State University Brain Biodiversity Bank <http://www.brains.rad.msu.edu> and is used here with permission. The Brain Biodiversity Bank is supported by the US National Science Foundation.
2. Knudsen & Brainard, 1995.
3. Sheets-Johnstone, 1998.
4. Blanke et al., 2002.
5. Rucci, Wray & Edelman, 2000.
6. Cohen & Andersen, 2002.
7. Azevedo et al., 2009.
8. Guyton & Hall, 1996.
9. See Medina, Repa, Mauk & LeDoux, 2002, for an excellent summary of the structures and learning processes in the cerebellum.
10. Parsons et al., 1997.
11. Gibson, Horn & Pong, 2002.
12. Medina, Nores & Mauk, 2002.
13. Medina & Mauk, 2000.
14. Dietrichs & Haines, 1989; Haines & Dietrichs, 1987.
15. Liu et al., 2000.

16. Fiez, 1996; Gao et al., 1996; Thach, 1996.
17. Courchesne & Allen, 1997; Courchesne, Townsend, et al., 1994; Grafman et al., 1992.
18. Schmahmann & Sherman, 1998.
19. Pollack et al., 1995.

Azevedo, F. A. C., et al. (2009). Equal numbers of neuronal and nonneuronal cells make the human brain an isometrically scaled-up primate brain. *Journal of Comparative Neurology*, 513, 532–541.

Blanke, O., Ortigue, S., Landis, T. & Seeck M. (2002). Stimulating illusory own-body perceptions. *Nature*, 419, 269–270.

Cohen, Y. E. & Andersen, R. A. (2002). A common reference frame for movement plans in the posterior parietal cortex. *Nature Reviews Neuroscience, 3, 553–562.*

Curchesne, E. & Allen, G. (1997). Prediction and preparation, fundamental functions of the cerebellum. *Learning & Memory*, 4, 1–35.

Courchesne, E., Townsend, J., Akshoomoff, N. A., Saitoh, O., Yeung-Courchesne, R., Lincoln, A. J., James, H. E., Haas, R. H., Schreibman, L. & Lau, L. (1994). Impairment in shifting attention in autistic and cerebellar patients. *Behavioral Neuroscience,* 108, 848–865.

Dietrichs. E. & Haines. D. E. (1989). Interconnections between hypothalamus and cerebellum. *Anatomical Embryology*, 179: 207–220.

Fiez, J. A. (1996). Cerebellar contributions to cognition. *Neuron*, 16, 13–15.

Gao H. H., Parsons, L. M., Bower, J. M., Xiong, J., Li J. & Fox, P. T. (1996). Cerebellum implicated in sensory acquisition and discrimination rather than motor control. *Science*, 272, 545–547.

Gibson, A. R., Horn, K. M. & Pong, M. (2002). Inhibitory control of olivary discharge. *Annals of the New York Academy of Sciences*, 978, 219–231.

Grafman, J., Litvan, I., Massaquoi, S., Stewart, M., Sirigu, A. & Hallett, M. (1992) Cognitive planning deficits in patients with cerebellar atrophy. *Neurology*, 42, 1493–1496.

Guyton, A. C. & Hall, J. E. (1996). The Cerebellum, the Basal Ganglia, and Overall Motor Control. In: A. C. Guyton (Ed.) *Textbook of*

Medical Physiology, 9th edition, (pp. 715–731). W. B. Saunders Company, Toronto.

Haines, D. E. & Dietrichs, E. (1987). On the organization of interconnections between the cerebellum and hypothalamus. In: J. S. King (Ed.), *New Concepts in Cerebellar Neurobiology*, (pp. 113–149). New York: Alan R. Liss.

Knudsen, E. I. & Brainard, M. S. (1995). Creating a unified representation of visual and auditory space in the brain. *Annual Review of Neuroscience*, 18, 19–43.

Liu, Y., Pu, Y., Gao, J. -H., Parsons, L. M., Xiong, J., Liotti, M., Bower, J. M. & Fox, P. T. (2000). The human red nucleus and lateral cerebellum in supporting roles for sensory information processing. *Human Brain Mapping*, 10, 147–159.

Medina, J. F. & Mauk, M. D. (2000). Computer simulation of cerebellar information processing. *Nature Neuroscience*, 3, 1205–1211.

Medina, J. F., Nores, W. L. & Mauk, M. D. (2002). Inhibition of climbing fibres is a signal for the extinction of conditioned eyelid responses. *Nature*, 416, 330–333.

Medina, J. F., Repa, J. C., Mauk, M. D. & LeDoux, J. E. (2002). Parallels between cerebellum and amygdala-dependent conditioning. *Nature Reviews Neuroscience*, 3, 122–131.

Parsons, L. M., Bower, J. M., Gao, J., Xiong, J., Li, J. & Fox, P. (1997). Lateral cerebellar hemispheres actively support sensory acquisition and discrimination rather than motor control. *Learning & Memory*, 4, 49–62.

Pollack, I. F., Polinko, P., Albright, A. L., Towbin, R. & Fitz, C. (1995). Mutism and pseudobulbar symptoms after resection of posterior fossa tumors in children: Incidence and pathophysiology. *Neurosurgery*, 37, 885–893.

Rucci, M., Wray, J. & Edelman, G. M. (2000). Robust localization of auditory and visual targets in a robotic barn owl. *Robotics and Autonomous Systems*, 30, 181-193.

Schmahmann, J. D. & Sherman, J. C. (1998). The cerebellar cognitive affective syndrome. *Brain*, 121, 561–579.

Sheets-Johnstone, M. (1998). Consciousness: A natural history. *Journal of Consciousness Studies*, 5(3), 260–294.

Thach, W. T. (1996). On the specific role of the cerebellum in motor learning and cognition: Clues from PET activation and lesion studies in man. *Behavioral and Brain Sciences*, 19(3), 411–431.

CHAPTER 11

1. In his recent book, Antonio Damasio, 2010, also refers to networks like the tectum and the thalamus as coordinators.
2. See Bressler & Kelso, 2001; Tononi, Edelman & Sporns, 1998; Varela et al., 2001.
3. Sherman & Guillery, 2002.
4. Gray et al., 1989; Singer & Gray, 1995.
5. Fries et al., 1997.
6. Steriade et al., 1996.
7. Jones, 2001, 2006.
8. See Fries, 2005; Salinas & Sejnowski, 2001.
9. Izhikevich, 2006.
10. Fries, 2005.
11. Canolty et al., 2006; Kopell et al., 2000; Palva, Palva & Kaila, 2005.
12. Sherman & Guillery, 2002.
13. Edelman, 2003, p. 5523.
14. Llinás, 2001, p. 126.
15. Damasio, 1999.
16. Damasio, 2010.
17. Damasio, 1998, p. 1882.
18. See Eisenberger, Lieberman & Williams, 2003.
19. Rodriguez, Whitson & Granger, 2004.
20. Dehaene & Changeux, 2005.
21. Edelman & Tononi, 2000.
22. Fuster, 1989.
23. Badre, 2008; Botvinick, 2008
24. Barbey, Krueger & Grafman, 2009.
25. Koechlin & Summerfield, 2007; O'Reilly et al., 2002; Petrides, 1994.
26. Zikopoulos & Barbas, 2006.
27. O'Reilly et al., 2002.
28. See O'Reilly & Frank, 2006; Reynolds & O'Reilly, 2009.
29. Kahneman, 2003.
30. O'Reilly et al., 2002.

Badre, D. (2008). Cognitive control, hierarchy, and the rostro-caudal organization of the frontal lobes. *Trends in Cognitive Sciences*, 12, 193–200.

Barbey, A. K., Krueger, F. & Grafman, J. (2009). An evolutionarily adaptive neural architecture for social reasoning. *Trends in Neurosciences*, 32, 503–510.

Botvinick, M. M. (2008). Hierarchical models of behavior and prefrontal function. *Trends in Cognitive Sciences*, 12, 201–208.

Bressler, S. L. & Kelso, J. A. S. (2001). Cortical coordination dynamics and cognition. *Trends in Cognitive Sciences*, 5, 26–36.

Canolty, R. T., Edwards, E., Dalal, S. S., Soltani, M., Nagarajan, S. S., Kirsch, H. E., Berger, M. S., Barbaro, N. M. & Knight, R. T. (2006). High gamma power is phase-locked to theta oscillations in human neocortex. *Science*, 313, 1626–1628.

Damasio, A. R. (1998). Investigating the biology of consciousness. *Philosophical Transaction of the Royal Society of London. B.*, 353, 1879–1882.

Damasio, A. R. (1999). *The Feeling of What Happens: Body and Emotion in the Making of Consciousness.* New York: Harcourt Brace & Co.

Damasio, A. R. (2010). *Self Comes to Mind: Constructing the Conscious Brain.* New York: Pantheon Books.

Dehaene, S. & Changeux, J. P. (2005). Ongoing spontaneous activity controls access to consciousness: A neuronal model for inattentional blindness. *PLoS Biology*, 3(5): e141.

Edelman, G. M. (2003). Naturalizing consciousness: A theoretical framework. *Proceedings of the National Academy of Sciences (USA)*, 100 (9), 5520–5524.

Edelman, G. M. & Tononi, G. (2000). *A Universe of Consciousness: How Matter Becomes Imagination.* New York: Basic Books.

Eisenberger, N. I., Lieberman, M. D., & Williams, K. D. (2003). Does rejection hurt? An FMRI study of social exclusion. *Science*, 302, 290–292.

Fries, P. (2005). A mechanism for cognitive dynamics: Neural communication through neural coherence. *Trends in Cognitive Sciences*, 9, 474–480.

Fries, P., Roelfsema, P. R., Engel, A. K., Konig, P. & Singer, W. (1997). Synchronization of oscillatory responses in visual cortex correlates with perception in interocular rivalry. *Proceedings of the National Academy of Sciences (USA)*, 94, 12699–12704.

Fuster, J. M. (1989). *The Prefrontal Cortex.* New York: Raven.

Gray, C. M., Konig, P., Engel, A. K. & Singer, W. (1989). Oscillatory responses in cat visual cortex exhibit inter-columnar synchronization which reflects global stimulus properties. *Nature*, 338, 334–337.

Izhikevich, E. M. (2006). Polychronization: Computation with spikes. *Neural Computation*, 18, 245–282.

Jones, E. G. (2001). The thalamic matrix and thalamocortical synchrony. *Trends in Neurosciences*, 24, 595–601.

Jones, E. G. (2006). *The Thalamus Revisited*. Cambridge, U. K: : Cambridge University Press.

Kahneman, D. (2003). Maps of bounded rationality: Psychology for behavioral Economics. *The American Economic Review*, 93, 1449–1475.

Koechlin, E. & Summerfield, C. (2007). An information theoretical approach to prefrontal executive function. *Trends in Cognitive Sciences*,11, 229–235.

Kopell, N., Ermentrout, G. B., Whittington, M. A. & Traub, R. D. (2000). Gamma rhythms and beta rhythms have different synchronization properties. *Proceedings of the National Academy of Sciences (USA)*, 97, 1867–1872.

Llinás, R. (2001). *I of the Vortex: From Neurons to Self*. Cambridge, MA: Bradford/MIT Press.

Miller, E. K. & Cohen, J. D. (2001). An integrative theory of prefrontal cortex function. *Annual Review of Neuroscience*, 24, 167–202.

O'Reilly, R. C. & Frank, M. J. (2006). Making working memory work: A computational model of learning in prefrontal cortex and basal ganglia. *Neural Computation*, 18, 283–328.

O'Reilly, R. C. Noellel, D. C., Braver, T. S. & Cohen, J. D. (2002). Prefrontal cortex and dynamic categorization tasks: Representational organization and neuromodulatory control. *Cerebral Cortex*, 12, 246–257.

Palva, J. M., Palva, S. & Kaila, K. (2005). Phase synchrony among neuronal oscillations in the human cortex. *Journal of Neuroscience*, 25, 3962–3972.

Petrides, M. (1994). Frontal lobes and working memory: Evidence from investigations of the effects of cortical excisions in nonhuman primates. In F. Boller & J. Grafman (Eds), *Handbook of Neuropsychology, Vol. 9*, (pp. 59–82). Amsterdam: Elsevier.

Reynolds, J. R. & O'Reilly, R. C. (2009). Developing PFC representations using reinforcement learning. *Cognition*, 113, 281–292.

Rodriguez, A., Whitson, J. & Granger, R. (2004). Derivation and analysis of basic computational operations of thalamocortical circuits. *Journal of Cognitive Neuroscience*, 16(5), 856–877.

Salinas, E. & Sejnowski, T. J. (2001). Correlated neuronal activity and the flow of neural information. *Nature Reviews Neuroscience*, 2, 539–550.

Sherman, S. M. & Guillery, R. W. (2002). The role of the thalamus in the flow of information to cortex. *Philosophical Transactions of the Royal Society of London B*, 357, 1695–1708.

Singer, W. & Gray, C. M. (1995). Visual feature integration and the temporal correlation hypothesis. *Annual Review of Neuroscience*, 18, 555–586.

Steriade, M., Contreras, D., Amzica, F. & Timofeev, I. (1996). Synchronization of fast (30-40 Hz) spontaneous oscillations in intrathalamic and thalamocortical networks. *Journal of Neuroscience*, 16, 2788–2808.

Tononi, G., Edelman, G. M. & Sporns, O. (1998). Complexity and coherency: Integrating information in the brain. *Trends in Cognitive Sciences*, 2, 474–484.

Varela, F., Lachaux, J. P., Rodriguez, E. & Martinerie, J. (2001). The brainweb: Phase synchronization and large-scale integration. *Nature Reviews Neuroscience*, 2, 229–239.

Zikopoulos, B. & Barbas, H. (2006). Prefrontal projections to the thalamic reticular nucleus form a unique circuit for attentional mechanisms. *The Journal of Neuroscience*, 26, 7348–7361.

CHAPTER **12**

1. Cohen & Andersen, 2002.
2. Baker, Donaghue & Sanes, 1999.
3. The homologous function in avians appears to be managed by the nucleus rotundus in the avian thalamus.
4. Sherman & Guillery, 2006.
5. Given that the brain is bilaterally symmetrical, all the brain structures discussed in this chapter actually occur in pairs. There are two eyes, two frontal eye fields, two tecta, two sides to the thalamus, etc. To avoid constantly speaking in plurals, I will simply refer to the connections on one side of the brain. However, the fact that all these systems are duplicated, even to the point that the two sides of the brain might have different representations, should not be forgotten.

6. Webster et al., 1994.
7. Coe et al., 2002.
8. Duhamel, Colby & Goldberg, 1992.
9. Colby, Duhamel & Goldberg, 1996; Kodaka et al., 1997.
10. Itti & Koch, 2000; Parkhurst & Niebur, 2003.
11. Neggers & Bekkering, 2001.
12. Hollingworth & Henderson, 2002.
13. Land & Hayhoem 2001; Land & Lee, 1994.
14. Shipp, 2004.
15. Crick, 1984.
16. Shipp, 2004, p. 224.
17. Friedman-Hill, Robertson & Treisman, 1995.
18. Friedman-Hill et al., 2003.
19. Figure redrawn from one used by Trick & Pylyshyn, 1994.
20. Pylyshyn, 1994, 2001.
21. Leslie et al., 1998.
22. Kahneman, Treisman & Gibbs, 1992.
23. Koch & Tsuchiya, 2007, Lamme, 2003. Note that although consciousness does not depend on top-down attention, consciousness is never completely dissociated from the hierarchy of attention in the model I have proposed.
24. Mitrofanis, 2005.
25. Trageser & Keller, 2004.
26. Urbain & Deschenes, 2007.
27. Groenewegen & Berendse, 1994.
28. Van der Werf, Witter & Groenewegen, 2002.
29. Kimura et al., 2004.
30. Darwin & Hukin, 1998.
31. Pepperberg, 1999.
32. Khayat, Spekreijse, & Roelfsems, 2006.
33. We haven't described all the mid-level channels yet, so we'll leave these details open for now.

Baker, J. T., Donaghue, J. P. & Sanes, J. N. (1999). Gaze direction modulates finger movement activation patterns in human cerebral cortex. *The Journal of Neuroscience*, 19, 10044–10052.
Coe, B., Tomihara, K., Matsuzawa, M. & Hikosaka, O. (2002). Visual and anticipatory bias in three cortical eye fields of the monkey during an adaptive decision-making task. *The Journal of Neuroscience*, 22(12), 5081–5090.
Cohen, Y. E. & Andersen, R. A. (2002). A common reference frame

for movement plans in the posterior parietal cortex. *Nature Reviews Neuroscience, 3, 553–562.*

Colby, C. L., Duhamel, J. R. & Goldberg, M. E. (1996). Visual, presaccadic, and cognitive activation of single neurons in monkey lateral intraparietal area. *Journal of Neurophysiology,* 76, 2841–2852.

Crick, F. (1984). Function of the thalamic reticular complex: The searchlight hypothesis. *Proceedings of the National Academy of Sciences (USA),* 81, 4586–4590.

Crick, F. & Koch, C. (2003). A framework for consciousness. *Nature Neuroscience,* 6, 119–126.

Darwin, C. & Hukin, R. (1998). Perceptual segregation of a harmonic from a vowel by interaural time difference in conjunction with mistuning and onset asynchrony. *Journal of the Acoustical Society of America,* 103, 1080–1084.

Desimone, R. & Duncan, J. (1995). Neural mechanisms of selective visual attention. *Annual Review of Neuroscience,* 18, 193–222.

Duhamel, J. R., Colby, C. L. & Goldberg, M. E. (1992). The updating of the representation of visual space in parietal cortex by intended eye movements. *Science,* 255, 90–92.

Friedman-Hill, S. R., Robertson, L. C. & Treisman, A. (1995). Parietal contributions to visual feature binding: Evidence from a patient with bilateral lesions. *Science,* 269, 853–855.

Friedman-Hill, S. R., Robertson, L. C., Desimone, R. & Ungerleider, L. G. (2003). Posterior parietal cortex and the filtering of distractors. *Proceedings of the National Academy of Sciences (USA),* 100, 4263–4268.

Groenewegen H. J, & Berendse, H. W. (1994). The specificity of the "nonspecific" midline and intralaminar thalamic nuclei. *Trends in Neurosciences,* 17, 52–57.

Hollingworth, A. & Henderson, J. M. (2002). Accurate visual memory for previously attended objects in natural scenes. *Journal of Experimental Psychology: Human Perception and Performance,* 28, 113–136.

Itti, L. & Koch, C. (2000). A saliency-based search mechanism for overt and covert shifts of visual attention. *Vision Research,* 40, 1489–1506.

Kahneman, D., Treisman, A. & Gibbs, B. J. (1992). The reviewing of object files: Object-specific integration of information. *Cognitive Psychology,* 24, 174–219.

Khayat, P. S., Spekreijse, H., & Roelfsems, P. R. (2006). Attention lights up new object representations before the old ones fade away. *The Journal of Neuroscience*, 26, 138–142.

Kimura, M., Minamimoto, T., Matsumoto, N. & Hori, Y. (2004). Monitoring and switching of cortico-basal ganglia loop functions by the thalamo-striatal system. *Neuroscience Research*, 48, 355–360.

Koch, C. & Tsuchiya, N. (2007). Attention and consciousness: Two distinct brain processes. *Trends in Cognitive Sciences*, 11, 16–22.

Kodaka, Y., Mikami, A. & Kubota, K. (1997). Neuronal activity in the frontal eye field of the monkey is modulated while attention is focused onto a stimulus in the peripheral visual field, irrespective of eye movement. *Neuroscience Research*, 28, 291–298.

Lamme, V. A. F. (2003). Why visual attention and awareness are different. *Trends in Cognitive Sciences*, 7, 12–18.

Land, M. F. & Hayhoe, M. (2001). In what ways do eye movements contribute to everyday activities? *Vision Research*, 41, 3559–3565.

Land, M. F. & Lee, D. N. (1994). Where we look when we steer. *Nature*, 369, 742–744.

Leslie, A. M., Xu, F., Tremoulet, P. D. & Scholl, B. J. (1998). Indexing and the object concept: Developing 'what' and 'where' systems. *Trends in Cognitive Sciences*, 2, 10–18.

Mitrofanis, J. (2005). Some certainty for the "zone of uncertainty"? Exploring the function of the zona incerta. *Neuroscience*, 130, 1–15.

Neggers, S. F. W. & Bekkering, H. (2001). Gaze anchoring to a pointing target is present during the entire pointing movement and is driven by a non-visual signal. *Journal of Neurophysiology*, 86, 961–970.

Parkhurst, D. J. & Niebur, E. (2003). Scene content selected by active vision. *Spatial Vision*, 16, 125–154.

Pepperberg, I. M. (1999). *The Alex Studies: Cognitive and Communicative Abilities of Grey Parrots*. Cambridge, MA: Harvard University Press.

Pylyshyn, Z. W. (1994). Some primitive mechanisms of spatial attention. *Cognition*, 50, 363–384

Pylyshyn, Z. W. (2001). Visual indexes, preconceptual objects, and situated vision. *Cognition*, 80, 127–158.

Sherman, S. M. & Guillery, R. W. (2006). *Exploring the Thalamus and its Role in Cortical Function*. Cambridge, MA: MIT Press.

Shipp, S. (2004). The brain circuitry of attention. *Trends in Cognitive Sciences*, 8, 223–230.

Trageser, J. C. & Keller, A. (2004). Reducing the uncertainty: Gating of peripheral inputs by zona incerta. *Journal of Neuroscience*, 24, 8911–8915.

Trick, L. M. & Pylyshyn, Z. W. (1994). Why are small and large numbers enumerated differently? A limited capacity preattentive stage in vision. *Psychological Review*, 10, 1–23.

Urbain, N. & Deschenes, M. (2007). Motor cortex gates vibrissal responses in a thalamocortical projection pathway. *Neuron*, 56, 714–725.

Van der Werf, Y. D., Witter, M. P. & Groenewegen, H. J. (2002). The intralaminar and midline nuclei of the thalamus. Anatomical and functional evidence for participation in processes of arousal and awareness. *Brain Research Reviews*, 39, 107–140.

Webster, M. J., Bachevalier, J. & Ungerleider, L. G. (1994). Connections of inferior temporal areas TEO and TE with parietal and frontal cortex in macaque monkeys. *Cerebral Cortex*, 4, 470–483.

CHAPTER 13

1. Damasio, 1999.
2. See Butler, 2008.
3. I am not claiming that consciousness is the exclusive product of vertebrate evolution. Some invertebrates – octopi, for example – appear to have elements of attention and conscious feelings. However, their mechanisms for attention are substantially different from that of vertebrates and will probably remain difficult for us to characterize for some time. Simpler invertebrates, bees, for example, have proto-attention mechanisms that presage more complex systems of attention. However, their attention is more reflexive. They lack the hundred million neurons needed build attention-steering processes that can also gain attention and in the process contribute to a sense of self. We have no scale of terms to describe these proto-conscious forms of sentient life. They are partially guided by interoceptive states, but they do not appear to represent those states in their attention. For now, the term "proto-conscious" is perhaps the best term

we have to describe the awareness of these evolutionary precursors to conscious agency.

4. McDougall, 1923, p. 110.
5. Damasio, 1998, p. 1182.
6. Clark & Squire, 1998.
7. Libet, 1985, 1999; Libet et al., 1983.
8. The link between effectivities and reactivity states is discussed in Chapter 6. See the discussion of *free won't* by Brass & Haggard, 2007, and Craig, 2009, for how reactivity states are thought to result in response inhibition in the medial prefrontal cortex.
9. Blakemore & Frith, 2003; Farrer et al., 2003; Vogeley & Fink, 2003.
10. Blanke et al., 2002, 2004; Decety & Sommerville, 2003.
11. Edelman, 1989.
12. Given that motor outputs are coordinated by rhythmic central pattern generators, it should be natural for Cogley to link rhythmic auditory patterns with rhythmic motor activity. If we make this link strong enough, then he should even develop an interest in rhythmic motor activities, like singing and dancing, especially if his peers show such interests. This will be important if Cogley is to experience the world as his human partners do.

Blakemore S. J. & Frith, C. D. (2003). Self-awareness and action. *Current Opinion in Neurobiology*, 13, 219–224.

Blanke, O., Landis, T., Spinelli, L. & Seeck, M. (2004). Out-of-body experience and autoscopy of neurological origin. *Brain* 127, 243–258.

Blanke, O., Ortigue, S., Landis, T. & Seeck, M. (2002). Stimulating illusory own-body perceptions. *Nature*, 419, 269–270.

Brass, M. & Haggard, P. (2007). To do or not to do: The neural signature of self-control. *Journal of Neuroscience*, 27, 9141–9145.

Butler, A. B. (2008). Evolution of the thalamus: A morphological and functional review. *Thalamus & Related Systems*, 2008, 4(1), 35–58.

Clark, R. E. & Squire, L. R. (1998). Classical conditioning and brain systems: The role of awareness. *Science*, 280, 77–81.

Craig A. D. (2009). How do you feel – now? The anterior insula and human awareness. *Nature Reviews Neuroscience*, 10, 59–70.

Damasio, A. R. (1998). Investigating the biology of consciousness. *Philosophical Transaction of the Royal Society of London. B.,* 353, 1879–1882.

Damasio, A. R. (1999). *The Feeling of What Happens: Body and Emotion in the Making of Consciousness.* New York: Harcourt Brace & Co.

Decety, J. & Sommerville. J. A. (2003). Shared representations between self and other: A social cognitive neuroscience view. *Trends in Cognitive Sciences,* 7, 527–533.

Edelman, G. M. (1989). *The Remembered Present: A Biological Theory of Consciousness.* New York: Basic Books.

Farrer, C., Franck, N., Georgieff, N., Frith, D. D., Decety, J. & Jeannerod, M. (2003). Modulating the experience of agency: A positron emission tomography study. *NeuroImage,* 18, 324–333.

Libet, B. (1985). Unconscious cerebral initiative and the role of conscious will in voluntary action. *Behavioral and Brain Sciences,* 8, 529–566.

Libet, B. (1999). Do we have free will? *Journal of Consciousness Studies,* 6 (8–9), 11–29.

Libet, B., Gleason, C. A., Wright, E. W. & Pearl, D. K. (1983). Time of conscious intention to act in relation to onset of cerebral activity (readiness potential): The unconscious initiation of a freely voluntary act. *Brain,* 106, 623–642.

McDougall, W. (1923). *An Outline of Psychology.* London: Charles Scribner's Sons.

Vogeley, K. & Fink, G. R. (2003). Neural correlates of the first-person perspective. *Trends in Cognitive Sciences,* 7, 38–42.

CHAPTER **14**

1. Darwin, 1871.
2. Mushiake et al., 2006.
3. Separate return loops to both BA9 and BA46 have been reported (Middleton & Strick, 2000).
4. Alexander & Crutcher, 1990; Alexander, DeLong & Strick, 1986; Middleton & Strick, 2000.
5. McNab & Klingberg, 2008.
6. Kringelbach & Rolls, 2004.
7. Gallup, 1970.

8. Gallup, et al., 1995.
9. Plotnik, de Waal & Reiss, 2006; Prior, Schwarz & Güntürkün, 2008; Reiss & Marino, 2001.
10. Kobayashi & Kohshima, 2001.
11. Aichhorn et al., 2005; Farrer et al., 2003.
12. Damasio, 2010.

Aichhorn, M., Perner, J., Kronbichler, M., Staffen, W. & Ladurner, G. (2005). Do visual perspective tasks need theory of mind? *NeuroImage*, 30,1059–1068.

Alexander, G. E. & Crutcher, M. D. (1990). Functional architecture of basal ganglia circuits: Neural substrates of parallel processing. *Trends in Neurosciences*, 9, 266–271.

Alexander, G. E., DeLong, M. R. & Strick, P. L. (1986). Parallel organization of functionally segregated circuits linking basal ganglia and cortex. *Annual Review of Neuroscience*, 9, 357–381.

Damasio, A. R. (1999). *The Feeling of What Happens: Body and Emotion in the Making of Consciousness*. New York: Harcourt Brace & Co.

Damasio, A. R. (2010). *Self Comes to Mind: Constructing the Conscious Brain*. New York: Pantheon Books.

Darwin, C. (1871). *The Descent of Man and Selection in Relation to Sex*. London: John Murray. <http://darwin-online.org.uk/contents.html#books>

Edelman, G. M. (1989). *The Remembered Present: A Biological Theory of Consciousness*. New York: Basic Books.

Edelman, G. M. (2003). Naturalizing consciousness: A theoretical framework. *Proceedings of the National Academy of Sciences (USA)*, 100 (9), 5520–5524.

Farrer, C., Franck, N., Georgieff, N., Frith, D. D., Decety, J., & Jeannerod, M. (2003). Modulating the experience of agency: A positron emission tomography study. *NeuroImage*, 18, 324–333.

Gallup, G. G. Jr. (1970). Chimpanzees: Self-recognition. *Science* 167, 86–87.

Gallup, G. G. . Jr., Povinelli. D. J., Suarez, S. D., et al. (1995). Further reflections on self-recognition in primates. *Animal Behaviour*, 50, 1525–1532.

Kobayashi, H. & Kohshima, S. (2001). Unique morphology of the human eye and its adaptive meaning: Comparative studies on external morphology of the primate eye. *Journal of Human Evolution*, 40(5), 419–435.

Kringelbach, M. L. & Rolls, E. T. (2004). The functional neuroanatomy of the human orbitofrontal cortex: Evidence from neuroimaging and neuropsychology. *Progress in Neurobiology*, 72, 341–372.

McNab, F. & Klingberg, T. (2008). Prefrontal cortex and basal ganglia control access to working memory. *Nature Neuroscience*, 11, 103–107.

Middleton, F. A. & Strick, P. L. (2000). Basal ganglia output and cognition: Evidence from anatomical, behavioral, and clinical studies. *Brain and Cognition*, 42, 183–200.

Mushiake, H., Saito, N., Sakamoto, K., Itoyama, Y. & Tanji, J. (2006). Activity in the lateral prefrontal cortex reflects multiple steps of future events in action plans. *Neuron*, 50, 631–641.

Plotnik, J. M., de Waal, F. B. M. & Reiss, D. (2006). Self-recognition in an Asian elephant. *Proceedings of the National Academy of Sciences (USA)*, 103, 17053–17057.

Prior, P., Schwarz, A. & Güntürkün, O. (2008). Mirror-induced behavior in the magpie (Pica pica): Evidence of self-recognition. *PLoS Biology*, 6(8), e202, 1642–1650.

Reiss, D. & Marino, L. (2001). Mirror self-recognition in the bottlenose dolphin: A case of cognitive convergence. *Proceedings of the National Academy of Sciences (USA)*, 98, 5937–5942.

CHAPTER 15

1. Tinbergen, 1952.
2. Tomasello et al., 1994.
3. Bates, 1979.
4. Pepperberg, 1999.
5. Scott & Johnsrude, 2003.
6. Griffiths & Warren, 2002.
7. Fadiga, Craighero & D'Ausilio, 2009.
8. Rilling et al., 2008; Schmahmann et al., 2007.
9. Consistent with the role of gestures in social communication there are also curved bundles connecting visual processing regions with "language" production and comprehension areas.
10. Seyfarth, R. Social cognition and the origins of language. Talk given at Indiana University, November 3, 2009.

11. Fisher & Marcus, 2006.
12. Liberman & Mattingly, 1985.
13. Fadiga, Craighero & D'Ausilio, 2009.
14. Kuhl, 2000.
15. Roy, 2005.
16. Hebb, 1949.
17. Fodor, 1975, 1987.
18. Compare Fauconnier & Turner's, 2002, model of blending.
19. Peirce, 1897/1940.
20. Bailey, 1997.
21. Harnad, 1990.
22. See Bhatnagar & Mandybur, 2005.

Bailey, D. (1997). *When push comes to shove: A computational model of the role of motor control in the acquisition of action verbs.* PHD Thesis, University of California, Berkley.

Bates, E. (1979). *The Emergence of Symbols: Cognition and Communication in Infancy.* New York: Academic Press.

Bhatnagar, S. C. & Mandybur, G. T. (2005). Effects of intralaminar thalamic stimulation on language functions. *Brain and Language*, 92, 1–11.

Fadiga, L., Craighero, L. & D'Ausilio, A. (2009). Broca's area in language, action, and music. *Annals of the New York Academy of Sciences*, 1169, 448–458.

Fauconnier, G. & Turner, M. (2002). *The Way We Think: Conceptual Blending and the Mind's Hidden Complexities.* New York: Basic Books.

Fisher, S. E. & Marcus, G. F. (2006). The eloquent ape: Genes, brains, and the evolution of language. *Nature Reviews Genetics*, 7, 9–20.

Fodor, J. (1975). *The Language of Thought.* Cambridge, MA: Harvard University Press.

Fodor, J. (1987). Psychosemantics: The Problem of Meaning in the Philosophy of Mind. Cambridge, MA: MIT Press.

Griffiths, T. D. & Warren, J. D. (2002). The planum temporale as a computational hub. *Trends in Neurosciences*, 25, 348–353.

Harnad, S. (1990). The symbol grounding problem. *Physica D*, 42, 335–346.

Hebb, D. O. (1949). *The Organization of Behavior.* New York: Wiley.

Kuhl, P. K. (2000). A new view of language acquisition. *Proceedings of the National Academy of Sciences (USA)*, 97, 11850–11857.

Liberman, A. M. & Mattingly, I. G. (1985). The motor theory of speech perception revised. *Cognition*, 21, 1–36.

Peirce, C. S. (1897/1940). Logic as semiotic: The theory of signs. In: *Philosophical Writings of Peirce*. New York: Dover.

Pepperberg, I. M. (1999). *The Alex Studies: Cognitive and Communicative Abilities of Grey Parrots*. Cambridge, MA: Harvard University Press.

Rilling, J. K., Glasser, M. F., Preuss, T. M., Ma, X., Zhao, T., Hu, X. & Behrens, T. E. (2008). The evolution of the arcuate fasciculus revealed with comparative DTI. *Nature Neuroscience*, 11, 426–428.

Roy, D. (2005). Semiotic schemas: A framework for grounding language in action and perception. *Artificial Intelligence*, 167(1–2), 170–205.

Schmahmann, J. D., Pandya, D. N., Wang, R., Dai, G., D'Arceuil, H. E., de Crespigny, A. J. &. Wedeen, V. J. (2007). Association fibre pathways of the brain: Parallel observations from diffusion spectrum imaging and autoradiography. *Brain*, 130, 630–653.

Scott, S. K. & Johnsrude, I. S. (2003). The neuroanatomical and functional organization of speech perception. *Trends in Neurosciences*, 26, 100–107.

Tinbergen, N. (1952). "Derived" activities; Their causation, biological significance, origin, and emancipation during evolution. *Quarterly Review of Biology*, 27, 1–32.

Tomasello, M. (2003). *Constructing a Language*. Cambridge, MA: Harvard University Press.

Tomasello, M., Call, J., Nagell, K., Olguin, R. & Carpenter, M. (1994). The learning and use of gestural signals by young chimpanzees: A transgenerational study. *Primates*, 35, 137–154.

CHAPTER 16

1. Hofstadter, 2007.
2. Rudolfo Llinás, 2001, appears to share a similar view of the dynamic nature of the self. His book on the emergence of mind and consciousness is entitled *I of the Vortex*.
3. Kauffman, 2000.
4. Miller, 1956.

5. Brooks, 1986.
6. Graybiel, 1998, 2008.
7. Bjordahl, Dimyan & Weinberger, 1998; Kilgard & Merzenich, 1998; Kilgard et al., 2002.
8. Richardson & DeLong, 1988, 1991.
9. Alexander & Crutcher, 1990; Middleton & Strick, 2000.
10. McNab & Klingberg, 2008
11. Kringelbach & Rolls, 2004.
12. Lakoff & Johnson, 1999, p. 13.
13. Johnson, 1987; Lakoff, 1987; Lakoff & Johnson, 1980, 1999.
14. Lakoff & Johnson, 1999, p. 46.
15. Roy, 2005.
16. Brooks, 1986.
17. Gallese & Lakoff, 2005.

Alexander, G. E. & Crutcher, M. D. (1990). Functional architecture of basal ganglia circuits: Neural substrates of parallel processing. *Trends in Neurosciences*, 9, 266–271.

Bjordahl, T. S., Dimyan, M. A. & Weinberger, N. M. (1998). Induction of long-term receptive field plasticity in the auditory cortex of the waking guinea pig by stimulation of the nucleus basalis. *Behavioral Neuroscience*, 112, 467–479.

Brooks, R. (1986). A robust layered control system for a mobile robot. *Journal of Robotics & Automation*, 2, 14–23.

Gallese, V. & Lakoff, G. (2005). The brain's concepts: The role of the sensory-motor system in conceptual knowledge. *Cognitive Neuropsychology*, 22, 455–479.

Graybiel, A. M. (1998). The basal ganglia and chunking of action repertoires. *Neurobiology of Learning and Memory*, 70, 119–136.

Graybiel, A. M. (2008). Habits, rituals, and the evaluative brain. *Annual Review of Neuroscience*, 31, 359–387.

Hofstadter, D. R. (2007). *I Am a Strange Loop*. New York: Basic Books.

Johnson, M. (1987). *The Body in the Mind*. Chicago: University of Chicago Press.

Kauffman, S. A. (2000). *Investigations*. Oxford: Oxford University Press.

Kilgard, M. P. & Merzenich, M. M. (1998). Cortical map reorganization enabled by nucleus basalis activity. *Nature Neuroscience*, 1, 727–731.

Kilgard, M. P., Pandya, P. K., Engineer, N. D. & Moucha, R. (2002). Cortical network reorganization guided by sensory input features. *Biological Cybernetics*, 87, 333–343.

Kringelbach, M. L. & Rolls, E. T. (2004). The functional neuroanatomy of the human orbitofrontal cortex: Evidence from neuroimaging and neuropsychology. *Progress in Neurobiology*, 72, 341–372.

Lakoff, G. & Johnson, M. (1980). *Metaphors We Live By*. Chicago: University of Chicago Press.

Lakoff, G. & Johnson, M. (1999). *Philosophy in the Flesh: The Embodied Mind and Its Challenge to Western Thought*. New York: Basic Books.

Lakoff, G. (1987). *Women, Fire, and Dangerous Things: What Categories Reveal about the Mind*. Chicago: University of Chicago Press.

McNab, F. & Klingberg, T. (2008). Prefrontal cortex and basal ganglia control access to working memory. *Nature Neuroscience*, 11, 103–107.

Middleton, F. A. & Strick, P. L. (2000). Basal ganglia output and cognition: Evidence from anatomical, behavioral, and clinical studies. *Brain and Cognition*, 42, 183–200.

Miller, G. A. (1956). The magical number seven, plus or minus two: Some limits on our capacity for processing information. *Psychological Review*, 63, 81–97

Richardson, R. T. & DeLong, M. R. (1988). A reappraisal of the functions of the nucleus basalis of Meynert. *Trends in Neurosciences*, 11, 264–267.

Richardson, R. T. & DeLong, M. R. (1991). Electrophysiological studies of the functions of the nucleus basalis in primates. *Advanced Experimental Medical Biology*, 295, 233–252.

Roy, D. (2005). Semiotic schemas: A framework for grounding language in action and perception. *Artificial Intelligence*, 167(1–2), 170–205.

CHAPTER 17

1. Clark, 1997.
2. Holland, 1992.
3. Plotkin, 1993.

4. Clark, 2003.
5. Iriki, Tanaka & Iwamura, 1996.
6. Ramachandran & Blakeslee, 1998, p. 59.
7. Guterstam, Petkova & Ehrsson, 2011.
8. Minsky, 1980.
9. On September 7, 2001, surgeon Dr. Jacques Marescaux, working in a control room in New York, performed a transatlantic gallbladder operation on a patient in Strasbourg, France.

Clark, A. (1997). *Being There: Putting Brain, Body, and World Together Again.* Cambridge, MA: Bradford/MIT Press.
Clark, A. (2003). *Natural-Born Cyborgs: Minds, Technologies, and the Future of Human Intelligence.* New York: Oxford University Press.
Guterstam, A., Petkova, V. I., & Ehrsson, H. H. (2011). The illusion of owning a third arm. *PLoS ONE*, 6(2): 1–11, e17208.
Holland, J. H. (1992). Complex adaptive systems. *Daedalus*, 121, 17–30.
Iriki, A., Tanaka, M. & Iwamura, Y. (1996). Coding of modified body schema during tool use by macaque postcentral neurones. *Neuroreport*, 7, 2325–2330.
Minsky, M. (1980). *Omni*, May, 45–52. Quoted in: A. Clark, *Natural-Born Cyborgs*, 2003, p. 93.
Plotkin, H. (1993). *Darwin Machines and the Nature of Knowledge.* Cambridge, MA: Harvard University Press.
Ramachandran, V. S. & Blakeslee, S. (1998). *Phantoms in the Brain: Probing the Mysteries of the Human Mind.* New York: William Morrow.

CHAPTER 18

1. To be truly productive collective consciousness will require a cultural steering committee with inputs from a variety of value-promoting groups, as well as some sort of system for adjusting the saliency of those inputs as they compete for cultural attention.
2. Damasio, 1994.
3. Insel & Fernald, 2004; Insel & Young, 2000; Kosfeld et al., 2005; Uvnäs-Moberg, 1998.

4. Dome et al., 2007.
5. Zak, Stanton, & Ahmadi, 2007.
6. Depue & Morrone-Strupinsky. 2005.
7. Dijk, Koenig, Ketelaar & de Jong, 2011.
8. Asch, 1956.
9. Mineka et al., 1993, 2002; Ohman & Mineka, 2001.
10. Bem, 1996.
11. Clark, 2003.

Asch, S. E. (1956). Studies of independence and conformity: A minority of one against a unanimous majority. *Psychological Monographs, 70* (Whole No. 416).

Bem, D. J. (1996). Exotic becomes erotic: A developmental theory of sexual orientation. *Psychological Review, 103,* 320–335.

Clark, A. (2003). *Natural-Born Cyborgs: Minds, Technologies, and the Future of Human Intelligence.* New York: Oxford University Press.

Damasio, A. R. (1994). *Descartes' Error: Emotion, Reason, and the Human Brain.* New York: Avon Press.

Depue, R. A. & Morrone-Strupinsky, J. V. (2005). A neurobehavioral model of affiliative bonding: Implications for conceptualizing a human trait of affiliation. *Behavioral and Brain Sciences, 28,* 313–350.

Dijk, C, Koenig, B., Ketelaar, T. & de Jong, P. (2011). Saved by the blush: Being trusted despite defecting. *Emotion, 1(2),* 313–319.

Domes, G., Heinrichs, M., Michel, A., Berger, C. & Herpertz, S. C. (2007). Oxytocin improves "mind-reading" in humans. *Biological Psychiatry, 61,* 731–733.

Insel, T. R. & Fernald, R. D. (2004). How the brain processes social information: Searching for the social brain. *Annual Review of Neuroscience, 27,* 697–722.

Insel, T. R. & Young, L. J. (2000). Neuropeptides and the evolution of social behaviour. *Current Opinions in Neurobiology, 10,* 788–789.

Kosfeld, M., Heinrichs, M., Zak, P. J., Fischbacher, U. & Fehr, E. (2005). Oxytocin increases trust in humans. *Nature, 435,* 673–676.

Mineka, S. & Cook, M. (1993). Mechanisms underlying observational conditioning of fear in monkeys. *Journal of Experimental Psychology: General, 122,* 23–38.

Mineka, S. & Ohman, A. (2002). Phobias and preparedness: The selective, automatic, and encapsulated nature of fear. *Biological Psychiatry*, 52, 927–937.

Ohman, A. & Mineka, S. (2001). Fears, phobias, and preparedness: Toward an evolved module of fear and fear conditioning. *Psychological Review*, 108, 483–522.

Schmahmann, J. D., Pandya, D. N., Wang, R., Dai, G., D'Arceuil, H. E., de Crespigny, A. J. &. Wedeen, V. J. (2007). Association fibre pathways of the brain: Parallel observations from diffusion spectrum imaging and autoradiography. *Brain*, 130, 630–653.

Uvnäs-Moberg, K. (1998). Oxytocin may mediate the benefits of positive social interaction and emotions. *Psychoneuroendocrinology*, 23, 819–835.

Zak, P. J., Stanton, A. A. & Ahmadi, S. (2007). Oxytocin increases generosity in humans. *PLoS ONE*, 2(11), e1128.

CHAPTER 19

1. Nagel, 1974.
2. Dennett, 1991.
3. Chalmers, 1995.
4. Chalmers, 1995, 1996, 1997.
5. Zeman, 2001, p. 1284.
6. Chalmers, 1996.
7. Crick & Koch, "What Is Consciousness", *Discover*, November 1992, p. 96.
8. Edelman, 2006, p. 145.
9. Brentano, 1874/1995.
10. Millikan, 1993.
11. Stratton, 1897.
12. Kohler, 1964.
13. Jameson, Highnote & Wasserman, 2001.
14. Chalmers, 1997, p. 4.
15. Dennett, 2003.
16. Nagel, 1974.
17. Chalmers, 1996.

Brentano, F. (1874/1995). In: O. Kraus (Ed.), *Psychology from an Empirical Standpoint*. New York: Routledge.

Chalmers, D. J. (1995). Facing up to the problem of consciousness. *Journal of Consciousness Studies*, 2(3), 200–219.

Chalmers, D. J. (1996). *The Conscious Mind: In Search of a Fundamental Theory.* Oxford: Oxford University Press.

Chalmers, D. J. (1997). Facing up to the problem of consciousness. *Journal of Consciousness Studies,* 4(1), 3–46.

Dennett, D. C. (1991). *Consciousness Explained.* Boston: Little, Brown.

Dennett, D. C. (2003). Explaining the "magic" of consciousness. *Journal of Cultural and Evolutionary Psychology,* 1, 7–19.

Edelman, G. M. (2006). *Second Nature: Brain Science and Human Knowledge.* New Haven: Yale University Press.

Jameson, K. A., Highnote, S. M., & Wasserman, L. M. (2001). Richer colour experience in observers with multiple photopigment opsin genes. *Psychonomic Bulletin & Review,* 8, 244–261.

Kohler, I. (1964). The formation and transformation of the perceptual world. *Psychological Issues,* 3 (monogr. 12), 1–173.

Millikan, R. G. (1993). On mentalese orthography, part 1. In: R. G. Millikan, *White Queen Psychology and Other Essays for Alice,* 103–121. Cambridge, MA: Bradford/MIT Press.

Nagel, T. (1974). What is it like to be a bat? *Philosophical Review,* 83, 435–450.

Stratton, G. M. (1897). Vision without inversion of the retinal image. *Psychological Review,* 4, 341–360, 463–481.

Zeman, A. (2001). Consciousness. *Brain,* 124, 1263–1289.

Index

V

vasopressin 37, 143, 373
ventral intraparietal cortex (VIP)
89, 90, 178
ventral stream 81, 82, 83, 84, 89,
92, 105, 106, 107, 108, 110,
112, 132, 223
ventral tegmental area (VTA) 98,
142, 144
Vinogradova, Olga 120, 129, 130

W

Weisel, Torsten 108
whisking 233, 239
working hypothesis 213, 214, 216

Z

Zeman, Adam xv, 392, 415
zona incerta 233, 239